Commerce in Space:
Infrastructures, Technologies, and Applications

Phillip Olla
Madonna University, USA

Information Science REFERENCE

INFORMATION SCIENCE REFERENCE

Hershey · New York

Acquisitions Editor: Kristin Klinger
Development Editor: Kristin Roth
Senior Managing Editor: Jennifer Neidig
Managing Editor: Sara Reed
Copy Editor: Lanette Ehrhardt
Typesetter: Jeff Ash
Cover Design: Lisa Tosheff
Printed at: Yurchak Printing Inc.

Published in the United States of America by
Information Science Reference (an imprint of IGI Global)
701 E. Chocolate Avenue, Suite 200
Hershey PA 17033
Tel: 717-533-8845
Fax: 717-533-8661
E-mail: cust@igi-pub.com
Web site: http://www.igi-pub.com/reference

and in the United Kingdom by
Information Science Reference (an imprint of IGI Global)
3 Henrietta Street
Covent Garden
London WC2E 8LU
Tel: 44 20 7240 0856
Fax: 44 20 7379 0609
Web site: http://www.eurospanonline.com

Library of Congress Cataloging-in-Publication Data

Commerce in space : infrastructures, technologies, and applications / Phillip Olla, editor.
p. cm.
Summary: "This book explains the role of earth observation satellite initiatives to meet information needs. It details the importance of the space infrastructure to deliver IT capabilities such as mobile broadband Internet and mobile communication connectivity; it also offers a review of how space technology can influence the future of IT architecture in health, education, logistics, business, and accounting"--Provided by publisher.
Includes bibliographical references and index.
ISBN-13: 978-1-59904-624-2 (hardcover)
ISBN-13: 978-1-59904-626-6 (ebook)
1. Artificial satellites in telecommunication. 2. Global positioning system. 3. Remote sensing. 4. Space industrialization. 5. Space tourism. 6. Astronautics and civilization. 7. Astronautics--Social aspects. I. Olla, Phillip.
TK5104.C59 2008
384.5'1--dc22
2007007282

British Cataloguing in Publication Data
A Cataloguing in Publication record for this book is available from the British Library.

Table of Contents

Detailed Table of Contents

This chapter summarizes the collective work of a team of students who participated in the 2004 International Space University Summer Session Program in Adelaide, Australia. The project is called STREAM, which stands for Space Technologies for the Research of Effective wAter Management. The work represented in this chapter was accomplished as part of the intensive space studies curriculum offered during the summer session. The team project focused on the importance of fresh water resource management and its impact on the surrounding communities. The team explored various space technologies and their current and future potential to enhance water resource management. A real world case study of Australia's Murray-Darling Basin (MDB) was performed to provide the central focus of the project. Based on the results of the case study, the team then extrapolated their results to other regions of the globe that are experiencing challenges to their fresh water supply. A significant space technology recommendation developed by the STREAM project team was to improve the soil moisture measurement capabilities in the MDB. The primary goal of the STREAM project team is that the recommendations outlined in the extensive final report (STREAM Team, 2004) will receive full attention from policy makers concerned with the water issues surrounding the MDB.

Deliberate exploitation of natural resources and excessive use of environmentally abhorrent materials have resulted in environmental disruptions threatening the life support systems. Human centric ap-

Section II
Satellite Internet and Navigational Technologies

This chapter introduces the concept of Satellite Navigation in the context of space infrastructures and technologies that can contribute for improvement of life on Earth. It includes a review of the motivations for developing a satellite navigation system, and the applications and services these systems have in daily life. Furthermore, currently existing Global Navigation Satellite Systems (GPS and GLONASS) and other GNSS systems under development (GALILEO) are described from different perspectives: from the technical and architectural aspects to the ways chosen to finance their development and operations. To round up this chapter, an analysis of the expected trends in GNSS systems is presented and potential scenarios for future evolution of global satellite navigation are discussed.

The aim of this chapter is to show the satellite Internet as a new quality, which was created thanks to the convergence of satellite communication and data networks. The chapter describes the development of satellite communication and satellite data networks, presents methods of Internet access via satellite and discuss the opportunities and challenges of building effective commercial services based on satellite Internet. The main advantages of the satellite Internet are high bandwidth, very good availability (in practice: anywhere in the world), and natural IP multicasting. Although getting broadband Internet access by satellite is considered very expensive, independence from the local infrastructure results in the satellite Internet being a good solution for both business communications (a corporate network or its fragments) and remote area communications (rural communications and services to isolated communities).

Nowadays it is possible to achieve low cost and short production times space missions using satellites with a mass below 10 kg. These small satellites are described as nanosatellites.
Current microelectronic technology makes it possible to develop nanosatellites for scientific experiments and relatively complex measurements (as well as for other applications) making it easy for universities and small research groups to have access to space science exploration and to exploit the new economic possibilities that emerge. This chapter, describes an experiment developed in Argentina at the Universidad Nacional del Comahue to design and construct a nanosatellite called Pehuensat-1.

Space technology has advanced rapidly in recent years. Nevertheless, a number of countries still lack the human, technical and financial resources required to conduct even the most basic space-related activities, such as meteorology, communications natural-resource management and education. The need to make the benefits of space technology available to all countries has thus grown more urgent with each passing year. This chapter proposes a two phased approach for using space technology to deliver Information Communication Technologies (ICT) to underserved areas. The first phase involves the definition and implementation of the Satellite Global infrastructure to provide connectivity to underserved regions. The second phase introduces the concept of a Coalition of Space Internet Providers (COSIP) model. The aim of this model is to encourage the diffusion of space technology delivered by the GBBS infrastructure to the grassroots level. The model defines how Internet capabilities should be introduced to rural under-privileged societies to provide health and educational services in a sustainable manner. This model is a reincarnation of the Local Information Utility (LIU) model that was successfully implemented over a decade ago, to aid the diffusion of the Internet to rural American communities. This chapter explains the technology at the foundation of the COSIP model and describes the actors required along with their roles and responsibilities.

This chapter examines the development and progress of European space policy from its beginnings over a decade ago up to today's perspectives for a European space policy. By outlining the institutional structures and responsibilities between the differing communities of the EU and ESA, it demonstrates the financial parameters behind the European space programmes and highlights accompanying structural difficulties between the institutions. Current European space efforts and the solutions adopted for cooperation are then highlighted within the background of the structures developing in the EU on security issues for Europe. The paper concludes with a prognosis and summary of Europe's efforts to create a strong European space policy.

This chapter introduces the potential of satellite Earth Observation (EO) as a tool for improving the implementation of Multilateral Environmental Agreements (MEAs). It provides the technical and legal

of the organization. For a space-faring organization, mission cadence is one of the key determinants of cost and risk. Mission cadence is also important as it determines the number of people in the organization with direct and relevant experience with space missions. Under a constrained budget, mission cadence can be increased by reducing the size and scope of the missions. Small spacecraft missions can afford to be innovative and thus create a culture in which new ideas are welcomed and/or sought. These smaller missions can preserve and generate knowledge by training the next generation of scientists, engineers and program managers.

This chapter introduces the ethical questioning in the field of space activities, especially space commerce. If the 1967 Outer Space Treaty defines space as the "property of all" and its exploration as the "province of all mankind", the future utilization of near-Earth (and tomorrow Greater Earth) space needs probably a new ethics (if ethics means not only legal applications but also and for example the application of the "rule of three Ps": protection, promotion and preparation). Orbital debris mitigation, the International Charter on Space and Major Disasters or, in the future, the safety of private astronauts crews offer lessons in realism and sources of prospective reflections. Space ethics is still in its infancy.

Space tourism is the term broadly applied to the concept of paying customers traveling beyond Earth's atmosphere. Operating reusable launch vehicles might be a first step to realize mass space tourism. Thus, the aim of this chapter is to investigate the potential hurdles along with other important aspects of space tourism flights utilizing reusable launch vehicles. The primary elements are social issues, e.g. "Is space tourism acceptable concerning ethical aspects?", institutional issues, e.g. "Is environmental pollution caused by space tourism harmful compared to other emission sources?" and financial issues, e.g. "Are there any potential investors interested in space tourism?".

The first part of this paper describes how the Space Elevator is expected to work, and the advantage of access to space via the SE versus using primarily rockets. A compendium of information from a variety of sources is included in order to explain how the Space Elevator would be designed, constructed, and how it could solve the problems of transporting cargo into Space easily, cheaply, and frequently. The Space Elevator is a relatively new topic in the area of realistic science concepts and was merely science fiction not too long ago. The Space Elevator ("SE") concept has only been in the spotlight in the last five years due to the work of Dr. Bradley Edwards of Carbon Designs Inc. Acceptance of the SE will be a

Foreword

There is no longer any doubt that the peaceful uses of space infrastructure provides a prevailing tool for expanding the well-being of humanity and the Earth's environment. Space architecture and applications are fundamental for providing important services to people on earth. Some of these services are intertwined into our daily lives such as the use weather forecasts, satellite TV and radio. More recently, we are seeing more innovative personal communications devices with integrated GPS technology. Space technology is currently being incorporated into Health care, learning, transport systems, disaster relief, and search-and-rescue operations.

Using space infrastructures provide opportunities in a wide range of public missions in a cost-effective way. In particular, space infrastructure can generate solutions for long-term societal needs such as climate monitoring, resource management, disaster relief, and digital inclusion. Unfortunately, these opportunities are not being exploited for a variety of reasons, which range include lack of information, technical problems, and the existence of bureaucratic rules that prevent the effective use of the infrastructures.

To realize the full potential of space infrastructure, there are some critical issues that need to be addressed: space infrastructure must be continuously improved and upgraded; there must be increased efforts to integrate space applications with terrestrial systems, ensuring infrastructure sustainability, address the digital divide, and mandating ethical use of space resources. In order to achieve these critical issues, governments and decision makers will need to consider the challenges from a technological, legal, economic, and regulatory dimension.

This book provides an insight into work that is being done around the globe to address these critical issues. It also provide, in only one document, a comprehensive review of how space technology can influence the future of people on earth across multiple domains such as health, education, disaster management, and communication. The book describes real world problems using case studies and opens at the same time a perspective towards the future by discussing the utilization of space technology at global, national and regional levels for resolving specific problems.

Preface

THE UNTOLD IMPACT OF SPACE TECHNOLOGY ON SOCIETY

Satellites and space technology play important roles relaying information to terrestrial systems for knowledge generation and decision-making. They are becoming more prominent in the data transfer, communication, navigation, and environmental observation markets. In developed nations, the use of space technology is deeply entrenched in modern applications (Cohendet, 2004). The areas that rely heavily on space infrastructures include meteorology, mobile communication systems, television broadcasting, natural resource management, navigation, health, environmental management, and disaster management, which consequently influence virtually every facet of human endeavor. It is therefore no surprise that this industry is anticipated to be a significant growth industry in the 21st century, leading to technological developments in several fields ranging from telecommunications, tele-health, tele-education, multimedia, opto-electronics, robotics, life sciences, energy, and nanotechnology (Hukill, Ono, & Vallath, 2000).

When discussing the advancement of space science and space technology most people instinctively think about deep space flights, lunar stations, and thrilling outer space adventures. The fact is that the majority of the human technology in space, which comprises of satellites, point toward Earth and most of the technology in space is used to provide services and fulfill the goals for people on planet Earth. The growing role of space technology is so profound it has almost become ubiquitous and prevalent in society.

In today's global society, it appears that economic prosperity is the most important human goal; however, the foremost goal of the human race should be to sustain a livable biosphere. The objective must be to improve the environment at least gradually over the next few centuries, a task "easier said than done." This is not something that can be done quickly. The planet is facing some fundamental challenges, which are expected to become much worse over the next couple of decades (Pelton, 2000). The problems that must be addressed extend over a spectrum of environmental, technological, and humanitarian domains. Currently, one of the most topical issues is the dilemma of global warming, which comprises of problems such as the carbon dioxide and methane build up and the disappearing ice caps.

The next set of challenges stems from global pollution and includes issues such as destruction of the rain forests, desertification, reduction of arable land, and the over reliance on dwindling petrochemical energy sources. Another series of problems relates to humanitarian issues that are compounded by the spiraling growth of the human population. There are inappropriate distributions of natural and agricultural resources to manage the growing population. About 1 billion people—one fifth of the world's population—live on less than $1 a day (World-Bank, 2006). Unfortunately, this is also reflected in the lack of universal access to information technology, global education, and health care, called the digital divide. Over the last decade, we have also increased the level of hatred between nations and races, fueled primarily by religious ideology rather than political allegiances, which has led to an increase in terrorism. Other random problems that we must overcome in order to survive the next century are new biological virus mutations such as bird flu and HIV AIDS, techno-terrorism, nuclear proliferation, and abuses from technologies such as bio-engineering and cloning.

Humans are not the only inhabitants on Earth who have ever faced environmental dilemmas. One could hypothesize the giant dinosaurs which once ruled the Earth would have been more fortunate if our sophisticated space technology was available. As Sir Arthur Clarke pointed out, "The dinosaurs failed to survive due to the lack of a space program." The most important message from this saying is not necessarily the creation of an International Space Program, but the ability to have the foresight to predict the future and plan accordingly.

The idea of using data and technologies from space infrastructure is not entirely new; however, the rate at which data is being integrated into terrestrial systems is experiencing colossal growth and acceptance, creating a phenomenon this book describes as space business (s-business). S-business relates to any venture performed by a group of diverse actors leading to the provision of goods or services involving financial, commercial, or humanitarian activity that is facilitated by the use of a space technological infrastructure in the Earth's orbit. The commercialization of space is creating opportunities for new types of information systems (IS) and information technology (IT) architectures.

Although the space technological infrastructure is primarily composed of the satellites in orbit, the supporting infrastructure is a collection of interconnected technological systems, social processes, and organizational elements that enable space data to be collected, processed, stored, and broadcast to devices or base stations on Earth. Once this data is received on Earth, it can be translated into meaningful information, leading to knowledge which can be used to aid the planning process and decision making. There are five established discernible space infrastructures: telecommunication, positioning and navigation, broadcasting, Earth observation, and micro gravity research (MGR) and tourism. Each space infrastructure has a specialized function; however, there is an increasing theme of convergence between these infrastructures, as can be seen from the descriptions as follows.

Telecommunications

It is currently the most important and the most dynamic market for space applications. It includes voice, data, Internet, and multimedia mobile services. Communication satellites are used to transmit voice and data services anywhere in the world. The main advantage over terrestrial communication systems is that satellites do not have to be connected to a ground network. Communication satellites have ubiquitous access and can reach people in remote villages, crews in the middle of the ocean, or explorers at the top of a mountain. They are ideal for situations in which the ground infrastructure is not available or has been temporarily damaged by natural disasters. In combination with ground-based networks, satellites can provide access to the World Wide Web. Satellite telecommunications have the potential to deliver information to rural and remote areas, and may also help countries kick start their economical development. Information and services delivered via satellites to technology deprived regions can also contribute to a nations sustainable development program, providing access to information and helping members of the public participate in decision-making, or more generally by improving education and health services and promoting favourable conditions for environmental protection.

The satellite telecommunications sector has faced some difficulties in recent years. Some of these difficulties are as follows:

- Reduction in demand for wireless satellite constellation development
- The proliferation of new actors in the field
- Larger commercial communication satellites lasting longer on their on-orbit stay times

There are numerous applications, such as domestic/international trunking, in-flight Internet service, broadband Internet, Internet backbone, roaming, wireless networks (VSAT), messaging, and asset management that make this multibillion dollar industry the most profitable space sector. The growth of the Internet, along with interactive multimedia applications, is causing an overload on terrestrial networks run by the telecommunications operators, creating an increase in the adoption of satellite communication.

Broadcasting

Organizations in the satellite broadcasting sector provide services such as digital TV, digital radio, multimedia, Internet content, and educational content. Satellite television and radio provide entertainment delivered by satellites as opposed to conventional terrestrial technologies. Satellite television has been around since the 1960s. This infrastructure was originally conceived with limited coverage and to serve a limited number of professional users. Satellite broadcasting has evolved to provide global coverage to a wide array of users.

The first satellite television signal was relayed from Europe to the Telstar satellite over North America in 1962, and the world's first commercial communication satellite, called Early Bird, was launched into synchronous orbit on April 6, 1965. In many areas of the world, satellite television services supplement older terrestrial signals, providing a wider range of channels and services, including subscription-only services. The recent additions to this technology include cutting-edge innovation, such as high definition, interactive programming features, mix channels, and pay-per-view.

Satellite radio is a digital radio that receives signals broadcast by communications satellite, which covers a much wider geographical range than terrestrial radio signals.

Satellite radio services are all commercial business entities. Accessing packages of channels requires a subscription to a commercial satellite for signal propagation. Currently, the main satellite radio providers are WorldSpace, XM, and Sirius. WorldSpace covers Europe, Asia and Africa, while XM Radio and Sirius both cover North America.

Earth Observation

Earth observation is vital for measuring and monitoring the world's climate and atmosphere, and for recording and mapping our valuable resources. Earth observation is a continuation of meteorology that is extending new domains, including agriculture, resource management, exploration, mapping and planning, hazard monitoring, and disaster assessment (landslides, earthquakes, volcanic eruptions, floods, and droughts) as well as security, defense, and the enforcement of international agreements (OECD, 2006).

Currently, there are millions of space-based sensors collecting data around the world by various countries and organizations providing the foundation for sound decision-making. Unfortunately, they operate independently with little integration. The concept of an integrated Earth information system requires an interdisciplinary focus, utilizing a wide array of technological sensors. There are sound social, economic, and scientific drivers that are dictating the need for building an integrated Earth information and data management system. Images retrieved from Earth observation satellites incorporated into geographic information systems (GIS) offer a wealth of vital information to policy makers, scientists, and the general public about the planet's variable environment. Satellite images provide information about:

- Land cover and land use
- Remote and difficult-to-access areas like dense forests, glaciated areas, deserts and, swamps
- Areas undergoing rapid environmental change, including loss or fragmentation of ecosystems and related

loss of biodiversity
- Effects of natural disasters such as floods, droughts, forest fires, and volcanic eruptions
- Wide-ranging impacts of pollution, from depletion of the ozone layer to tracing oil spills and photochemical smog
- War-torn regions and the environmental impacts of armed conflicts

Positioning and Navigation

The uses of satellites for localization and navigation activities are rapidly expanding. The implementation of satellite positioning constellations have created a growing number of applications such as air transport, maritime transport, land transport, and localization of isolated individuals, and provides a universal referential time and location standard for a number of systems. Satellite navigation is achieved by using a global network of satellites that transmit radio signals from approximately 11,000 miles in high Earth orbit.

The technology is accurate enough to pinpoint locations anywhere in the world, 24 hours a day, and can operate in any weather. These constellations of satellites are referred to as global navigation satellite systems (GNSSs). There are currently two GNSS systems in operation, the navigation satellite timing and ranging system (NAVSTAR), commonly referred to as the global positioning system (GPS) owned by the United States of America, and GLONASS (Global'naya Navigatsivannaya Sputnikovaya Sistema) of the Russian Federation. A third system called GALILEO is under development by the European Community (EC) countries, which will be interoperable with the existing systems. The United States and Russia have offered the international communities free use of their respective systems.

The business model for GALILEO will be similar to GPS for basic users; however, not all applications will be free, as some applications that require a high quality of service will incur a charge. GNSS is revolutionizing and revitalizing a variety of application markets such as aviation, maritime, land transportation, mapping and surveying, precision agriculture, power and telecommunications, urban gaming, and disaster monitoring.

Micro Gravity Research and Space Tourism

The initial stage of micro gravity activities is currently under way at the International Space Station (ISS). If the costs of space flights fall significantly below $1 million per passenger, a growing number of companies will find it profitable to finance commercial activities in space (Collins, 1990). Initially, these will potentially include high-technology activities such as microgravity research, advanced materials development, bio-technology, and solid-state physics research. Some researchers believe that the demand for launches could exceed 100 passengers per day from the Pacific Rim alone within 20 years (Yamanaka & Nagatomo, 1986).

Involvement to date by these companies in the space industry has been restricted to the use of telecommunications satellites for transmission and broadcasting of programming material, including concerts, sports events, films, and television programmers[6]. When the launch of personnel into orbit is relatively low-cost, safe, and routine, it is likely they will make extensive use of the unique possibilities of zero gravity to record programming material in orbit. As the cost of a flight to orbit falls below $100,000 per passenger, the demand from individuals for recreational space travel is expected to grow rapidly. Several unique attractions (Collins, 1990) of a short orbital visit include: observation of the Earth, astronomical observation, zero gravity phenomena, zero gravity flying, and zero gravity water sports.

Synopsis of Chapters

This book contains work from scientists, educators, lawyers, and policy analysts from 12 countries. The book is divided into three sections, each containing between three and four chapters. The sections reflect the current trends that are emerging from the space arena.

- Section I: Space Technology for Managing Resources
- Section II: Satellite Internet and Navigational Technologies.
- Section III: Space Policy and Economics
- Section IV: Space and Society

The first section, "Space Technology for Managing Resources" contains four chapters.
The first chapter in this section is titled "Space Technologies for Research of Effective Water Management: A Case Study," written by Angie Bukley and Dr. Olga Zhdanovich. This chapter summarizes findings from a project called *STREAM*, which stands for Space Technologies for the Research of Effective wAter Management. The team project focused on the importance of fresh water resource management and its impact on the surrounding communities. A real world case study of Australia's Murray-Darling Basin (MDB) was performed. Based on the results of the case study, the team then extrapolated their results to other regions of the globe that are experiencing challenges to their fresh water supply.

The second chapter in this section is titled "Using Space Technology for Natural Resource Management," by N. Raghavendra Rao. This chapter recommends a model for the creation of knowledge-based systems for natural resources management. It also describes techniques for utilizing unique capabilities of remote sensing satellites for conserving natural resources and managing natural disasters. This chapter was written exclusively for the people who are not familiar with the technology and who are given the task of framing policies.

The third chapter in this section is titled "Using Space Technology for Disaster Monitoring, Mitigation, and Damage Assessment," by Pasquale Pace, Gianluca Aloi, and Luigi Boccia, from Italy. This chapter presents information on how space technologies and satellite applications can mitigate the impact of natural and man-made disasters. The objective is to provide the reader with an overview of the most important space technologies for both monitoring and telecommunications, and to highlight the main issues in managing a disaster response.

The final chapter in this section is titled "Cospas-Sarsat Satellite System for Search and Rescue," by James V. King. This chapter outlines the development and evolution of the Cospas-Sarsat system, describes the principle of operation, presents the current status, and looks at the future of the system. Cospas-Sarsat, an international satellite system for search and rescue, started operating in 1982 and has been credited with saving many thousands of lives since then. More than a million aviators, mariners, and land users worldwide are equipped with Cospas-Sarsat distress beacons that could help save their lives in emergency situations anywhere in the world.

The second section of this book contains a combination of social and technical chapters that investigate using next generation satellite more effectively to deliver Internet and location services. This section is called "Satellite Internet and Navigational Technologies." The first chapter in this section is titled "Global Navigation Satellite Systems and Services," by Justo Alcázar Díaz and Tirso Velasco. This chapter provides a detailed analysis of global navigation satellite systems (GNSS), which refers collectively to the worldwide positioning, navigation, and timing (PNT) determination capabilities available from satellite constellations. This chapter provide a summary of the future and existing systems along with the business models, limitations, and capabilities.

The second chapter in this section is titled "The Satellite Internet: The Convergence of Communication and Data Networks," written by Agnieszka Chodorek and Robert R. Chodorek. This chapter investigates the development of satellite communication and satellite data networks. A brief overview of the Internet technology, along with a convergence of the Internet and space technologies, is presented. This chapter discusses the importance of broadband Internet, which is available anywhere—in almost all locations on the Earth—and for millions people.

The third chapter in this section is titled "The Era of Nanosatellites: Pehuensat Development Status," authored by Juan Jorge Quiroga, Roberto Fernández, and Jorge Lassig. This chapter describes the technical elements of the development of a nanosatelite by an educational institution in Argentina.

The final chapter in this section is titled "Digital Bridges: Extending ICT to Rural Communities Using Space Technologies," by Phillip Olla. This chapter describes how a satellite technology can be used to bridge the digital divide. This chapter proposes a two-phased approach. The first phase involves the definition and implementation of the satellite global infrastructure to provide connectivity to underserved regions. The second phase introduces the concept of a coalition of space Internet providers (COSIP) model. The aim of this model is to encourage the diffusion of space technology infrastructure at the grassroots level.

The first chapter in the "Space Policy and Economics" section was written by Lesley Jane Smith and Kay-Uwe Hörl. The chapter is titled "Constructing the European Space Policy: Past, Present, and Future." This chapter assesses today's European space landscape, with a focused view on the evolution of the European Space Policy as well as on the ongoing and projected space applications. This chapter also sheds some light on the historical parameters that framed Europe's first steps into the space arena, and examines the current European motivations to implement policy and develop space programs in a particular way.

The next chapter in this section is titled "Application of Satellite Earth Observation for Improving the Implementation of Multilateral Environmental Agreements," written by Ikuko Kuriyama from Japan's Aerospace Exploration Agency (JAXA). This chapter provides a broad overview of the issues related to the application of EO for effective implementation of multilateral environmental agreements (MEA). The chapter reviews the characteristics of EO technology, and examines its potential to support effective implementation of and compliance with MEAs. The chapter proposes possible solutions and future steps to facilitate actual application of EO in the MEA process, and provides an insight on this issue for readers in the IT community.

The third chapter in this section is titled "Extraterrestrial Space Regimes and Macroprojects: A Review of Socioeconomic and Political Issues," by Dimitris J. Kraniou. The author examines macroprojects to be deployed in outer space. A feasibility study is being used to analyze the deployment of such projects in extraterrestrial realms.

The fourth chapter in this section is titled "Commercialisation of Space Technology for Tomorrow's Space Missions," by Stella Tkatchova and Michel van Pelt. The chapter discusses the direct and indirect benefits for the different stakeholders from commercialization of space technologies for future moon and Mars missions. These include NASA's Moon and Mars vision and the ESA Aurora program.

The final chapter in this section, by Larry J. Paxton, is titled "Challenges in Knowledge Management: Maintaining Capabilities Through Innovative Space Missions." This chapter discusses how organizations improve and maintain performance by generating, maintaining, and sharing knowledge. This chapter uses NASA as an example. High tech organizations are much more dependent on knowledge as a commodity than those in the manufacturing sector. NASA certainly is the epitome of a high tech organization. It faces complex and deep challenges, not the least of which is how to address the loss of knowledge as the workforce ages and retires.

The first contribution in the "Space and Society" section is by Jacques Arnould, from the French Space Agency (CNES). This chapter asks some critical questions of the space industry, which must be addressed at some point in the future. These critical questions include who will dispose of satellites at the end of their life? What resources should governments be prepared to commit to tracking such debris, for the benefit of commercial business? What is the use of satellite-based observation, communication, and positioning systems? How will the safety of private crews be guaranteed? If they are no longer "envoys of humanity" in the accepted modern legal sense, what duties will governments have toward them on Earth and in space?

The second chapter is titled "Commerce in Space: Aspects of Space Tourism," by Robert A. Goehlich. Space tourism is the term broadly applied to the concept of paying customers traveling beyond Earth's atmosphere. Operating reusable launch vehicles might be a first step to realize mass space tourism. Thus, the aim of this chapter is to investigate any potential hurdles and other aspects of importance to space tourism flights by using reusable launch vehicles. The primary ones are social issues, for example, "Is space tourism acceptable concerning ethical aspects?," institutional issues, for example, "Is environmental pollution caused by space tourism harmful compared to other emission sources?," and financial issues, for example, "Are there any potential investors interested in space tourism?"

The third chapter in this section was written by Paul E. Nelson and is titled "Space Elevator: Generating Interest in the Future of Space Access." The first part of this chapter describes how the space elevator is expected to work, along with the benefits over using primarily rockets. A compendium of information from a variety of sources is included in order to explain how the space elevator would be designed and constructed, and how it could solve the problems of transporting cargo into space easily, cheaply, and frequently.

The final chapter in this section is "Commerce in Space: Infrastructures, Technologies, and Applications," by Dr. Chris Robinson. This chapter attempts to track the history and meaning of space art. Space Art holds such widely varied meanings for differing constituencies and, compared with other disciplines, has diverse participation, but little formal history in space development. There are two major constituencies that are worlds apart, the usually consistent formal and popular visual documentation of the development of space and the intermittent and reluctant interest of the fine or academic arts.

CONCLUSION

Over the past few years, space business has seen substantial investment in the established space infrastructure, with technological improvements to launch equipment, satellites, and user devices. This is creating an increase in the capabilities of downstream market applications and is lowering the cost to use space infrastructure. Over the next decade, significant technological and policy advancements are planned to each of the five space infrastructures, instigating new opportunities for information technology.

This book will explore some of these new opportunities and highlight the impact that the data from space infrastructure is having on our society. Due to our great space programs around the world, we posses the ability to predict the future; however, what we seem to lack is the capability to plan accordingly. There seems to be an endless array of challenges ahead that threaten the existence of human race as we know it today, yet space technology has been shown to hold the key to combating most of the challenges listed above. The true value of the space infrastructure on the environment is the ability to provide us with the information we need to plan for the future. If we are unable to process this information or incapable of understanding the vast amount of information, or we simply choose to ignore the information, then eventually we will deservedly share a similar fate as the dinosaurs. We hope you enjoy reading these diverse and interesting chapters as much as we enjoyed pulling them together.

REFERENCES

Cohendet, P. (2004). *Space systems to bridge the digital divide: A European socio-economic perspective.* Paper presented at the 55th International Astronautical Congress 2004, Vancouver, Canada.

Collins, P. Q. (1990). The coming space industry revolution and its potential global impact. *Journal of Space Technology and Science, 6*(2), 21-33.

Hukill, M., Ono, R., & Vallath, C. (Eds.). (2000). *Electronic communication convergence: Policy challenges in Asia*. London.

OECD. (2006). *The commercialisation of space and the development of space infrastructure.* Organisation for Economic Co-operation and Development. Retrieved February 15, 2007, from http://www.oecd.org/document/20/0,2340,en_2649_34815_1837844_1_1_1_1,00.html

Pelton, J. N. (2000). *The next billion years and the significance of the emerging global brain.* Retrieved February 15, 2007, from http://www.clarkefoundation.org/archives/2000.php

World-Bank. (2006). *World Bank figures on poverty.* Retrieved February 15, 2007, from http://web.worldbank.org/WBSITE/EXTERNAL/NEWS/0,,contentMDK:20040961~pagePK:64257043~piPK:437376~theSitePK:4607,00.html

Yamanaka, T., & Nagatomo, M. (1986). Spaceports and new industrialized areas in the Pacific basin. *Space Police, 2*(4), 342-354.

Phillip Olla, Editor
Madonna University, USA
June 2007

Acknowledgment

First of all, I would like to thank all the chapter authors for their excellent contributions. I would also like to recognize the publisher for the opportunity to publish this book. The editorial assistance provided was excellent. The staff at IGI Global, especially Meg Stocking and Kristin Roth were excellent in keeping the project on track and helping out with the difficult issues.

Next, I would like to acknowledge the role of the reviewers and editorial team Ish Jumani, Claire Jolly, Rachel Demetrio, Eric Williams, and Janet Kotecha for their continuous support. I have also been very fortunate to have the excellent support from HUDDSNET group. Without them, the world would be a lonely place.

Finally, I would like to thank my wife. This project would not have happened without the persistent work of the editorial assistant Venus Olla. Venus was instrumental in proofreading and pulling the work together.

... *For the strength of the pack is the wolf, and the strength of the wolf is the pack.* — Rudyard Kipling

Phillip Olla, Editor
Madonna University, USA
June 2007

Section I
Space Technology for Managing Resources

Chapter I
Space Technologies for the Research of Effective Water Management: A Case Study

Angie Bukley
Ohio University, USA

Olga Zhdanovich
Analytical Centre for Economics & Management in Aerospace Technologies, Russia

ABSTRACT

This chapter summarizes the collective work of a team of students who participated in the 2004 International Space University summer session program in Adelaide, Australia. The project is called space technologies for the research of effective water management (STREAM). The work represented in this chapter was accomplished as part of the intensive space studies curriculum offered during the summer session. The team project focused on the importance of fresh water resource management and its impact on the surrounding communities. The team explored various space technologies and their current and future potential to enhance water resource management. A real world case study of Australia's Murray-Darling basin (MDB) was performed to provide the central focus of the project. Based on the results of the case study, the team then extrapolated their results to other regions of the globe that are experiencing challenges to their fresh water supply. A significant space technology recommendation developed by the STREAM project team was to improve the soil moisture measurement capabilities in the MDB. The primary goal of the STREAM project team is that the recommendations outlined in the extensive final report (STREAM Team, 2004) will receive full attention from policy makers concerned with the water issues surrounding the MDB.

BACKGROUND

Fresh water is one of the most valuable resources on the planet. Although three quarters of the Earth's surface is covered by water, less than 1% of the water is suitable for meeting human needs (Young, Dooge, & Rodda, 1994). Some parts of the globe suffer from droughts or the inaccessibility of fresh drinkable water, while torrential floods plague other parts of the world. The demand for fresh water does not end simply with human consumption. Every facet of our lives depends on the adequate supply of fresh water, from irrigating agricultural lands to raising livestock, manufacturing commercial goods, and preserving the health of our wetlands. To manage and maintain an equitable distribution of fresh water for all parties concerned, it is necessary to understand the flow of water through various tributaries and the impacts on the environment and various communities when the movements are altered or, in some cases, cut off. Space remote sensing technology offers scientists and environmentalists an avenue to see an area in its entirety. Aside from spatial information, other space technologies can offer temporal and spectral data of the land, sea, and atmosphere.

As the world population grows, both competition for available fresh water resources and degradation of those resources is increasing. The effective management of water collection and distribution is essential for the sustainable development of populated areas, particularly in Australia, one of the driest continents in the world. The available supply of fresh water must be balanced with demand in order to properly manage the equitable distribution of water for all those who need it.

Space offers an ideal vantage point for synoptic hydrological and climate studies, especially if coverage over large areas is required over an extended period of time. However, to date, the most widely used remote sensing tools are not space borne, but airborne sensors and ground networks, especially for studies on regional and local scales (Young et al., 1994). Consequently, proponents of space-based remote sensing must focus on providing an end-to-end strategy to provide products that meet the needs of the user community.

The MDB covers roughly one seventh of Australia's land mass and is home to approximately 10% of its population (Oliver, 2003). The basin contributes between 30 and 40% of Australia's total production from resource-based industries, and it generates about 50% of the nation's gross value of agricultural production. This number reflects the importance of the MDB system in Australia. Growing water-related problems not only have severe affects on the MDB ecosystem but on the lives of all the Australians who depend on the MDB for their sustenance or their livelihood.

Of all the parameters that are measured within the MDB, soil moisture is currently the only one that is not readily available (Murray-Darling Basin Commission, 2004). To better understand the MDB water cycle, more soil moisture data with increased accuracy and reliability is required. Soil moisture is a required parameter both for the study of the global water cycle using general circulation models, and for water management issues like floods and drought predictions.

THE GLOBAL WATER CYCLE

The global water cycle (GWC) is a continuous renewal process that recycles and circulates the water on the planet. It is also the principal mechanism whereby fresh water is produced and distributed to different ecosystems around the world.

Fundamentally, the GWC involves the movement and transformation of water through the processes of evaporation (transpiration), condensation, and precipitation. The evaporation of water takes place primarily from the surface of the ocean, while transpiration is the transfer of mois-

Figure 1. The water cycle (Courtesy of U.S. Department of the Interior)

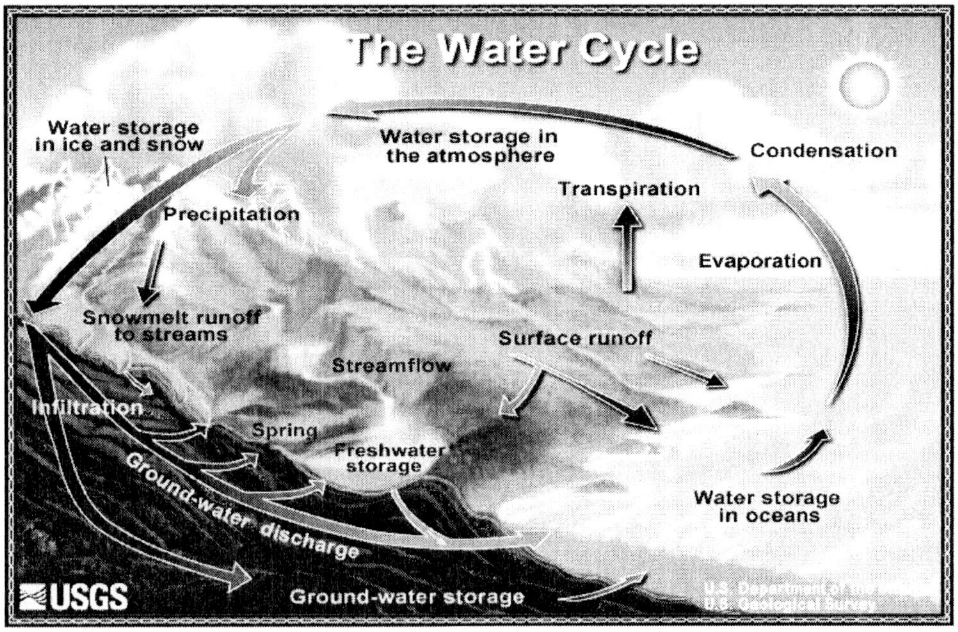

ture to the atmosphere by plants. Transpiration is required to cool the leaves and bring nutrients to different parts of the plant. The energy needed for evaporation originates from the sun. Incoming solar radiation is absorbed by the Earth and converted into heat energy. Evaporation is a cooling process; therefore, heat energy is absorbed during the evaporation of water. As the moist air rises, it begins to cool. The humid air, along with the stored heat energy, can be transported to different parts of the world via wind movements.

When the air reaches a certain altitude, the moisture begins to condense into water vapor and forms clouds. The water droplets continue to increase in size as more and more moisture condenses. Because condensation is a warming process, the stored heat energy is released into the atmosphere. When the atmosphere can no longer support the weight of the water vapor, precipitation occurs in the form of rain or snow. Because evaporation, condensation, and wind movement alter the heat distributions of the planet, the GWC also regulates the heat energy around the Earth and helps balance the overall heat distribution around the planet. When the precipitation reaches the Earth, a portion of it is absorbed into the ground and accumulates into underground fresh water or deep aquifers. Another portion evaporates back into the atmosphere. The remaining water forms overland runoff that enters streams, rivers, and lakes and eventually returns back to the ocean.

REMOTE SENSING AND THE GLOBAL WATER CYCLE

Space-based instruments can provide a global view of general activity in the atmosphere as well as on the surfaces of both landmasses and bodies of water. However, many characteristics of the water cycle are difficult to derive from space-based remote sensing data alone. To obtain the full picture of the water cycle processes, data from satellites must be fused with data obtained from airborne and ground-based sensors as well as with data from in-situ measurements. Using multisource data is a key issue in hydrological data analysis as the information provided by each single sensor might be incomplete, inconsistent, or imprecise because of limitations in data collection, storage, and manipulation. Efficient data processing and knowledge discovery methods have become an important issue for the remote sensing community. The use of data processing methods and geographical information systems (GIS) has greatly improved the accessibility and ease of analysis of remote sensing data in many different application contexts, providing the end-users with an effective tool for water management.

The development of technology should enable the use of satellite data for monitoring smaller water bodies and other parameters of the water cycle, such as water quality, salinity, and soil moisture. Of these, soil moisture is particularly important in helping determine global and local water balances.

Space Technology for Monitoring Soil Moisture

Australia is a dry continent that faces drought regularly. This happens when the moisture stored in the soil falls below an adequate limit. As the climate changes, uncertainties in rainfall patterns increase. Australia is likely to become even drier as global warming continues. All fresh water comes from precipitation and groundwater. However, the drought that plagues the MDB is a multifaceted problem. Rainfall, evapotranspiration, and runoff leave water in the soil available for plants. Even now the hydrologic balance of the MDB is being altered as a result of extensive land degradation, which has been caused by clearing the land for raising crops and grazing cattle. The use of ground water for irrigation has further exacerbated the problem. Other issues affected by soil moisture levels are crop forecasting, flooding, and fire risk.

Current space-based systems provide good estimations of the surface soil wetness. Nevertheless, they are not capable of performing a direct measurement of the moisture through the soil profile below the thin surface layer. Current practice is to derive soil moisture using correlation techniques applied to a variety of measurements from all forms of soil moisture space remote sensors.

Current space remote sensing assets that may be used for soil moisture measurements include:

- **Precipitation radar (PR):** This radar is used primarily for detecting rainfall in the tropics, but its data may be applicable to soil moisture estimation as well. Current research is aimed at developing methods to derive soil moisture estimates using PR data.
- **Passive microwave radiometers:** These instruments detect microwave emissions from the Earth's surface and atmosphere. Algorithms for determining soil moisture from this data are being developed.
- **Synthetic aperture radar (SAR):** This is a radar technique used to detect small changes in topography, but can also be used to monitor deforestation and surface hydrological states. Raw measurements do not correspond directly to soil moisture, and therefore processing the SAR data with other variables is required.

Space technology can also be used to gather ground based measurement data by employing telemetry. To determine soil moisture, ground based monitoring is essential to ascertain the moisture profile in the various soil layers. Ground probes are easy to implement with the added benefit of being low cost. Ground probe measurements can be linked by communication satellites to a data processing center, where they can then be used in conjunction with other remote sensing information to derive soil moisture estimates. Obtaining accurate and reliable measurements of soil moisture with considerable resolution will improve the understanding of the MDB water cycle and enable water managers to enact policies for promoting improvement of water distribution and application.

To create effective models for the MDB, the quantification of certain parameters needs to be improved. In the future, sensors on the following satellites can provide soil moisture measurements to Australia, with varying degrees of accuracy and resolution:

- **The soil moisture and ocean salinity (SMOS)** satellite will be launched in 2007 by the European Space Agency (ESA). SMOS will map soil moisture and ocean salinity, which are two crucial variables for understanding weather and climate. It will also monitor the water content in vegetation, snow, and ice cover.
- **The hydrosphere state (HYDROS)** mission will be launched between 2009 and 2010 by NASA. The main objective is to monitor daily soil moisture and freeze-thaw cycles globally from space. The HYDROS remote sensing data will be used for weather and climate prediction, hydrosphere modeling, and water resource availability prediction. The long-term goal of HYDROS is to enable scientists to better understand the global water, energy, and carbon cycles.
- **The advanced land observing satellite (ALOS)** was launched by JAXA in January of 2006. The estimation of soil moisture distribution is one of the general goals for the science program of ALOS, which has a phased array type L-band synthetic aperture radar (PALSAR) as one of its three land observing sensors.

Soil Moisture Monitoring for the Murray-Darling Basin

Soil moisture studies have not received much attention in the MDB as of yet. Some direct measurements using ground probes have been made. Indirect solutions have also been developed, using other variables (remote sensing or ground data) and correlation algorithms between these variables (temperature and precipitation) and soil moisture. But these solutions remain inaccurate because the sparse number of ground stations, requiring interpolation of data, coupled with the fact that cloud cover affects the remote sensing of surface temperature by the advanced very high resolution radiometer (AVHRR). To improve the understanding of the MDB water cycle, more soil moisture data with increased accuracy and reliability are required. Measurements of soil moisture are important both for the study of the global water cycle using general circulation models and for more local water management issues like flood and drought prediction. Soil moisture measurements should be acquired:

- Over large areas (the whole MDB)
- Over long periods of time (continuous information), which will necessitate effective data storage and management strategies
- With a high level of repetitiveness for change analysis and variability studies

Soil moisture information derived from these data should still be made available to users, not only to scientists doing research studies but also to policy makers and any water-dependent users. Obviously, scientists and policy makers do not need the same level of information detail. While scientists may use remote sensing data or ground data of several variables, policy makers and other nonscientists users require processed data and integrated information readily available in user-friendly form.

Direct measurements of soil moisture are available, but the ground-based methods used to collect these data, which are based on contact measurement using the thermostat method, do not allow the accumulation of information on spatial distribution of soil moisture over the measured area. Thus, the development of methods for the measurement of soil moisture is based on an integrated approach using ground observation and remotely sensed data is required. The number of ground stations distributed throughout the MDB is indeed sparse and assumptions are used to interpolate information in the regions without stations. An error of at least 15% is seen when determining soil moisture, even when highly sophisticated models are used. Since 1980, there have been approximately 80 meteorological ground stations recording data that measure air temperature and daily rainfall. Of these 80 sites, 50 have complete data and can be used with confidence. The other 30 miss numerous periods.

Satellite data in the MDB are retrieved from the AVHRR sensor. The five bands of the AVHRR measure visible, near infrared, and three thermal bands of radiance. This allows the average radiant temperature of the MDB's surface over approximately 1 km pixel to be estimated at basically regular day and night passes. AVHRR data are archived in real time using the Australian standard data archive (ASDA) format developed by the Bureau of Meteorology and the Commonwealth Scientific and Research Organization (CSIRO) and are also split into their components (Australian Government, 2004). The AVHRR digital data acquired by the US National Oceanographic and Atmospheric Administration (NOAA) satellites are transmitted to ground stations and can be used to reconstitute an image of the Earth's surface not too dissimilar from an aerial photograph. Data are received through the Australian Center for Remote Sensing (ACRES) antennae at the data acquisition facilities at Alice Springs, enabling coverage of the Australian landmass. The recorded data are sent via a high-speed communication link daily to the data processing facility in Canberra where they are catalogued and archived for 7 days. Data are recorded and archived by orbit or swath. If the region of interest extends across two adjacent paths, the appropriate data set from each path will need to be extracted. Because of the orbital variables of the satellite, adjacent paths are not acquired sequentially and information from each path may differ according to the time delay between the two passes. The data transmitted to Earth from an Earth observation satellite are in a form unsuitable for use by customers, so ACRES processes this raw data in varying degrees to produce products suitable for use by clients. ACRES processes AVHRR data within 12 hours of acquisition and places it on the ACRES Web pages for ftp download for a limited time. Historical AVHRR data may be accessed from the CSIRO Earth Observation Center (Australian Government, 2004).

Data are processed into man-computer interactive data access system (McIDAS) area files (up to about 120 Mbytes per orbit). Data files may be downloaded from Geoscience Australia's Web site at http://www.ga.gov.au/acres/prod_ser/noaaprice.htm. The data are subject to Copyright and a license agreement is required. Data are stored and available in digital satellite imagery (Australian Government, 2004).

Integration of various data, including satellite data, has only been possible since 1986, when the first AVHRR sensor flew, and a series of issues are still present. First, the cloud cover determines the accuracy of the AVHRR data over the sensed area. The varying cloud cover results in a temporal gap in the series of the surface temperature images. Furthermore, the number of ground meteorological stations is rare and the distribution over the territory is irregular. This means that interpolations are necessary in order to retrieve data for the locations where there are no ground stations. The range of observations at the ground stations is also comparatively short and heterogeneous in respect to the frequency of the remote sensing data.

STREAM SOLUTION FOR SOIL MOISTURE STUDIES IN THE MDB

A 5-year soil moisture management strategy for the MDB Commission (MDBC), which builds on existing practices within the MDB, is proposed. The objective of this strategy is to improve integrated water management through system-wide analysis of soil moisture data in the MDB. This strategy consists of two parts:

1. Develop a hybrid system of space, airborne, and ground-based sensor systems for soil moisture, and
2. Establish a central processing center for matching data sets to better support agencies involved in soil moisture and related studies and water management decision-making.

Figure 2 provides a visual impression of the hybrid solution (space, airborne, and ground sensors) proposed by the STREAM team for gathering soil moisture data in the MDB.

The hybrid solution consists of installing many additional remote ground sensors and processing the data obtained along with space-based and airborne data at the central data processing center. The ground probes can be installed in the short term. These sensors are low cost, meaning that this step can be easily implemented. The ground probes, linked to a central office, possibly by communication satellites, will provide reliable data of the ground moisture at multiple depths and in real-time. The overall quality of soil moisture data will be improved with the use of in-situ, space-based, and other remote sensing or ground-measured data. As the quality and quantity of the data improve, the development of operational data post-processing techniques also improves. In addition, an increase in space system-derived data is expected with the operation of SMOS, HYDROS, ALOS, and other space sensor satellite programs within the next 5 years. The central data processing centre needs to be enhanced to handle the constantly increasing data volume.

Figure 2. Soil moisture data-gathering elements in the MDB

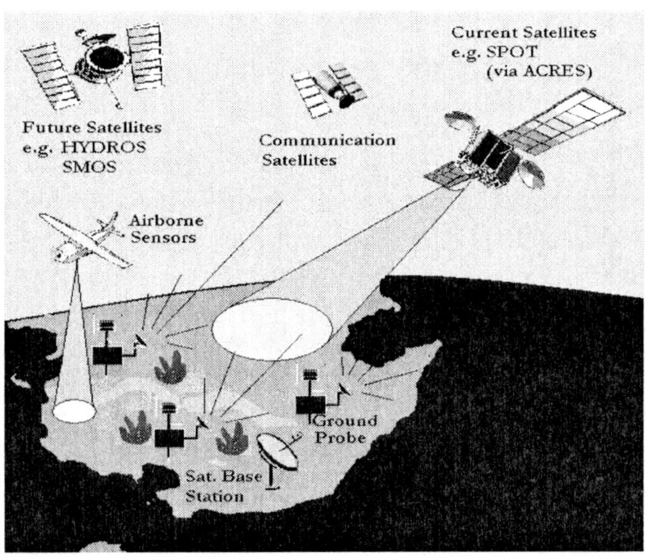

The combination of the two essential components of the strategy mentioned above enables the development of an integrated and enhanced soil moisture data flow to the end user. Processing and analyzing of the space-based data with other information must also be a high priority. A conceptual approach suggested by the STREAM team is detailed in Figure 3, which builds on the existing arrangements and practices of the MDBC and other organizations involved. The figure shows three distinctive functional levels as follows:

- **Source level:** Covers data sourcing, from space-based remote sensors to meteorological and other sensors; for soil moisture, satellite data could be supplemented with a telemetric ground system, which will enable a hybrid of ground- and space-based data.
- **Processing level:** An organizational construct achieving an enhanced means to associate or match all available records soon after the time of observation; involves library function for long-term referrals and some standard processing functions.
- **Outputs level:** Capability to provide standard reports and requested data sets to customers; two prime customers are envisaged: water managers (policy makers and water-dependent users) and scientists (who have research objectives)

Data Source Level

Currently, the MDBC and its agencies receive some remote sensing imagery and metrological data on a routine basis. This report suggests that the MDBC might wish to establish a supplementary data gathering system, via in-ground remotely reporting soil moisture sensors, as shown in Figure 3. The probes would be positioned at selected sites of interest, for example, where a long-term study of areas adjacent to wetlands might be required. The STREAM concept is to use in-situ data to complement the space-based sensor image data, to arrive at a comprehensive understanding of the subject environment. This multisensor system (space-based, airborne, and ground sensors) is a STREAM hybrid approach.

As an example, the Australian company Measurement Engineering Australia (MEA) supports remote monitoring or telemetry for in-situ weather stations and soil moisture stations (both for water content and soil water instruments). The system uses a short-haul radio link connection (up to 5 km) to bring field soil moisture and climate sensor information back to a central data logger. Remote access to the data logger is then possible via cell phone or landline connections.

In the near future, the SMOS and HYDROS missions will deliver global coverage images, which can be used for the MDB. Furthermore, SMOS will permit scientists to measure evapotranspiration. The repetition period of these satellites is between 2 and 5 days. Scientists would be able to retrieve soil moisture content and evapotranspiration data if the rainfall information is available. The success rate of the measurements performed by HYDROS and SMOS will be close to 100% because cloud cover does not affect the instruments. The instruments of these two missions can determine the soil moisture up to a depth of five centimeters.

Data Processing Level

The MDBC currently has arrangements in place to receive and store satellite images. The STREAM Soil Moisture Management Strategy develops a library of consistent data sources. It is suggested that the central library center might be undertaken as a function of CSIRO. The center would match satellite data records with other data records from a similar time of observation. This library would significantly enhance future historical trend analysis activities. Some standard post-processing of data would also be applied at the library center, including processing through basic models.

Figure 3. Conceptual arrangement: Enhanced organization for data flow and processing

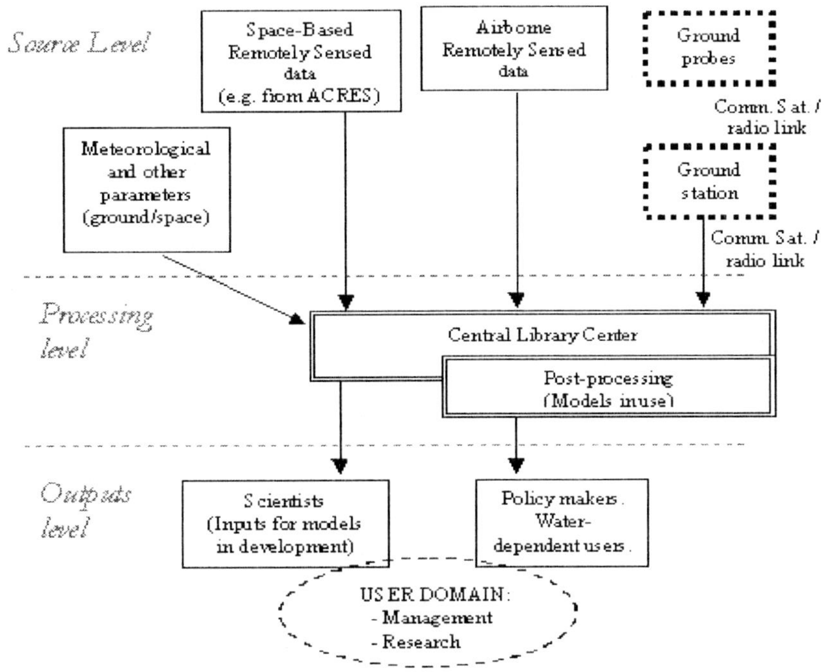

The library center would also produce standard reports, including those associated with image change detection techniques.

Data Output Level

Within the soil moisture management strategy, the outputs level is designed to specifically meet basic requirements for the major user groups (policy makers, scientists, irrigators, farmers, and urban users). The water managers are likely to require automatically generated reports, while the experts may require raw data sets, in particular, formats. The scientific community undertakes development of new models. Overall, scientists and experts should be able to more easily identify, obtain, and use required data sets for their research.

User Domain Data Processing and Integration

With the existing and coming technology, the trend for obtaining good and useful data is increasing. The end user receives data from a library; the library may provide some small-scale data processing. The majority of available data will need to be further integrated in the user domain. This will be feasible thanks to:

- The development of new tools, that is, GIS and data mining
- Observational techniques with new opportunities to collect and transmit data from in-situ systems
- The better understanding of specific processes, thanks to field studies

To achieve suitable results in integrated soil moisture management, better communication between the users and suppliers of soil moisture data is necessary. First, the users have to clearly define their needs. It is also important to have a well-defined strategy for observations with well-planned in-situ, airborne, and space-based technology, based on an established cooperation between the users and the airborne and space missions' planning groups. Today, there is a vital need for information that will facilitate the mapping of soil moisture in the MDB. There is also a necessity for a comprehensive database that provides historical data as well as mathematical models that forecast the soil moisture in the Murray-Darling Basin.

Soil Moisture Data Management Strategy for the MDB

The STREAM report suggests a management strategy for enhancing the gathering and processing of soil moisture data for consideration by the MDBC. The basic steps for a 5-year program are:

- Confirm strategy and areas of interest through a feasibility study (3 month activity)
- Establish ground probe systems over small regions every year for a period of 5 years
- Prepare for, and begin, receipt of data from the SMOS and HYDROS research satellites (liaison activities with Australian scientists commence in 2006)
- Establish additional personnel and capabilities to assist with data matching and correlation between space and ground data sets, and to carry out cataloguing, storage, and distribution functions in the central library center (2005)
- Contribute data handling concepts and modeling developments to the global scientific community through partnerships with countries associated with delivering space-based data (2006-2010)
- Publish annually a water management plan for areas of high interest (for the hybrid sensor program)

Areas of land adjacent to major wetlands were selected as the subject of a cost analysis for the soil moisture management strategy. These areas are under consideration for controlled flooding by the MDBC. The wetlands require a full flood cycle about every 10 years for regeneration processes, and it is important to determine how much water should be released in the most efficient and effective manner for wetlands regeneration. The soil moisture of the areas that are close to the wetlands must be studied in order to develop an effective flood plan. As the scientific and management strategies for these regions mature, a master wetlands restoration controlled flood plan for effective water management in these areas can be developed.

Soil Moisture Studies Summary

Soil moisture has been identified as a key parameter both for improving understanding of the water cycle and for improving operational monitoring and prediction techniques for water management. Soil moisture has not, however, been adequately addressed in the MDB as yet, mainly because most current, comprehensive, direct measurement systems are expensive.

The STREAM study proposes a soil moisture management strategy for the MDB. The strategy consists of two parts: a hybrid system of space, airborne, and ground sensors, and a central library center.

The planned SMOS and HYDROS remote sensing missions (scheduled for launch in 2007 and 2009, respectively) will provide integrated solutions for soil moisture monitoring. In the meantime, this report emphasizes the use of ground probes as a short-term, low cost, and easily implemented solution. These probes, linked to a central library center, possibly by communication satellites, will provide reliable soil moisture data at multiple depths and in real-time. These data may be used in conjunction with other remote sensing or ground-measured parameters in order to address the most urgent water management and water cycle understanding issues.

The function of the central library center would be to gather, centralize, and store data and to offer users a library of temporally matched data sets. This structure could be built on the current practices of the MDBC. The central library center would process data to provide information to end users at a level that takes into account the different requirements of each user; for example, raw data would be provided to scientists for research models, whereas policy makers would be supplied with post-processed and analyzed information. This central library center could also include data forecasting and basic modeling capabilities in order to provide direct recommendations and integrated information to policy makers and any water-dependent users.

GLOBAL RELEVANCE OF THIS STUDY

The previous sections contain discussions of potential technologies that would assist in the management of the water resources in the Australian Murray-Darling Basin. They review current and planned remote sensing solutions that measure the properties of water resources and factors that directly, or indirectly, affect them, with a focus on soil moisture content. These solutions apply not only to the MDB but to other regions across the globe as well. We now focus on providing a starting point for applications of space technologies, such as remote sensing, that could be used by national space or environmental agencies to enhance water resource management.

The following sections present a discussion of the various issues that are being experienced throughout the world. Several themes can be found throughout the text and include the impact of drought and human consumption on water resources, pollution, and increasing water salinity. The issues and solutions for each area have the potential of filling volumes. This section is intended to highlight some of the issues in water basins around the world and then recommend possible directions to take in the search for a global solution. Most importantly, this chapter recommends the elaboration of a United Nations (UN) charter for water management.

Asia

Asia contains 36% of the global water runoff and about 60% of the world population. Yet it has the lowest ranking in the world of annual water resources per capita. The most prominent of the Asian issues include flood disasters in the south and southeast and drought conditions in the remainder of the area. In addition, water bodies near cities of the region's developing countries received domestic sewage, industrial effluents, chemicals, and solid waste pollution. Basins within China represent a cross section of the problems with Asia due to the simultaneous increases in the local economy and population. China is listed by the United Nations as one of 13 countries with a water deficit; however, the variation in precipitation levels across the country means China also has frequent floods. China possesses 2,800 km^3 of water resources, ranking the sixth in the world, but the water volume per capita is only 2,300 m^3 which amounts to only a quarter of the per capita amount around the world, and water distribution country-wide is uneven. According to Chinese government statistics, it is predicted that China will require 698.8 km^3 of water in 2010, while the water supply is only 667 km^3, leaving a deficit of 31.8 km^3. Continuing and accelerating growth of population and industry over the past century has resulted in increasingly severe problems related to freshwater shortage. The situation is most serious in the northern part of China, which includes the

Yellow/Huanghe, Huaihe, and Haihe River basins, otherwise known as the 3H basin. The increasing water shortage problem in the 3H basins has been making continued growth increasingly difficult, to the extent that the government recognizes the need for prompt critical analysis to formulate an action plan.

The lack of water availability was dramatically demonstrated during the drought of 2001, where approximately 22 million people in rural areas were temporarily without an adequate drinking water supply. The Yellow River has nearly reached the level of complete water resource exploitation and other major rivers within the 3H basin are basically fully developed, leaving little or no water flowing to the sea.

Chemical and physical pollution of water bodies in Yangtze-3H River basin have also become serious concerns. The problem of the 3H basin extends to groundwater as well. A water quality assessment of 2,015 wells in the Haihe basin showed that two-thirds of the investigated wells did not meet quality standards for drinking. The pollution levels in the Yellow River are causing serious problems for water supply and the environment. Two other important factors are the unmeasured wastewater discharge from industry in rural township and village enterprises and unmeasured pollution sources from agriculture. At points in 1993, some areas in the 3H basin received water of such low quality that it could not even be used for agriculture irrigation.

Two-thirds of agriculture in China is in the north while four-fifths of the water is in the south. The uneven distribution of water resources in the Yangtze-3H basin have been illustrated by increasing periods of no water flow in the lower river while the Yangtze River basin faces increasing flood disasters and surplus rain. With frequent floods, droughts, and shortage of water resources, the water issue is vitally important to the regions sustained growth (Kim, 2003).

It has been recognized that the traditional and conventional ways of managing the water resources must be modified in order to support continuing and sustainable growth in China. At the national level, changes took place in the 1990s that began to impact water management in the Yangtze-3H basin. In 1982, the new Chinese constitution caused a shift toward legal methods for guiding action and decision-making and also reiterated state ownership of water resources. The major legal landmark was the 1988 water law, which provided the basic framework and principles for water management. China has begun to implement a cross regional canal system, known as the south-to-north water transfer project (SNWT), in an attempt to equalize the water distribution within the country. Additional concepts such as evaluating water prices, rights, and markets have also been discussed and tested within China (Giordano & Zhu, 2004).

With an independent space and communication infrastructure, China has implemented an integrated water management system focusing on the Yangtze-3H basin. As part of their policy, China has recognized the advantages of using remote sensing technology. Remote sensing is routinely applied for the measurement and investigation of surface water resources, vegetation (and corresponding ecology), salinity, regional desertification, drought and flood monitoring, and planning of water diversion projects between basins. Space remote sensing and the national water observation system in China are forcing important changes in China's agricultural economy. Clearly, there will be less water available for irrigation purposes in the future. Therefore, in north China, more dry-land crops may be planted rather than the traditional high water use crops such as corn and rice. In addition, new regulations for water distribution and water market mechanisms instead of traditional coarse water management should be used in future (Li, 2002).

Middle East

The Middle East is the area where Africa, Europe, and Asia meet. For the purposes of this discussion, the region includes countries as far west as Israel and Turkey and as east as Pakistan. This region contains its own set of issues that are different than those found in other areas of the world. The area contains four major rivers and basins, including the Tigris, Euphrates, Arru Darya, and the Indus. But while these basins are the largest, there are several smaller systems, such as the Jordan River, that also warrant attention. Because the region is mostly arid or semiarid, the primary water resource issue tends to be low water levels. The natural cycle of winter rain and summer drought results in irrigation and human consumption putting a strain on the system. This is compounded by the fact that the population in the area is growing rapidly, resulting in even higher demand and undesirable future impacts. In response, some countries within the Middle East have instituted methods of conserving water resources. For example, Israel has instituted "reclamation" projects so that water used for irrigation is actually a combination of fresh water and processed grey water (Dale, 2001). However, even programs such as this are only a start and much more drastic reductions will be required to maintain pace with the population growth.

To date, there have been no major conflicts due to the availability, or lack thereof, of water resources (Alan, 1998). Whether this condition will remain will have to be verified in the future. Political conflict is an important part of the overall status of water management within the Middle East region; however, it is outside the scope of this work and will not be discussed to any great detail. The fact that water is a scarce resource in the region, coupled with the rapid increase in population and the possibility of long term droughts means that it is crucial that water management and conservation become high priorities.

North America

North America covers a wide range in latitude and has a climate varying from arctic to tropical to desert. Most of the fresh water in the continent comes from mountains covered with dense vegetation. These areas accumulate water during wet periods or snowmelt and release it slowly during the dry season (Sedell, Bennett, Steedman, Foster, Ortuno, Campbell, & Achouri, 2002). In northern watersheds, 80% of the drinking water is derived from the Great Lakes and water shortage is not an issue. In eastern Canada, the St. Lawrence-Great Lakes Water basin suffers mostly from pollution (airborne, urban, and agricultural), erosion caused by deforestation of the riverside, and the effects of a growing population and riverside infrastructures. In central Canada, the main issues of the Mackenzie basin are related to sudden floods of water and ice that occur in the spring due to the melting of permafrost. This phenomenon changes the navigation channels significantly every year. The Fraser River in western Canada travels through the Rocky Mountains where it collects a large amount of sediments. The gravel and sand deposited at the bottom of the mountains and the rapid modification of the flood plains are a main concern for human activities. On the other hand, they are quite beneficial to the natural habitats of the local wildlife.

In the Midwestern United States, droughts are a main concern. In order to mitigate those risks, the U.S. government built a large number of dams in the Colorado and Columbia watersheds, including the famous Hoover Dam. The construction of these dams and reservoirs has changed the landscape and surrounding habitats quite significantly. The growing population in the surrounding large cities is a concern because the demand for water will only increase with time (Garono, McFall, Burke, & Sutherland, 2003).

The Mississippi water basin dominates eastern United States. The main concern in this basin relates to pollution caused by the combination of an increasing population, constant use of chemicals for agriculture, and the addition of numerous industrial complexes on the riverside. Salinity issues in areas of diversion are also a major concern (Caffrey, Coreil, & Demcheck, 2002).

In Mexico, the largest water basin is the Rio Grande, which runs along the U.S-Mexican border and empties into the Gulf of Mexico. The primary use of water from the Rio Grande is for irrigation. Droughts are a concern in this region and, because there are no major tributaries to replace the extracted water, the Rio Grande can dry up. In addition, human activities are cause for concern over water quality (i.e., salinity, nutrients, and fecal coliform bacteria) (U.S. Environmental Protection Agency, 2001).

South America

Contrary to the Murray-Darling basin in Australia, there is no shortage of water in South America. In fact, it has been estimated that this continent holds 13,400 km^3 in renewable water supplies, of which 21% flows annually in the rivers (McClain, 2002). The natural environment of South America is extremely diverse, ranging from the tropical rain forests of the Amazon basin to the heavy populated urban centers of La Plata basin and from the rich wetlands of the Pantanal to the high-altitude lakes of Bolivia and the semidesert regions of Brazil, Argentina, and Peru. Although each watershed is unique in its own right, the problems and issues associated with them are similar in some respects.

Approximately 300 million people inhabit the continent, distributed among 13 countries originally established as Spanish, Portuguese, French, Dutch, and British colonies. The potential water availability per capita in South America is estimated at 38,200 m3/yr, which represents twice that available to Australia and nearly 10 times as much than that in Asia (Shiklomanov, 2000). South American watersheds are a source for drinking water, food, transportation, and recreation, as well as waste disposal. The annual withdrawals of water from the environment were estimated to be at 167 km^3 in 1995 and are forecasted to rise to 213 km^3 by 2010 (Shiklomanov, 2000). Some methods of water withdrawal are unsustainable and, when combined with the development of the South American economy, increases in population and predicted changes in climatic conditions, the stress on the water systems of the continent will increase.

Eleven major water basins cover South America, of which the Amazon, Paraná, and Orinoco watersheds are the largest. The Paraná River drains the plains and grassland of the Gran Chaco, as well as the coastal hills of central Brazil, while the Amazon and Orinoco rivers drain into large portions of rain forest and tropical savannas. The Chubut, Colorado, and Parnaiba water basins are large geographically, draining the more arid parts of South America; however, their discharges are small when compared to other rivers in South America. Finally, the Uruguay River drains the humid plains between the Paraná and the Atlantic coast, and the Magdalena River drains the moist central valleys of the northern Andes. The Lake Titicaca basin is located at 3,800 meters above sea level and represents the only watershed on the arid Bolivian high plains.

Discussion and Recommendations

Table 1 presents a high level summary of the major water management issues for the basins studied in this section. Issues range from direct water level measurements (e.g., droughts, floods) and water quality to land related problems (e.g., erosion, soil moisture). Although the compared regions seem to differ greatly among themselves, the range of problems related to water show commonalities. As an example, water quality issues between the MDB and the Mississippi watersheds differ, but salinity issues observed in both basins show commonalities. The table also provides insight

Table 1. Summary of existing major issues within studies water basins

Water Basin	Drought	Water Level Fluctuation	River Flow Rate	Flood	Precipitation	Ice Pack Topography and Thickness	Water Quality	Fresh Water Salinity	Encroachment of Sea	Soil Moisture	Erosion	Sedimentation
	Major Water Issues											
Murray-Darling (Australia)	●	◖	◖	●	●		O	O	●	◖		◖
Yangtze-3H (Asia)	●	◖	◖	●	●	●	O			◖	●	◖
Indus River (Middle East)	●	◖	◖	●	●	●	O		●	◖		◖
Great Lakes-St. Lawrence River (North America)		◖	◖		●	●	O			◖	●	◖
Overall South America	●		◖		●	●	O			◖	●	◖

Legend:
● Issue exists / Current space technologies available for monitoring
O Issue exists / Few space technologies are available for monitoring
◖ Issue exists / Space technologies can provide some form of monitoring
Blank indicates issue is not a major concern

into the areas where space-based technologies are already employed or could play a more significant role.

Table 1 highlights some elements such as erosion, ice, flood, and precipitation monitoring that are globally important and could, or are currently, using data from current space-based earth observation missions. While some countries have already established space-based monitoring programs, other nations have not yet developed the necessary infrastructure. These programs contribute some information for water management and hydrological sciences.

- **Recommendation 1:** Implement a global data-sharing infrastructure for remote sensing data related to the water cycle.

The above recommendation can be extrapolated even further so that the information collected by space remote sensing techniques can be included in such endeavors as the global water system project (DIVERSITAS, 2004). The goal of this project is to merge data from various sources and create a global water system information base; however, the current framework has not incorporated space technologies to the fullest extent.

- **Recommendation 2:** Integrate the global remote sensing data on the water cycle with the natural and social science global information base.

Measurement of water level and flow are also very important for hydrologic modeling. This data is used in understanding and forecasting the behavior of water basins. Even though the water flow is not necessarily a water management issue in many watersheds, it is important data to have. Space technology can be used to perform water level and flow monitoring, but ground-based methods are bound to be more effective and certainly more cost beneficial.

There is also a global interest in acquiring data to monitor soil moisture. Soil moisture plays a significant role in the evaporation part of the water cycle and is very important for hydrologic and atmospheric modeling. No current satellite technology exists to provide this very important data to scientists.

- **Recommendation 3:** Invest in the development of future missions to acquire practical temporal and spatial soil moisture data.

Water quality remains a very important issue that is not being addressed fully by space-based remote sensing. As can be seen in Table 1, there is no space technology available to monitor water quality per se, but some space technology can be used to monitor some water quality-related elements like water temperature, suspended solids and phytoplankton, algae, and chlorophyll levels. The significance of water quality will grow with the increasing demand for fresh water as the global population increases. Hyperspectral remote sensing, which has the capability to detect and identify individual pollutants and sediments, represents the most promising technology for space-based water quality monitoring.

- **Recommendation 4:** Invest in the development of space-based hyperspectral technologies for future global water quality monitoring missions.

Water flow does not stop at national borders. Many watersheds span multiple countries (e.g., Rio Grande, Amazon, Indus). Issues also arise when watersheds span multiple states or provinces within the same country. In addition, considering that water could be the source of disputes and possibly conflicts, there is a need at an international level for a charter with a global focus on water management.

- **Recommendation 5:** Implement a United Nations (UN) charter of water management. This charter shall focus on the following items:
    - Harmonization of water management policies between nations
    - Provide a conduit for water related data sharing
    - Plan and launch a global water monitoring system in which the data is owned by the UN

CONCLUSION

This chapter addressed water management issues in the MDB and for various regions of the world. Although problems vary according to regional climate and geographical differences, common threads can be found where space technologies could provide significant benefits in the management of water. Among these, some elements such as erosion, ice, and flood and precipitation monitoring, have already or could be addressed by existing space-based remote sensing missions. Additional elements that could benefit global water management have been identified under five recommendations. Most importantly, we have recommended the instigation of a United Nations (UN) charter for water management.

ACKNOWLEDGMENT

The authors of this chapter would like to extend their greatest appreciation to the members of the International Space University 2004 Summer Session STREAM Team. The members of the team are D. Bédard (Canada), G. Grillmayer

(Austria), R. Riegerbauer (Sweden), M. Tetlow (Australia), M. Loweth (United Kingdom), R. Annes (Netherlands), H. Arid (Morocco), D. Bates (United States), M. Bonnet (France), Y.-A. Bordier (France), J. Cain (Canada), M. Ciotola (United States), K. Clayfield (Australia), R. Condessa (Portugal), F. Dell'Endice (Italy), U. Demšar (Slovenia), M. Ejova (Russia), M. Eskelinen (Finland), M. Falcolini (Italy), A. Galvez (Spain), J. Gill (United States), H. Goldsmith (United States), S. Gürol (Turkey), Ø. Helleren (Norway), J. Hermann (Australia), W. Kondo (Japan), W. Krauser (United States), C. Kulik (Sweden), P. Li (China), U. Livoff (Denmark), A. Nayar (Canada), S. Ng (Canada), M. Oelke (United States), R. Patricio (Portugal), S. Ruel (Canada), Y. Seki (Japan), F. Trolio (Belgium), and A. Vallgren (Sweden). Without the hard work and dedication of this team, this project would not have been realized.

REFERENCES

Allan, T. (1998). Watersheds and problemsheds: Explaining the absence of armed conflict over water in the Middle East. *MERIA Journal, 2*(1), 49-51.

Australian Government (2004). *Longest rivers: Geoscience Australia*. Retrieved February 15, 2007, from http://www.ga.gov.au/education/facts/landforms/longrive.htm

Caffrey, R. H., Coreil, P., & Demcheck, D. (2002). *Mississippi River water quality: Implications for coastal restoration*. Interpretive Topic Series on Coastal Wetland Restoration in Louisiana (National Sea Grant Library No. LSU-G-02-002).

Dale, W. N. (2001). *Middle East water problems*. Retrieved February 15, 2007, from http://www.unc.edu/depts/diplomat/archives_roll/2001_07-09/dale_water/dale_water.html

DIVERSITAS (2004, May 20). *The global water system project: Scientific framework and implementation activities*. Retrieved February 15, 2007, from http://www.gwsp.org/downloads/GWSP.gws-framework-20-05-04.pdf

Garono, J. R., McFall, K., Burke, J., & Sutherland, B. (2003) *Critical habitat on the lower Columbia River*. Geospatial Solutions. Retrieved February 15, 2007, from http://www.geospatial-online.com/geospatialsolutions/article/articleDetail.jsp?id=76575

Giordano, M., & Zhu, Z. (2004). *Water management in the Yellow River Basin: Background, current critical issues and future research needs*. Retrieved February 15, 2007, from http://www.iwmi.cgiar.org/Assessment/files/pdf/publications/ResearchReports/CARR3.pdf

Kim, H. S. (2003). *Sustainable development and the South-to-North Water Transfer Project in China*. Retrieved February 15, 2007, from http://fred.ccsu.edu:8000/archive/00000054/01/etd-2003-12.pdf

Li, J. (2002). *Application of remote sensing to water resources management in arid regions of China*. Retrieved February 15, 2007, from http://www.gisdevelopment.net/aars/acrs/2002/pos3/283.pdf

McClain, M. E. (Ed.). (2002). *The ecohydrology of South American rivers and wetlands* (IAHS Special Publication No. 6). Wallingford, Oxfordshire, UK: International Association of Hydrological Sciences.

Murray-Darling Basin Commission (2004). Retrieved February 15, 2007, from http://www.mdbc.gov.au

Oliver, M. (2003). River Murray: The big picture. River Murray Urban Users Local Action Planning Committee, 2003. Retrieved February 15, 2007, from http://www.murrayusers.sa.gov.au

Sedell, J. R., Bennett, K., Steedman, R., Foster, N., Ortuno, V., Campbell, S., & Achouri M. (2002, October). Integrated watershed management issues in North America. In *Proceedings of the 21st Session of the North American Forestry Commission, Food and Agriculture Organization of the United Nations*, Hawaii, USA.

Shiklomanov, I. A. (2000). *Appraisal and assessment of world water resources. Water International, 274*(1), 11-32.

STREAM Team (2004). *Space technologies for the research of effective water management* (ISU Team Project Report). Adelaide, Australia: International Space University.

US Environmental Protection Agency (2001, November). American heritage rivers progress report.

Young, G. J., Dooge, J. C. I., & Rodda, J.C. (1994). *Global water resource issues.* UK: Cambridge University Press.

Chapter II
Using Space Technology for Natural Resource Management

N. Raghavendra Rao
SSN School of Management & Computer Applications, India

ABSTRACT

Deliberate exploitation of natural resources and excessive use of environmentally abhorrent materials have resulted in environmental disruptions threatening the life support systems. A human centric approach of development has already damaged nature to a large extent. This has attracted the attention of environmental specialists and policy makers. It has also led to discussions at various national and international conventions. The objective of protecting natural resources cannot be achieved without the involvement of professionals from multidisciplinary areas. This chapter recommends a model for the creation of knowledge-based systems for natural resources management. Further, it describes making use of unique capabilities of remote sensing satellites for conserving natural resources and managing natural disasters. It is exclusively for the people who are not familiar with the technology and who are given the task of framing policies.

INTRODUCTION

Rapid changes in global economy have led to heavy consumption of natural resources. It is said that the global environmental crisis is due to the mismanagement of natural resources. A human centric approach of economic development has damaged the natural resources to a considerable extent. While explaining eco-management Chatterji (2002, p. 9) says: "Change and development are essential to human progress and exploiting nature in this process is not a new phenomenon. In this march of progress, we have over exploited nature and caused harm to its environmental quality."

The problems related to environmental issues have become the main theme at seminars and conventions organized by environmental specialists. The topics such as "sustainable development," "global warming," and "disaster management" are being discussed at these seminars and conventions. However, it has been said that development is also possible by making peace with nature. This approach is known as "sustainable development." Nonadherence to this approach results in climatic variability and unpredictability. While explaining why we should be concerned about global warming, Sir John Houghton (2004, p. 200) says:

Firstly, there is our responsibility to future generations. It is a basic instinct that we wish to see our children and grandchildren well set up in the world and wish to pass on to them some of our most treasured possessions. A similar desire would be that they inherit from us the earth which has been looked after and which does not pose to them more difficult problems than those we have had to face.

This chapter explains the components of natural resources and the impact of human activities on natural resources. Further, it recommends a model for making use of space and information technologies to create a knowledge-based system for natural resources. This model will mainly be useful to the various agencies which are involved in the management of natural resources and environmental issues. It also suggests a model for handling the damage caused by natural disasters.

Environmental scientists and biotechnology experts will clearly explain the composition and functions of natural resources on planet Earth. They will provide information related to environmental issues. Professionals from information and space technologies will consider their inputs for creating this model. Integration of the services of these professionals is an important base for creating this model. Figure 1 indicates the components required to create a knowledge-based system for natural resources management.

Figure 1. Components for a knowledge-based system for natural resources management

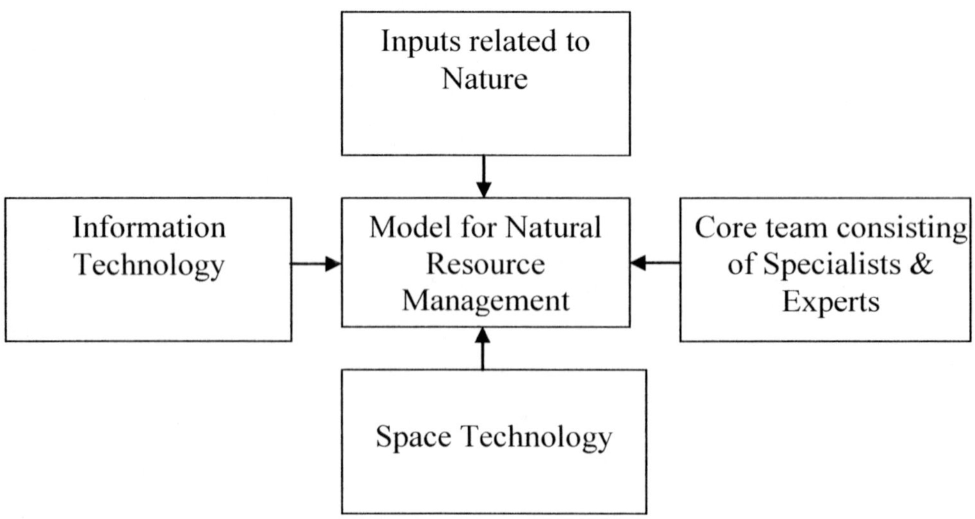

INPUTS RELATED TO NATURE COMPONENTS OF NATURAL RESOURCES

The structural composition and functions of the various components of natural resources play an important role as life supporting systems. Life on this planet Earth depends on the above systems. The components of life supporting systems are mentioned in Figure 2.

Nature belongs to all and is thus important. Everyone should know the basic components of nature.

It is very important to use resources of nature intelligently so that they do not get exhausted. Natural ecosystems operate by themselves under natural conditions without any interference by human beings. If their services are properly utilized, human beings will be benefited. The misuse of natural resources will affect human beings on the planet Earth.

Functionality of each component in resource of nature and misuse of these components is explained in Table 1.

Kraak and Ormeling (2004, p. 51) rightly say:

Environmental maps are created to gain a better understanding of the earth's natural resources. Some of these maps are inventories related to vegetation, soil, hydrology, geology, geophysics and forestry. Others are related to the use and misuse of these resources, such as maps showing water, air or soil pollution.

Change and development are essential to human progress. Exploiting nature in this process is not a new phenomenon. The over exploitation of nature has been causing harm to environmental quality. People make use of resources of nature for creating economic activities. Aggressive approach of economic activities results in generating greenhouse gases. The overexploitation of resources of

Figure 2. Components of life supporting systems

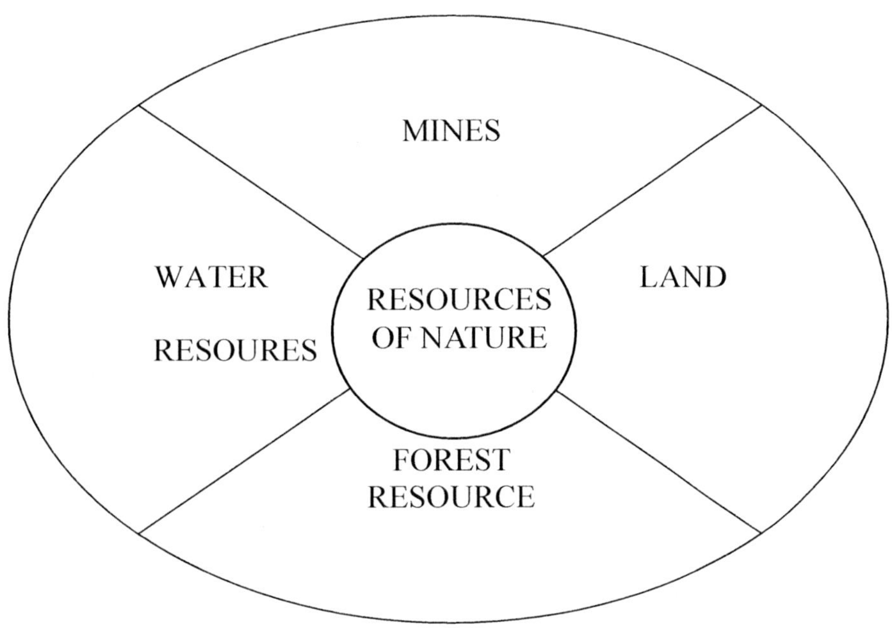

Table 1. Resources and functionality

TYPE OF RESOURCE	FUNCTIONAL ATTRIBUTES	MAJOR USE	MISUSE
FOREST	• PRODUCTION OF OXYGEN • ABSORPTION OF TOXIC GASES • ABSORPTION OF RAIN WATER • SOIL CONSERVATION • WILD LIFE HABITAT	• REDUCING GLOBAL WARMING • POLLUTION MODERATORS	• DEFORESTATION
WATER RESOURCE	• PRODUCTION OF COOLING EFFECT AS IT EVAPORATES • CARRIER OF NUTRIENTS • SUSTAIN AQUATIC ORGANISMS IN EXTREME COLD	• IRRIGATION • DOMESTIC CONSUMPTION • INDUSTRIAL CONSUMPTION	• EXCESS CONSUMPTION • DISCHARGE OF WASTES AND CHEMICALS
MINES	• METALLIC MINERALS • NON METALLIC MINERALS • ENERGY GENERATING MATERIALS	• COMMERCIAL USE • FERLILIZERS • MEDICINES • COAL, LIGNITE • URANIUM	• EXPLOITATION TO MAXIMUM EXTENT
LAND	• HOLDING ABOVE AND BELOW LAND LEVEL OF THE RESOURCES	• USED FOR HUMAN BEINGS EXISTENCE	• EXPLOITING TO MAXIMUM EXTENT

nature will cause changes in environment. The problems related to natural resources have to be addressed by making use of the advancements in information and space technologies. Figure 3 shows the effects of human activities on natural resources.

It would be apt to note the observations of Oliver and Hidore (2002, p. 300):

Earth is the only planet that we are sure has water in all three physical states—solid, liquid and gas. The balance among the three is a delicate one. Any global change in temperature will alter the balance among the different forms.

Nature has been very kind to human beings. Ever since human beings appeared on the planet Earth, they have been dependent on nature for their subsistence. In return, it is unfortunate that human beings are creating stress to the planet Earth. While explaining the need for biotechnology Thakur (2006, p. 9) says:

Environmental movements arise largely due to the clash between economic development and exploitation of natural resources.

CORE TEAM: ENVIRONMENTAL SPECIALISTS AND TECHNOLOGY EXPERTS

Management of any resource needs a detailed inventory. Natural resource management is no exception to this. Assessing the existing resources of nature and the present utilization of natural

Figure 3. Causes and effects on natural resources

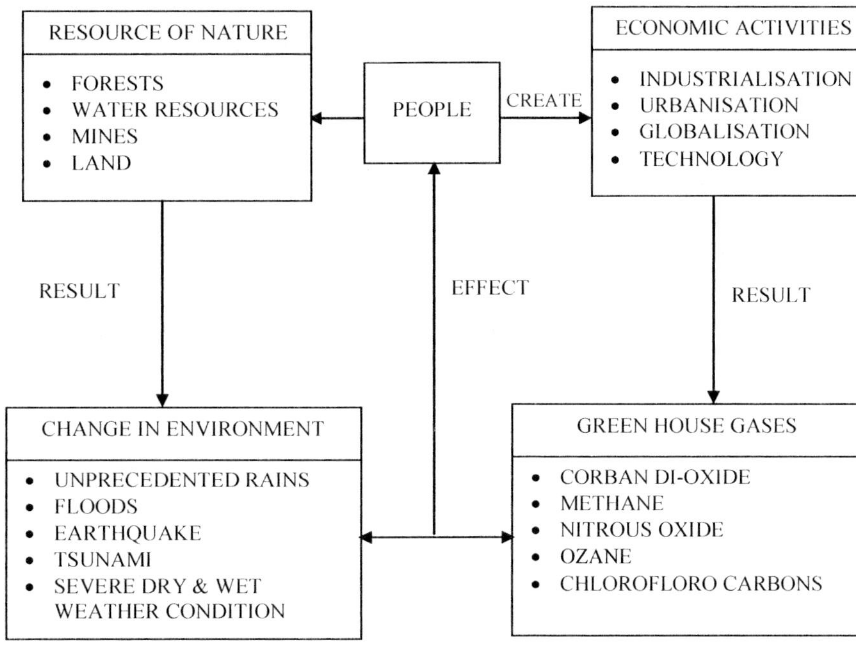

resources is the important basis for getting the information. The information obtained for this purpose can be used for planning the consumption by the present generation and conserving the resources for future generations also. Studying this type of inventory requires an enormous amount of gathering data, compilation, analysis, and modeling. The advanced concepts in space, information, and communication technologies can be used for creating knowledge-based systems for natural resource management. Natural resource management is interdisciplinary, where coordination is required among the various government agencies. It is advisable to form a core team consisting of the representatives of the above agencies, along with environmental and biotechnology experts. Professionals from space, information, and communication technologies will provide their expertise to the core team for creation of knowledge-based systems for resource of nature. The core team can analyze and draw conclusions from the knowledge-based system. This system will help them frame policies pertaining to usage of natural resources. While explaining environmental studies—a multidisciplinary subject—Kaushik and Kaushik (2006, p. 3) quote a Chinese proverb: "If you plan for one year, plant rice, if you plan for 10 years, plant tree and if you plan for 100 years, educate people." Further they observe that if we want to manage our planet Earth, we have to make all the persons environmentally educated.

INFORMATION TECHNOLOGY SOFTWARE AND SOFTWARE TOOLS

The information from the components of natural resources and causes and effects on natural resources can be classified into three types of databases. The graphical information pertaining to the different areas can be stored in "GIS database." The textual information can be stored in "text database," and the quantitative information can be stored in "data warehouse." "Data mining" is a tool used for analysis of quantitative data stored in "data warehouse." This tool helps one to know the relationships and patterns between data elements. "Text mining" is also a tool used for analyzing textual data in "text database." Like data mining, it identifies relationships among the vast amount of text data. The graphical referenced data in "GIS database" will be useful for analysis along with textual and quantitative data by the core team. The type of information required for the purpose of analysis and taking decisions for an environmental management varies from place to place. It also depends on the resources available and usage in the respective areas in a particular country. A well-structured database can help the people who are given the task for planning and framing the policies for managing the resources and handling the environmental issues of the different areas under their control.

SPACE TECHNOLOGY

It would be apt to recall the observations of Raja Rao (2005, p. 1) on space and satellite:

Thanks to Arthur C. Clarke, a fiction writer who envisaged the importance of satellite in 1945. Though the distance communication has been there since centuries, space communication has come to stay as the main means of distance communication particularly in unapproachable regions of the globe from the last two decades or so. One could classify satellites as nodes (active or passive) for the space communication.

Sophisticated sensors are used in spacecrafts and satellites for gathering information about the Earth's surface. Further, the developments in the supporting tools, such as image processing, photogrammetry, and geographic information systems, with the advancements in ICT, have pushed the application of remote sensing in satellites to hitherto unexplored areas. A remote sensor has two kinds of features for transmitting images. They are active and passive systems. Passive systems utilize sun-rays. Active systems utilize the source of energy emitted by sensors. In both the systems, a remote sensing sensor records reflections from the Earth's surface. A sensor in the satellite transmits the images to the Earth through a ground receiving station. The data received by antenna dish are recorded on a magnetic tape. If the ground receiving station is nearby at the time of data collection, then the data can be recorded virtually at the same time as they are transmitted. If there is no ground receiving station in line of sight from the satellite, the signals may be transmitted to the ground via a relay satellite in an orbit higher than that of the transmitting satellite. Figure 4 gives a macro view of how a satellite works.

It is interesting to note the observation of Chang (2003, p. 7) on satellite images: Using satellites in space as reference points, a GPS receiver can determine its precise position on the Earth's surface, which can then be used to determine the location and shape of spatial features."

Table 2 explains the classification of satellites. Earth is considered as a primary objective for these satellites, and altitude is decided from the Earth's surface.

Orbits generally fall into three categories, low, medium, and geosynchronous Earth orbit. Low Earth orbit (LEO) is used mostly for scientific and military purposes. Middle Earth orbit (MEO) is used in global positioning system (GPS) satellites. geosynchronous Earth orbit (GEO) is used for communications.

Selection of Satellite Data Products

The choice for selecting data from the satellite for a required application is wide. There are a number of satellites in orbit having sensors with coarse and fine resolutions. These satellites have the capability to revisit an area at various frequencies.

Figure 4. Macro view of working of a satellite

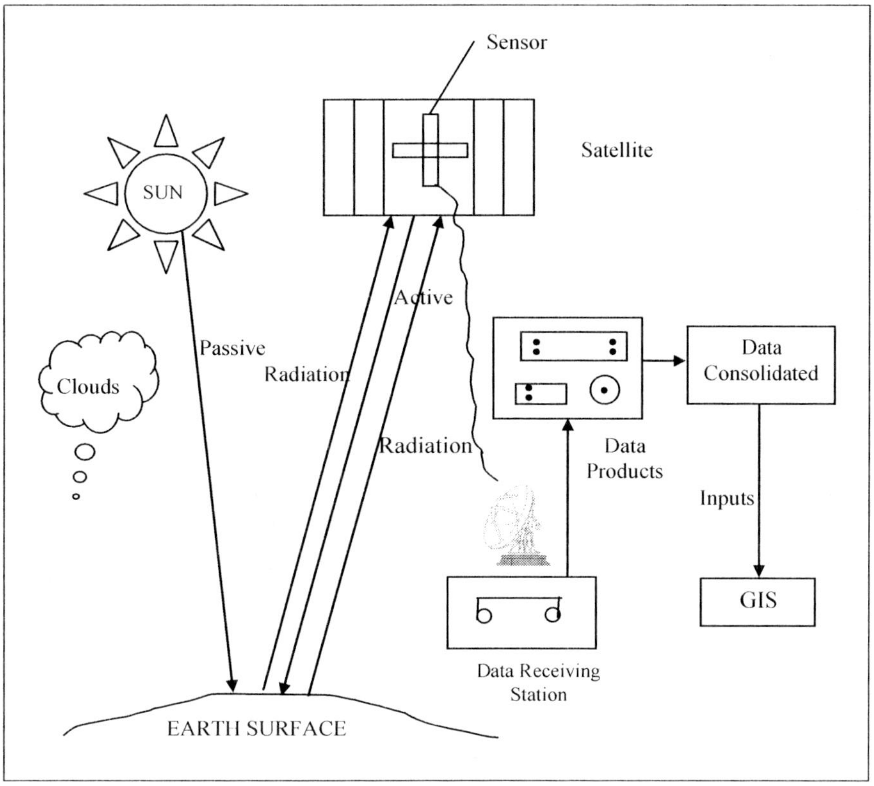

Table 2. Classification of satellites

Satellites in orbits	Altitude from earth's surface
LEO	100 – 1000 KM
MEO	5000 – 12000 KM
GEO	35000 KM

Selection of satellite products are to be considered on the following basis:

1. Project theme
2. Type of application: Geographic, climate, and scale of operation
3. Characteristics of sensors: Spectral and spatial
4. Combination of images in spectral, spatial, and temporal

A remote sensing satellite is designed according to its application. A satellite carrying sensors with medium to high spatial resolution and a slow repeat cycle is used for studying Earth resources. A satellite carrying sensors with low spatial resolution and a fast repeat cycle is used for meteorology.

Application Areas in Remote Sensing

The concept of remote sensing is being used in areas such as agriculture, forestry, water resources, mapping of land usage and land cover, and monitoring environmental hazards. Importance of remote sensing can be understood from the following examples.

Agriculture

Remote sensing technology can be useful in the understanding of land utilization in the field of agriculture. This will facilitate in knowing patterns of crop, fallow lands, waste lands, and surface water bodies. Timely detection of pests and diseases and assessment of crop condition are other processes where remote sensing can play an important role.

Forestry

The remote sensing data from aerial and satellite platforms play a significant role in forest resource survey, monitoring forest cover, evaluating eco systems, and studying wildlife habitat. The mapping of the forest type or vegetation is essential for forest resource surveys. The contrast is to bring out the difference between single crop fields and the mixture of many tree species occupying a forest land. The identification of tree species can be done with prior knowledge of the area.

Water Resources

Knowledge of ground water location is important for the supply of water and management of water resources. The remote sensing data can provide useful information on the factors controlling the occurrence and movement of ground water (e.g., geology, geomorphology soils, and land cover). A systematic study of these features helps in a better delineation of the prospective ground water zones in a region.

Mapping of Land Use and Land Cover

The remote sensing data both from air and space have become popular as information sources, particularly for mapping land use and land cover. Information generally covers the features present on the Earth such as vegetation, rocks, and buildings. Aerial photographs provide a wealth of information. It needs a specialized knowledge for interpretations of the features.

Monitoring of Environmental Hazards

The images from an environmental satellite, in conjunction with the conventional data, will help in monitoring convective situations such as cyclones and storms. The visible, infrared, and enhanced infrared images are utilized to estimate rainfall in sensitive areas. In an area prone to intense convection, such images are of significant value for assessing the related hazards and mitigating and averting the associated disasters.

Compatibility between Remote Sensing and GIS

Remote sensing systems have a very important feature for the collection and classification of the spatial data. GIS is used for managing and analyzing the spatial data. Thus, both systems are complementary to each other. Many advancements are taking place in satellite sensors of very high resolution. The link between a remote sensing system and GIS is facilitating in solving problems related to natural resources management.

The importance of linkage between remote sensing and GIS can be observed from the statement of Panda (2005, p. 208):

With the development in satellite sensors of very high resolution of the order of a meter (for example, the Ikonos data), the compatibility of geographic information system and remote sensing system has become stronger day by day to tackle problems of any desired scale.

Remote Sensing Inputs to Geographic Information Systems

Remote sensing output is expressed in polygons. Geographical information systems require accurate, digital, polygonal, point, or network data sets as its inputs to be processed further, depending on the type of problems.

GIS Output to Remote Sensing System

Remote sensing data analysis requires collateral and ancillary data for development and improvement of its classification of data. This information is available in GIS in usable form for the remote sensing systems. It may be noted that reverse flow of data from GIS to remote sensing improves the classification of remote sensing imagery.

It would be apt to note the methods suggested by Guha (2003, p. 70) on data interpretation and analysis. The various methods of extracting information from the remote sensing data are as follows: (a) manual interpretation of standard photographic products using very simple inexpensive optical instruments, (b) manual interpretation aided by photographic enhancements and employing costly optical instruments, (c) manual interpretation of special digitally-enhanced photographic products using costly optical instruments, and (d) manual interpretation of the computer output resulting from the digital analysis of computer-compatible tapes.

While explaining image interpretation keys, Chandran and Ghosh (2006, p. 77) say:

Image interpretation keys are valuable aids for summarizing complex information. Such keys serve either or both of the two purposes: (1) A means of training inexperienced personnel in the interpretation of complex or unfamiliar topics (2) A reference aid for experienced interpreters to organize information and examples pertaining to topics.

GRAPHIC INFORMATION SYSTEM (GIS)

There seems to be an impression among many people that computer assisted drafting (CAD), computer-assisted cartography (CAC), and geographic information system (GIS) are all the same because the graphic display from each of these systems is almost the same except for minor differences. CAC system is used to create maps from geographical objects and their descriptive attributes. It does not have analytical features. CAD system helps to generate graphic images. This system is not linked to external descriptive data files. It also does not have analytical features. The remotely sensed satellite data are the source for GIS. While stressing the importance of the use of applications of remote sensing for Earth resource management, Joseph (2005, p. 389) says:

The final solution to a problem requires integration of many spatial and aspatial information. That is, we have to relate information from different resources. Integration of data from different sources can help to resolve 'Conflicts' of interest on a scientific basis providing different trade-offs that exists for planning its pros and cons including the costs involved. Currently this is very conveniently carried out by using geographic information system (GIS).

GIS is a tool for converting the spatial data into information. This information can be utilized to manage natural resources. The term geographic in GIS indicates the data items related to natural resources located in latitude or longitude of a country. These locations can either be known or calculated. Data in a GIS will be available in two dimensions or three dimensions. The data in a GIS provide in the form of colored maps, images, graphics, and information tables.

CASE ILLUSTRATION

Generally, economic development and natural resources management are considered mutually antagonistic. Promotion of one would inevitably mean damage to the other.

In the present globalization scenario, it has become a necessity for integrating natural resources concerns into economic development activities. If we concentrate on urban development, we will face the risk of losing natural resources, because the former will take over the latter. The stress on the Earth's surface requires careful assessment. A new natural resources management agenda is needed, especially in the developing countries, for reducing the stress on natural resources and managing environmental problems in urban areas. The type of information required for the purpose of analysis and framing policies for natural resources management varies from country to country. It also depends on the resources available and their usage in the respective areas in a country. A well-structured system can help the authorities who are involved in planning and framing the policies for managing the resources of the area under their control. Developing countries need a model that helps them manage their country's resources judiciously. A core team consisting of multidisciplinary experts to develop a model for natural resources management is required. The components of information and space technologies will be the backbone for this model. This model can be referred as a knowledge-based system for natural resources management (KBSNRM).

Knowledge-Based System for Natural Resources Management (KBSNRM)

The KBSNRM model has the following five stages:

- **Stage 1: Geographical data for a country:** GIS software is to be used for creating geographical data of a country. This is the starting point of this model.
- **Stage 2: Quantitative and textual contents of geographical data:** Geographical data is converted into quantitative and textual data. This data will be useful for analysis and framing policies related to natural resources and issues related to natural resources.
- **Stage 3: Data for environmental management:** The type of data required for analysis of environmental issues related to urban area is explained. Depending on the requirements, any aspect of data can be taken from the KBSNRM model for analysis. KBSNRM is a knowledge repository system.
- **Stage 4: Knowledge-based system for natural resources management:** The design of the KBSNRM model is explained at macro level.
- **Stage 5: Disaster management system:** Places prone to natural disasters can be

selected from the KBSNRM model. The information from disaster management will be useful for managing the natural disasters.

GEOGRAPHICAL DATA FOR A COUNTRY

The government departments in a country have the details of natural resources, population, location of industries, business houses, educational institutions, residential areas, and other related information pertaining to their countries resources. In most of the cases, either it will be available partially in manual records or stand alone computer systems. It means that they are not available in an integrated system for using them. It is high time that the geographical information system (GIS) is used to know the inventory of natural resources and the exact location of these resources in a country. GIS applications are unlimited. GIS in the KBSNRM model will be useful in knowing the location and extension of resources such as agricultural land and forest, and the area covered by water and mountains in a country. Further, it provides information pertaining to the location of industries, business, and service units.

Geographically referenced data can be used for making boundaries and classifications. Figure 5 gives an idea of the geographical data of a particular area in a country.

QUANTITATIVE AND TEXTUAL CONTENTS OF GEOGRAPHICAL DATA

The components of natural resources can be classified as renewable resources. Renewable resources have the inherent ability to reappear or replenish themselves by recycling, reproduction, or replacement. These renewable sources are

Figure 5. Geographical data

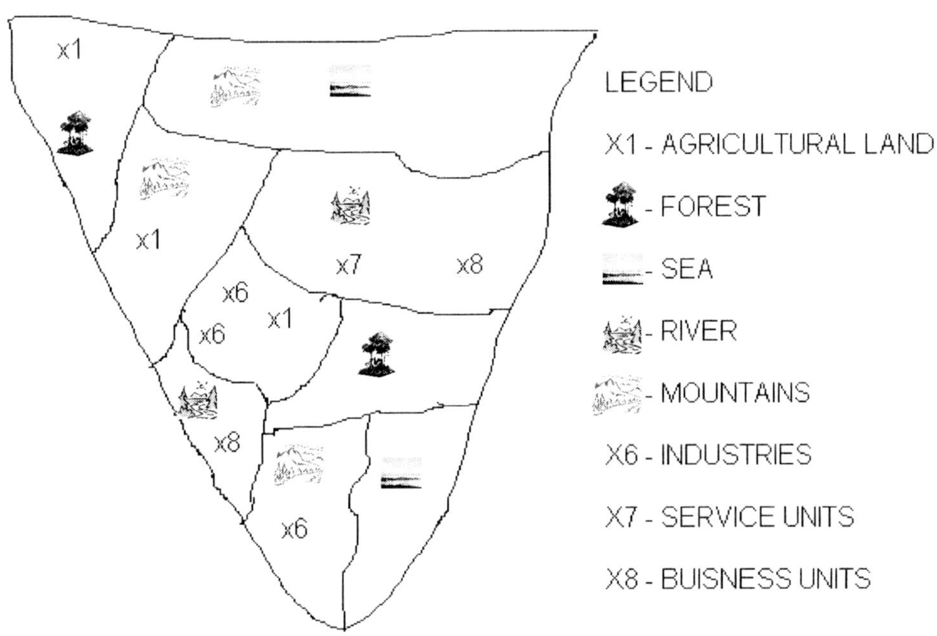

water, plants, animals, soil, and living organisms. Nonrenewable resources are the Earth's geologic endowments, that is, minerals, fossil fuels, non-mineral resources and other materials which are present in fixed amounts in the environment. Every member of the core team is expected to be aware of this.

The quantitative and textual data in reference to geographical data can be classified and stored in the KBSNRM model. Figure 6 explains the classification of data in a module in KBSNRM. The macro level contents in the module are shown in Table 3.

Natural Resources

- **Forests:** Data related to moist tropical forests, dry tropical forests, montane sub-tropical forests, montane temporal forests, subalpine forests, and alpine scrubs are identified under this category. Classification of forests varies from country to country.
- **Water resources:** Data related to fresh water lakes, saline lakes, rivers, ground water, and oceans are stored under this category.
- **Minerals:** Data related to metallic and non-metallic are stored under this category.
- **Agricultural land:** The data related to cultivable and noncultivable land will be stored under this category.
- **Industrial areas:** This will contain data pertaining to various industries such as automobile, textiles, chemicals, pharmaceuticals, consumer durables, and consumer-related products. This list is not exhaustive.
- **Urban areas:** Data related to residential, roads, transport infrastructure, and utility services will be available in this module.
- **Rural areas:** Data related to the areas which do not fall under the category of urban areas are stored in this module.

Figure 6. Classification of data in quantitative and textual contents in a database

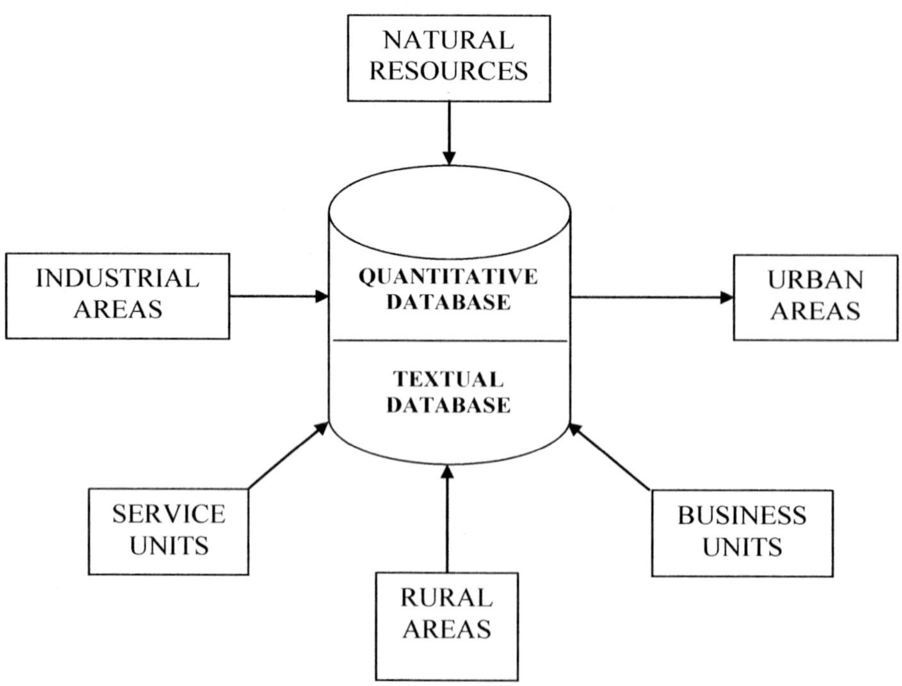

- **Service units:** Data related to educational institutions, health care units, and other related units fall under this category and will be stored in this module.
- **Business units:** Data related to trading organizations, financial institutions, hospitality units, and other units falling under this category will be stored in this module.

Data in the various modules in KBSNRM will be helpful for framing policies to manage natural resources of a country.

DATA FOR ENVIRONMENTAL MANAGEMENT

Industrialization and urbanization has become a worldwide phenomenon. Because requirements of people living in urban areas are increasing, providing and handling their requirements need to be assessed properly. Urbanization raises many environmental issues.

The relationship between the availability of natural resources and ecosystem and consumption of resources can be established from the quantitative and textual database in any particular area. This helps one to understand how to minimize environmental hazards and avoid the depletion of resources. The details pertaining to water usage, solid waste produced, disposition of package materials, and the services related to utilities can be taken from this system. This will be most useful for environmental management. Figure 7 explains the type of data required for analysis.

The Macro level contents for analysis for environmental management are shown in Table 4.

Remote Sensing Application in Urban Areas

Planned development does not seem to take place in the developing countries in respect to urbanization. Information in respect to cities and suburban infrastructure is needed constantly. Remote sensing systems can be used in the urban areas for getting data related to buildings, transport and utility infrastructure, and environment assessment. The details pertaining to water usage, disposal of packing materials, and services related to utilities can be ascertained and stored in the KBSNRM model.

Table 3. Modules and contents in quantitative and textual database

MODULES	CONTENTS
Natural Resources	Details of agricultural land, mines, rivers, sea, forest and other related resources.
Industrial Areas	Details of industrial belt.
Urban Areas	Details of various urban areas.
Rural Areas	Details of various rural areas.
Service Units	Details of various service units.
Business Units	Details of various business units.

Figure 7. Data for environmental management analysis

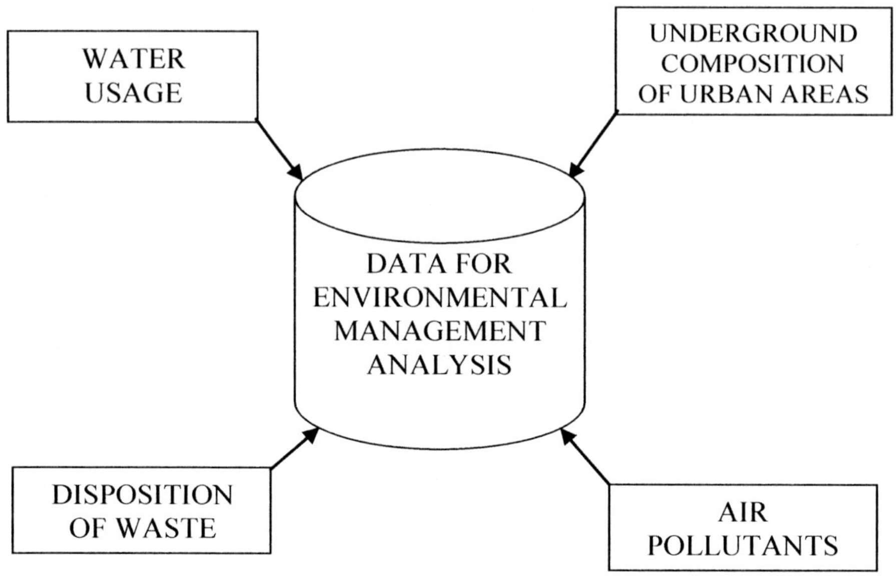

Water Usage

The data related to water consumption by residents and industrial, service, and business sectors can be ascertained and stored in the model. This will be useful in assessing the requirements of consumption of water by these sectors. A contingency plan at the time of shortage of water can be prepared on the basis of the information in this module.

Disposal of Wastage

The advancement in the package industry has created a good scope of marketing of the products manufactured across the globe. Demand for packed products is on rise in the developing countries. Disposal of every kind of packing materials referred to in Table 4 need to be carried out by civic authorities. Statistics of discarded packing material is required for allocation in cities and suburban areas. This module will provide this information.

Air Pollutants

Oxygen and nitrogen are the major constituents of the atmosphere. Coal, fuel oil, and gasoline used by us emit carbon monoxide. This human activity is contributing to changes in the atmosphere. The largest single source of this emission from automobile pollution in air is from the above factors. The health of human beings is affected by this pollution.

Underground Composition of Urban Areas

The Earth is being dug up in the developing countries very frequently for the purpose of laying lines for communication, electricity, water mains, sanitary, and sewage lines. Generally, it is not well-planned activity in developing countries. The details of the information pertaining to every area in cities and suburban parts are required for developing urban areas.

It is interesting to note the statement of Jensen (2004, p. 460) on remote sensing the urban

Table 4. Modules and contents in data for environmental management analysis

MODULES	CONTENTS
Water Usage	1. Residents 2. Industrial Sector 3. Service Sector 4. Business Sector
Disposal Of Waste	PACKAGING MATERIALS 1. Textiles- Jute, Sacks, Bags 2. Multi metals- Milk & Juice Cartons 3. Glass- Bottles 4. Aluminum cans, and steel strapping 5. Paper-Corrugated cartons and boxes 6. Wood Pallets, Boxes and crates SOLID WASTE 1. Industrial Waste 2. Municipal Waste
Air Pollutants	1. Particles 2. Sulphur dioxide 3. Nitrogen Oxide 4. Hydrocarbons 5. Carbon Monoxide
Underground Composition Of Urban Areas	1. Communication Lines 2. Electricity Lines 3. Water Main Lines 4. Sanitary Sewage Lines 5. Storm Water Lines

landscape: urban/suburban environments are enormous consumers of electric power, natural gas, telephone service and potable water. The removal of storm water from urban impervious surfaces is also a serious problem.

KNOWLEDGE-BASED SYSTEM FOR NATURAL RESOURCES MANAGEMENT

The importance of the services of hardcore software professionals and space technology experts cannot be underestimated because they are the backbone of the system along with the other members of core team. The core team's analysis and solutions can be stored in the knowledge-based system for natural resources. The data and information can also be made available to executives of the various governmental agencies who are associated with the natural resources management and environmental issues with an access through mobile communication.

The KBSNRM model gives an idea of how the concepts of space and information technologies can be used for creating the inventory of the natural resources of a country.

The remarkable developments in space-borne remote sensing technology and its applications during the last few decades have firmly established its immense and potential mapping of various natural resources. The capabilities of remote sensing and GIS have to be taken advantage of for pursuing better natural resources and environmental management. Information technology can be used in the development and application of computational tools to acquire, store, organize, achieve, analyze, and visualize satellite and biological data. The type and level of information extracted depends on the expertise of analysts and the area of application of the information. The concepts of space technology will be useful for this system. Executives of the various governmental agencies located at different locations in a country will be making use of this system. Global system for mobile communication (GSM) architecture is suggested for the reason that it will be successful in wireless technology in the present scenario. The macro level design of the system is explained in Figure 8 in knowledge-based systems for natural resources management.

The KBSNRM model is created by the joint efforts of the core team. Data from legacy system, that is, geographical, quantitative, and textual databases, will be the base for creating graphical database, data warehouse, and textual database. Knowledge repository bases for natural resources contain the data at a country, state, region, city, and village level. This will provide a good scope for analyzing the various aspects of the natural resources, usage, and consumption of them.

Analysis of data in knowledge repository bases can be carried out by using the software tools such as GIS analysis, data mining, and text mining. The data analyzed by the specialists identified by the core team will be available in knowledge-based systems for natural resources. The analyzed data can be made available to the end users at different locations in a country with the devices, such as mobile hand sets and mobile laptops or notebooks.

SIGNIFICANCE OF THE KBSNRM MODEL

The major problem in the developing countries is in the identification of the effects of the mismanagement of natural resources. Assessment of mismanagement of natural resources can be carried out only when one is clear about the functionality of each component of natural resources. The KBSNRM model stresses the importance of a systematic approach in creating a knowledge-based system for natural resources of a country. This model facilitates to hold data pertaining to a country on the basis of two parameters. They are quantitative and qualitative. Data in this model helps the government agencies to analyze framing policies for conserving the resources of a country and managing the issues related to the environment. Technology is the major stimulus for change in any sphere of activity on the planet Earth. The exponential increase in computing power in information technology and the rapid advancements in space technology have led to the development of this model.

Importance of Space Technology in Natural Disasters

Global warming causes climatic changes and natural disasters such as unprecedented rains, floods, earthquake, tsunami, and severely dry and wet weather. A proper "disaster management" plan is also required for providing services to the victims of natural disasters whenever they occur. The stress caused on planet Earth needs a careful assessment of the use of natural resources.

Atmospheric changes and disturbances can be observed through satellite systems. High resolution remote sensing satellites integrated with GPS systems provide the inputs at the various phases of natural disasters for preparedness, prevention, mitigation, and disaster management. Communication satellites will be helpful for sending warning signals to people in remote and inaccessible areas. Figure 9 gives an overview of space technology in natural disaster management. While stressing the role of space technology in disaster management Rao (1996, p. 387) says:

Space systems, which derive basic advantage from the altitude of their operation, have unambiguously demonstrated their capability in providing vital information and services towards all the three major

Figure 8. Knowledge-based system for natural resources management

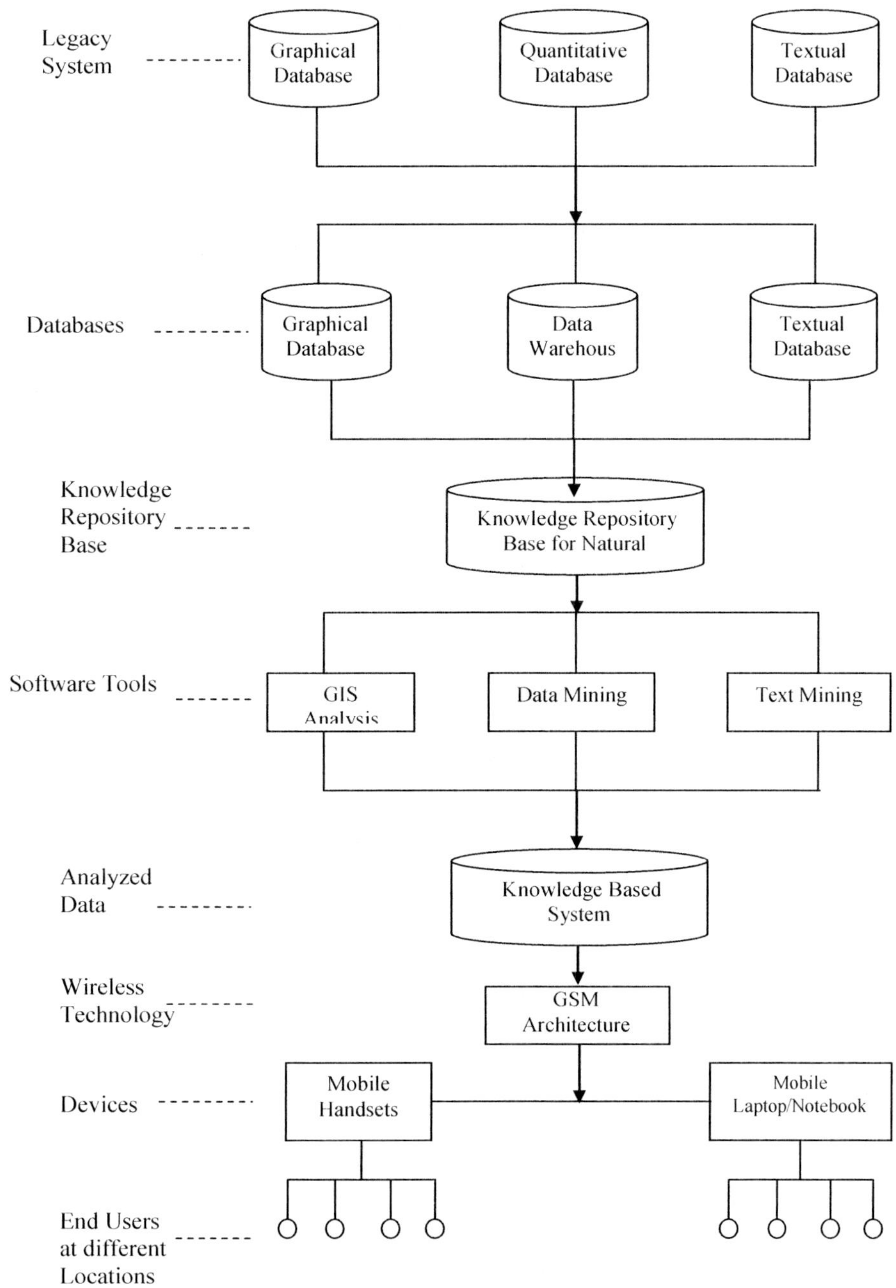

aspects of disaster management, namely—prevention, preparedness and mitigation.

Satellite technology can help in disaster preparedness by providing repetitive and synoptic up to date information on the locally available resources. Disaster prevention measures can be improved through satellite technology in three ways. They are: (1) mapping the disaster prone areas, (2) forecasting of impending areas, and (3) disaster affected areas. Geostationary satellite data is capable of providing information every half an hour and are useful in monitoring short term disasters like cyclones and tornadoes. It is felt that a combination of high spatial, temporal, and spectral resolution data would certainly be beneficial in disaster management.

Some of the disasters that occur world wide on a regular basis are earthquakes, floods, cyclones, avalanches, landslides, tsunami, drought, and forest fires. The time scale indicating their general nature of occurrence and duration of the disaster is given in Table 5.

DISASTER MANAGEMENT

In times of natural disaster governmental agencies, which are entrusted with disaster management, get severe criticism, as the expectations of the victims are high and they demand immediate response. Timely action by the former generally is not as quick as one would want it to be. Natural disasters are likely to occur more often due to global warming. Information pertaining to the type of help and services to be rendered during this period is required to be stored in an

Figure 9. Role of space technology in natural disaster

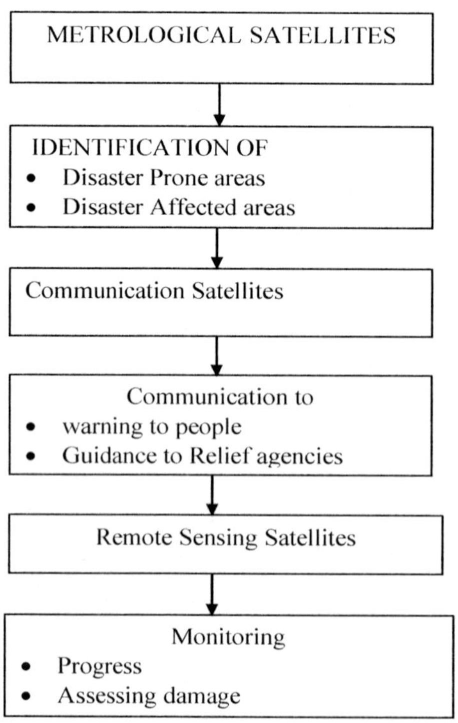

Table 5. Space technology in natural disaster management (Adopted from Rao, 1996)

S.	Disaster	Severity Index on a relative scale of 10	Time scale	Duration	Features derivable from remote sensing satellites	Application areas
1	Cyclones	9	Not sudden	Days	Cloud dynamics	Geostationary, polar orbiting satellites and cyclone warning radars to monitor development and progress of cyclones
2	Hailstorms, Thunderstorms, Tornadoes, Dust storms	7	Sudden	Hours	Cloud structure, cloud types and atmospheric monitoring	Temporal (every 5-30min) imagery to monitor the development of the events
3	Lightning	7	Very Sudden	Seconds	Energy and intensity	Lightning detectors on geostationary satellite
4	Landslides	7	Sudden	Hours to days	Digital elevation models of terrains prone to landslides, and fault zones from polar orbiting satellites	Intensity of rainfall together with information on slope, gradient and loose soil to identify potential sites of occurance
5	Floods	10	Varies from sudden (flash floods) to not so sudden (regular floods)	Hours to days	Terrain characteristics, vegetation cover, rainfall, snow melt run off from polar orbiting satellites	Flood damage assessment and delineation of high risk zones and their treatment to reduce the risk
6	Drought	9	Very Slow	Months	Crop vigour from vegetation index, surface water bodies and their temporal changes from polar orbiting satellites	Comparison with the time profile of vegetation index in a normal year combined with rainfall information for prediction of extent and severity of drought
7	Forest fires	7	Not sudden	Hours	Extent of forest cover affected by fire, from polar orbiting satellites	Sequential imagery show progress of the event and extent of damage
8	Heat and cold waves	3	Slow	Weeks	Surface temperature, humidity, winds (geosynchronous satellites)	Sequential data (twice a day) help formulating procedures for locating vulnerable areas

continued on following page

Table 5. continued

9	Crop pests and diseases	9	Slow	Weeks	Crop types, crop vigour. Soil characteristics from polar orbiting satellites	Spectral and tonal differences at critical stages of crops for identification of disease affected areas
10	Earthquakes	10	Very sudden	Minutes	Mapping of faults, lineaments and use of GPS system at appropriate locations	Detection of crustal movement, tilt and creep along the faults
11	Accidents	6	Very sudden	Minutes	Urban and land cover changes	Planned development of urban growth

information system. This will be more useful for taking action in an emergency. The kinds of services needed are medical services, transporting people from the affected areas to safe places, and organizing food and provisions to the people in distress. Figure 10 explains the basic modules in a disaster management system.

The macro level data and information in the disaster management system are shown in Table 6.

The experiences of the people who have been directly involved in providing services to the victims of disaster will be the inputs for creating this database.

Resource Allocation Module

This system will have the list of basic supplies such as food items, clothes, raincoats, and umbrellas required to help the victims of natural disaster. Addresses of voluntary rescue teams and medical doctors are to be stored in the database.

Transport Module

The details of heavy transport vehicles that can wade through water, boats, and aerial survey aircrafts should be available in the system.

Evacuation Module

Vulnerable areas and safe regions' exact locations are to be stored in the system by using GIS Application.

Shelter Supplies Module

This module will contain the details of resources of various shelter locations.

Deprivation Module

This module will contain the details of short supply of food items in the earlier disaster-affected areas and statistics of number of people infected with diseases. This information will be useful for avoiding such type of situations.

Disaster Module

This module will contain the details of handling the various situations at various disaster management operations. Success and failure of the rescue operations can be derived from the information from this module. It will be useful for the future operations.

Table 6. Modules and contents in disaster management system

MODULES	CONTENTS
Resource Allocation Module	Basic Supplies, Equipments, Transport, Rescue Teams and Medical Teams
Transportation Module	Types of vehicles required to various consignments
Evacuation Module	Vulnerable areas and safer regions
Shelter Supplies Module	Deploying the allocated resources to various shelters
Deprivation Module	Computation of shortage matrix, number of people deprived of food, likely to be infected with diseases, dead and recovered from injury
Disaster Module	Computes and categories the damages, evaluating the success or failure of the support systems

Figure 10. Modules in disaster management system

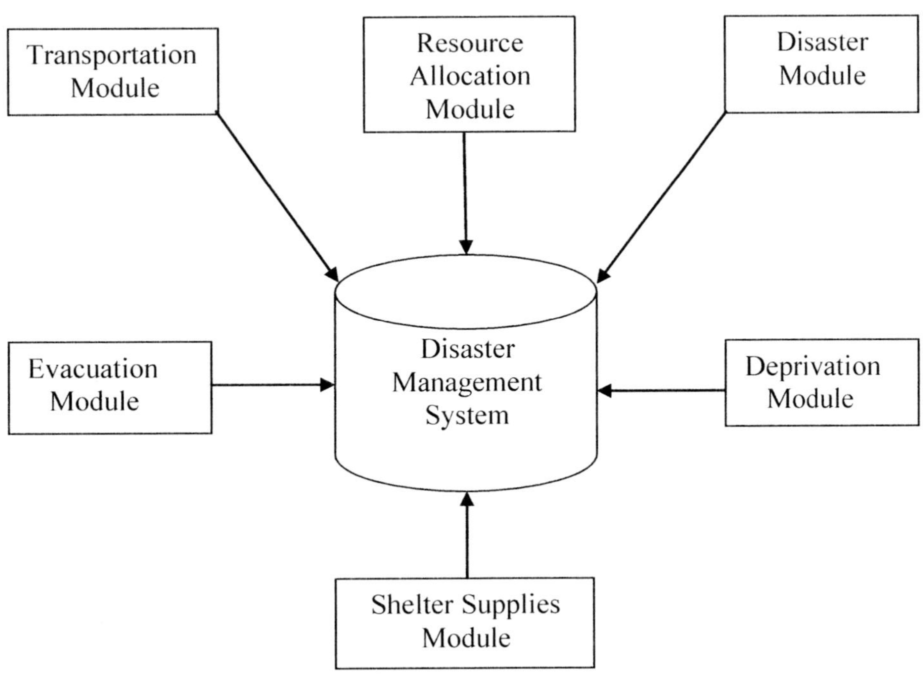

Future Trends

The policy of many countries is to encourage space-related programs. The advancements in space technology have provided the means to acquire large volumes of data related to planet Earth. The concept of a neural network can be applied for generalizing the relation between the evidence (e.g., remote sensing data) and the conclusion (e.g., land cover classification) without any mathematical models. There is scope to pursue research with this concept in the Geographical Information System.

CONCLUSION

Economic development has helped to raise the standard of living and has also led to mismanagement of natural resources. This has resulted in environmental issues. Wisdom lies in maintaining a balance between the needs of human beings and supplies from natural resources so that the delicate ecological balance is not disrupted. In their zeal to go ahead with ambitious plans of development, the integration of knowledge relating to environmental sciences, economics, and information and space technologies has escaped the attention of these people. A knowledge-based system for managing natural resources is explained by using the above in the integration of knowledge. The KBSNRM model gives an idea, using GIS application and space technology, for obtaining the inventory of natural resources of a country. This model further explains the type of information needed to handle issues related to urbanization. Information required for managing natural disasters is discussed. Features derivable from remote sensing satellites are provided.

The problems of managing resources of nature are diverse and approaches toward their solution to mitigation depend on the information, monitoring, and pressure for action which is acute. Knowledge-based natural resource management will help the various government agencies to change their practices for using the natural resources and addressing the problems of depletion of nonrenewable resources, handling environmental issues, and managing disasters. Natural resource management becomes easier through the components of space and information and communication technology.

REFERENCES

Chandra, A. H., & Ghosh, S. K. (2006). *Image interpretation, remote sensing and geographical information system* (p. 77). New Delhi: Narosa Publishing House.

Chang, K.-T. (2003). *Introduction to geographic information systems* (p. 7). New Delhi: Tata McGraw-Hill Publishing.

Chatterji, A. K. (2002). *Ecosystem approach in environmental management: Introduction to environmental biotechnology* (p. 9). New Delhi: Prentice Hall of India.

Guha, P. K. (2003). *Interpretation of remote sensing data: Remote sensing for the beginner* (p. 70). New Delhi: Affiliated East—West Press.

Houghton, J.(2004). *Why should we be concerned? Global warming, the complete briefing* (p. 200). Cambridge, UK: Cambridge University Press.

Jensen, J. R. (2004). *Remote sensing the urban landscape: Remote sensing of the environment* (p. 460). New Delhi: Pearson Education (Singapore).

Joseph, G. (2005). *Applications of remote sensing for Earth resources management: Fundamentals of remote sensing* (p. 389). Hyderabad:University Press.

Kaushik, A., & Kaushik, C. P. (2006). *Environmental studies—A multidisciplinary subject: Perspectives in environmental studies* (p. 3). New Delhi: New Age International Publishers.

Kraak, M.-J., & Orheling, F. (2004). *GIS applications: Which map to use? In Cartography: Visualization of geospatial data* (p. 51). Delhi: Pearson Education (Singapore).

Oliver, J. E., & Hidore, J. J. (2002). *The warming of planet Earth: Climatology* (p. 300). Delhi: Pearson Education (Singapore).

Panda, B. C.(2005). *Principles of geographic information system (GIS): Remote sensing principles and applications* (p. 208). New Delhi: Viva Books.

Raja Rao, K. N. (2005). *An overview of space and satellite: Fundamentals of satellite communication* (p. 1). New Delhi: Prentice Hall of India.

Rao, U. R. (1996). *Space and disaster management: Space technology for sustainable development* (p. 387). New Delhi: Tata McGraw-Hill Publishing.

Thakur, I. S. (2006). *Introduction: Environmental biotechnology* (p. 9). New Delhi: I.K. International.

Chapter III
Using Space Technology for Disaster Monitoring, Mitigation, and Damage Assessment

Pasquale Pace
University of Calabria, Italy

Gianluca Aloi
University of Calabria, Italy

Luigi Boccia
University of Calabria, Italy

ABSTRACT

The theme of this chapter is how space technologies and satellite applications can mitigate the impact of natural and man-made disasters. The objective is to provide the reader with an overview of the most important space technologies for both monitoring and telecommunications and to show the main issues in managing a disaster response. The chapter is divided into three parts. Firstly, the potential of remote sensing satellites related to natural disasters is described. Secondly, the strength and the weakness of space-based telecommunication architectures for the emergency and recovery phase are outlined. Finally, international policies currently applied for emergency management and disaster recovery will be described, while trying to individuate the needs for an optimal provision of information and accessibility of space-related services and coordination of existing in-orbit assets in case of disaster.

INTRODUCTION

In the last few years, natural and man-made disasters have been affecting an increasing number of people (Figure 1), causing significant social and economic damages (Figures 2-3) (EM-DAT, 2006). This rising trend is mainly due to the augmented vulnerability of the world's population, which can be attributed to several factors, including population pressure, declining resource base, and degradation of the environment. In general, the effects of catastrophic events can be limited, reducing the various elements of vulnerability. However, an effective strategy aiming at a reduction of the disasters' impact cannot be developed without considering a pervasive application of modern technologies. In particular, space-based applications play a fundamental role in all the stages of a catastrophe. Earth observation and meteorological satellites are in fact an essential tool for collecting the information needed to mitigate the human and economic losses due to a disaster. Besides, most of the communications and data exchange occurring during the emergency response and recovery make use of satellite communication systems even as a complement of that available from other ground-based sources. Furthermore, a disaster remains one of the most critical and severe challenges to public emergency services and to the national governments. Disaster management, in fact, requires not only an efficient coordination at several hierarchical levels of all the national and international organizations involved in the recovery phase, but also the adoption of a correct policy for an optimal provision of the information concerning the situation. This results in the need of an international policy for disaster management, but it requires also well-based communication and data exchange procedures aiming at the creation of a global disaster monitoring and information network.

Figure 1. Total number of people affected by natural or man made disasters in the world (1900-2004)

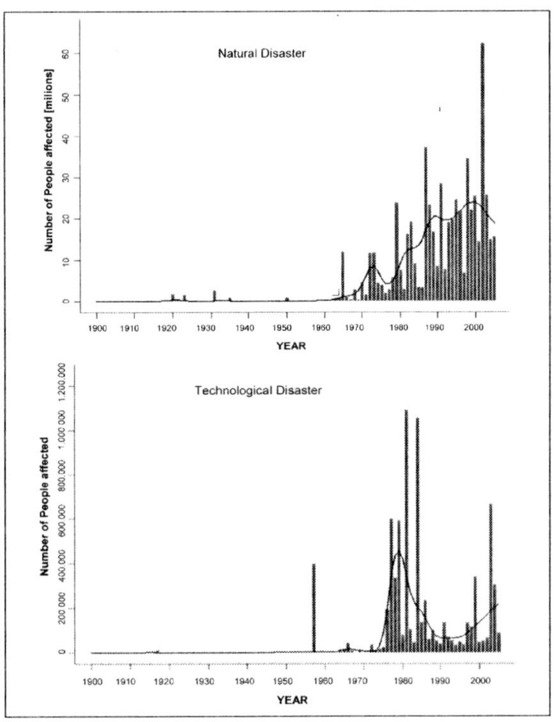

Figure 2. Total amount of damages caused by natural or man made disasters in the world (1900-2004)

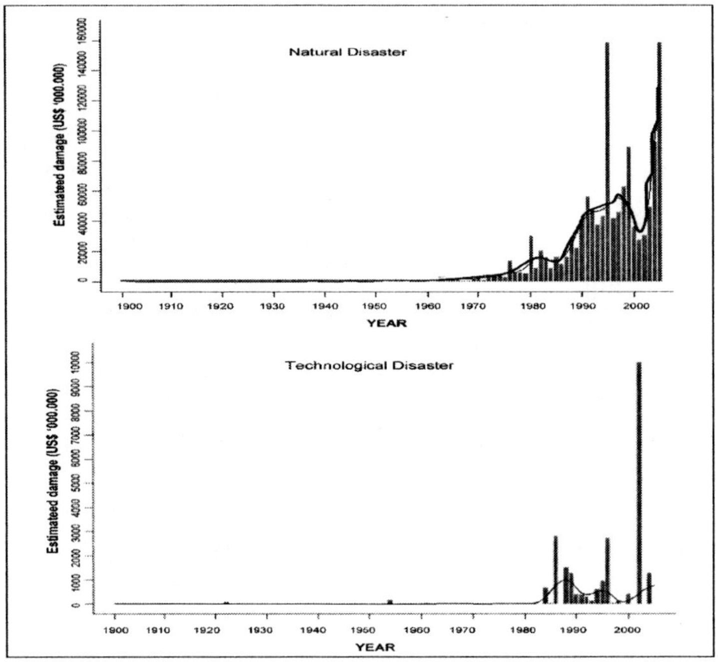

Figure 3. Top 25 countries: Total amount of economic damages reported 1994-2004 (2003 US$ billion)

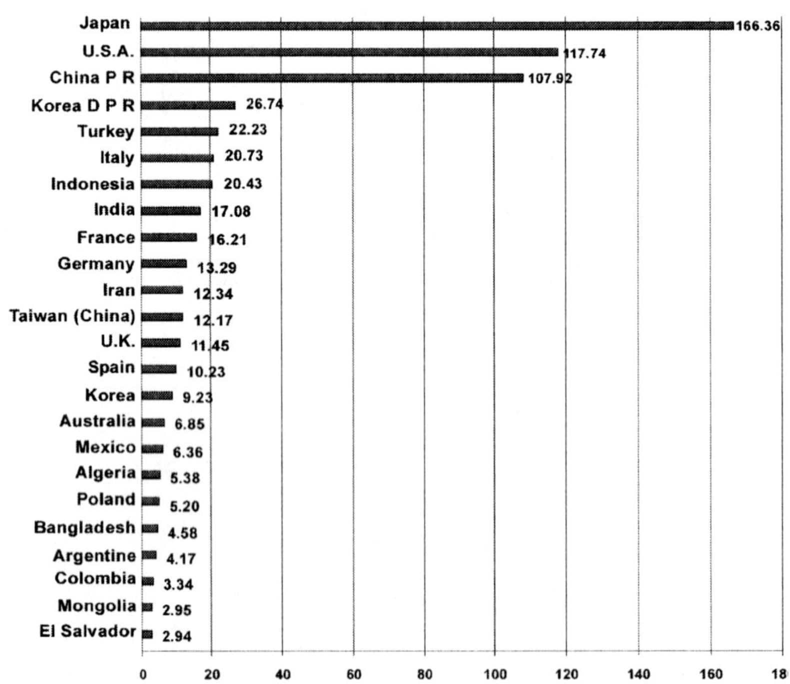

DEFINITIONS, PROCESSES, AND PROCEDURES FOR DISASTER MANAGEMENT

The occurrence of a natural or man-made disaster is contingent on many complex geophysical conditions and social circumstances. The causes predisposing a natural hazard can be classified on the basis of their origin, thus defining (Matar, 2005) hydro-meteorological, geophysical, and bio-chemical disasters. Floods, droughts, hurricanes, tropical cyclones, typhoons, or wild fires can be categorized as hydro-meteorological catastrophes, while geophysical disasters are the outcome of Earth processes such as earthquakes, volcanic eruptions, or landslides. A typical example of bio-chemical disasters are epidemics, pollutions, or insect infestation. Besides, man-made disasters are those catastrophic events whose causes can be ascribed to human behaviour.

In general, natural disasters cannot be prevented, but their deleterious effects and losses can be limited if appropriate planning and management methodologies are applied. Typically, management activities related to a catastrophic event can be grouped into three phases:

- **Predisaster:** This phase comprises all the activities aimed at the avoidance or reduction of risks, including disaster knowledge, preparedness, and prevention. In this phase are also included the actions taken in response to an ongoing or impending hazard (e.g., hazard forecasting, warning, and prediction).
- **Emergency response:** The actions undertaken immediately after the onset of a hazard are classified as emergency responses. They include rescue and relief operations, assessment of the extent and severity of damage, and implementation of remedial measures during the emergency.
- **Recovery and reconstruction:** Reconstruction and rehabilitation of the area affected by a disaster

In the following, a common disaster scenario is described, highlighting a conceptualization of the involved subsystems, and explaining the role of each subsystem and the connections between them.

As shown in Figure 4, a generic disaster scenario is composed of three different subsystems: data collection and detection, management, and response subsystems. These subsystems are strictly related through an integrated communication channel, which is conceived to develop and maintain sound relationships among these subsystems during an emergency; moreover, the integrated communication channel needs to have a very high reliability because it has to be working in a hazard situation.

The *data collection and detection subsystem* focuses on the relatively routine monitoring of the natural, technological, and civil environments that could induce an emergency. It collects, collates, assesses, and analyzes information about those environments and, when warranted, makes a prediction about the potential occurrence of an emergency. As shown in Figure 4, data are derived using remote sensing satellites and, in some cases, in situ sensors. The collected information is then distributed using the integrated communication channel. The data collection and detection subsystem is largely the domain of scientific organizations for natural hazards. Even if national monitoring systems exist, an increasing number of international charters are being created with the aim to optimize the provision of information on different phases of a disaster. This would be of great benefit for those countries lacking remote sensing satellites.

The *management subsystem* is focused on integrating the risk information received from the data collection and detection subsystem and warning the public when warranted. This subsystem is composed largely of local emergency management officials. After receiving information from the detection subsystem, these managers must interpret that information in terms of potential losses

Figure 4. Subsystems involved in a generic disaster scenario

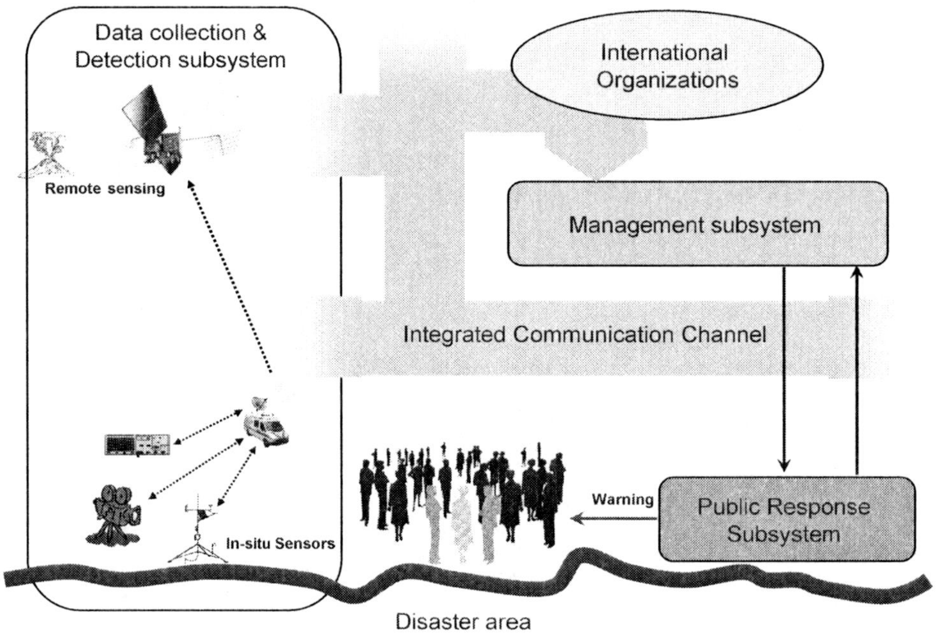

(e.g., loss of life and property) and then decide if the risk warrants a public warning. In making such decisions, managers use specified or ad hoc criteria; moreover, international charters can be used in order to exploit the satellite resources of different countries in a more effective way for the benefit of the international community. The management subsystem, which is also devoted to the preparation of emergency plans, is typically the domain of local government at various hierarchical levels. For example, a mayor or county executive is usually responsible for issuing evacuation advisements for floods. Occasionally, warning the public is the responsibility of the governor as, for example, in the case of nuclear power plant accidents in some states.

The public response subsystem implements all the actions and plans to be taken immediately before, during, and after a catastrophe occurs. The public response comprises both humanitarian organizations and national government forces involved in relief, rescue, and rehabilitation activities. This subsystem is also devoted to disseminate the warning messages received from the management subsystem. This specific scope requires the following particular structural characteristics. First, comprehensive and multiple communication channels to the public should be adopted. Second, warning messages have to be appropriate and provide the public with exhaustive emergency response plans. Third, people response should be monitored, also providing feedbacks with the management subsystem in order to make adjustments when needed.

SPACE TECHNOLOGY FOR EARTH MONITORING AND DISASTER PREDICTION

The role of space technologies is indispensable in all the phases of a disaster management programme providing a new and unique opportunity to access a series of precise and complementary tools essential for understanding the Earth's phenomena and features. Indeed, data derived from Earth observation satellites are fundamental to providing detailed information on geology, tectonics, seismicity, regimes of rivers and basins, local meteorological conditions, terrain, and topography, and thus helping in the prevention and reaction to a natural or man-made catastrophe. Furthermore, development of new sensors multiplied by increased temporal, spectral, and spatial resolution have made these data as an ideal means not only to monitor different surface and environmental phenomena, but also to define the parameters involved, to delineate the areas affected and, finally, to extract information required for planning and management.

The use of space-derived technology for Earth observation related to natural disasters is a very broad topic, which is briefly presented in this section from two different points of views. Firstly, a technological review of the most common classes of satellite instruments for Earth observation will be proposed. Then, the beneficial effects that can derive from space systems will be exposed, while considering different hazard scenarios.

Space Technology for Disaster Mitigation

In all disaster scenarios, the data collected from space are the fundamental layer over which all other actions and responses are structured. The provision of information on the Earth's phenomena from space is implemented using a variety of instruments flown on various space platforms and based on different measurement techniques. There are currently in operation (EOH, 2005) about 70 Earth observation satellites, but the sustained investment of all the world's civil space agencies will ensure that in the next 15 years over 100 missions with more than 300 instruments will be launched. However, the data collected by Earth observation satellites can be efficiently employed for an effective mitigation of a catastrophic event only when sufficient satellite coverage and appropriate continuity of the space-based measurements is ensured. For this reason, integrated satellite observations and international cooperation is essential when a natural disaster occurs.

Meteorology

Meteorology is certainly one of the most established Earth observation disciplines and it is likely to be the one most intensely focusing the public awareness. Weather forecasts use sophisticated models of the atmosphere, whose status is described by temperature, wind speed, humidity, and pressure (Holton, 1992). The temporal evolution of the atmosphere's status is thus obtained, solving the model equations where the current status is known. Observations are thereby essential inputs for any numerical weather prediction (NWP) model. Even if networks of in situ meteorological sensors still have a primary role, the function to collect data on the atmosphere is essentially provided by meteorological satellites which are generally geostationary (GEO) or polar orbital satellites providing continuous coverage of much of the globe. These satellites allow for both visible (VIS) and infrared (IR) images, which are then used to determine the movement of the atmosphere, thus providing detailed information on the winds. Images in the IR region have also proven to be very effective in the determination of top cloud temperatures, which in turn can be used to estimate the rainfall in tropical areas. The humidity and temperature vertical profiles of the atmosphere can be determined using atmospheric sounder instruments, which are generally based on radio occultation systems and on passive IR or microwave measurements. Other important parameters being measured by weather satellites are the sea surface temperature, precipitation, and liquid water and identification of the winds on the sea surface. In particular, the estimation of the intensity, direction, and possible circulation patterns of the surface winds is essential to classify interseasonal climate variation and to help storm centres.

Earth Observation

A large number of the satellite missions planned for the next few years have Earth observation as a main objective. The increasing interest and demand for satellite data is mainly due to the improved satellite and sensor technology. Another significant element that contributes to the increasing expansion of Earth observation systems is related to the use of small satellite platforms that can provide a consistent cost advantage in meeting the needs for higher temporal resolution by means of an affordable constellation of EO satellites. For these reasons, the use of remote sensing and of other geo-information technologies (GIT) has become an integrated, well developed, and successful tool in all the phases of a disaster management. Moreover, the increased availability of space-based data and the improved functionalities of geographical information systems (GIS), combined with the potential of other related applications, such as satellite-based positioning systems, have created new opportunities for a more detailed and rapid analysis of the parameters characterizing a natural hazard.

The correct and rapid analysis of a disastrous event requires a wide range of sensors, including thermal-IR, multispectral and hyper-spectral sensors, high-resolution panchromatic sensors, and synthetic aperture radars (SAR). The integrated and combined use of these enabling technologies is the core of a comprehensive disaster management system. However, the necessity to retrieve information simultaneously from several systems and the accessibility to EO data, together with their temporal resolution, are the two key parameters that should be considered for appropriate disaster monitoring, modeling, early warning management, evacuation planning, rescues operation management, and postdisaster assessment.

A detailed discussion of all the different classes of instruments employed for Earth observation in a disaster scenario is beyond the scope of this book. However, for ease of discussion a short overview of the most common instrument categories is reported in the following classification.

Atmospheric Chemistry Instruments

This class of instruments provides a measurement of composition of the atmosphere performed using different parts of the electromagnetic spectrum. Each atmospheric gas is in fact characterized by particular interaction with the electromagnetic waves transmitted by a satellite. This interaction is typical of each molecule and it is expressed by the absorption and emission spectra, which in turn are exploited by remote sensing instruments to identify the atmospheric composition.

Even if the earliest atmospheric chemistry instruments have been developed to individuate the stratospheric ozone depletion, this type of instrument is currently used (ACE, 2006) also to find traces of different gases and for pollution monitoring and climatology.

High Resolution Optical Images

High resolution optical images can provide a detailed image of the Earth based on panchromatic and multispectral sensors which simultaneously record data in the visible and IR spectrums. Typically, these instruments can supply images with a resolution in the range from 10 to 100m, although the last generation of instruments can reach a resolution in the range of 1-5m (Figure 5) (LandSat, 2006). Unfortunately, these sensors are limited by adverse meteorological conditions and they can operate only in the daytime. These high resolution optical images are very useful in natural hazards for damage assessment and for environmental monitoring.

Figure 5. Researchers at Michigan State University and the University of Hawaii have discovered the first ever image of waves from a Tsunami captured by satellite LandSat7 (LandSat, 2006) on Dec 26, 2004. In the image, two breaking waves can be seen along Devi Point on the eastern shoreline of India. The inset shows that each breaking wave is composed of several smaller undulating waves. These images coincide with the arrival on the Indian coast of the second tsunami wave crest, as measured by the ocean surface topography satellite Jason-1. Data available from U.S. Geological Survey/EROS, Sioux Falls, SD.

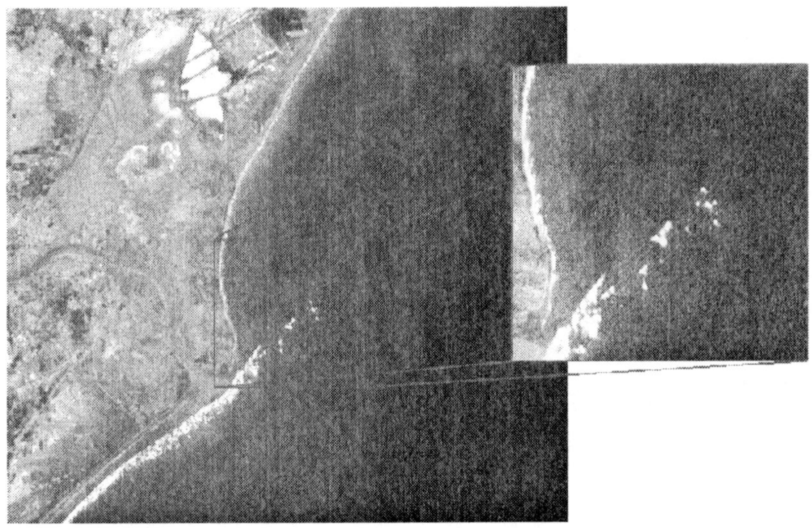

Imaging Microwave Radars

Microwave sensors play a key role in the remote sensing of the planetary surface. These instruments, which have been developed in the late 1970s and early 1980s, produce high resolution images using a radar operating in the range from 1 to 10GHz. Both synthetic aperture radars (SAR) and real aperture radars can reconstruct high resolution microwave images of the Earth through the analysis of signals backscattered by the Earth's surface. The main feature of these systems is their ability to retrieve images with any weather condition and on night/day basis. Most recent multipolarized SAR instruments have the capability to retrieve quantitative data on biophysical parameters such as soil moisture and biomass.

Land microwave images are currently used as an input for the prevention and assessment of many different natural disasters. In particular, SAR data allow the identification of oil slicks and, in case of near real-time applications, the detection of anomalous ocean waves. Another important application is the generation of 3-D topographical images and, when differential SAR interferometry is employed, it is possible to identify ground movements with a millimetre and even submillimetre accuracy.

HOW SPACE-BASED MONITORING DATA CAN BE HELPFUL IN DIFFERENT HAZARD SCENARIOS

The data collected from space can have different impacts in the reduction of the social and economic losses generated by a calamity. In fact, the capabilities of Earth observation have proven to be significantly beneficial for many different disaster scenarios, although the potential of satellite technologies depends on the characteristics and peculiarities of each catastrophic event.

Cyclone Detection

One of the most common applications of Earth observation satellites for disaster prevention is the identification of storms and the capability to comprehend their evolution. NOSS, ENVISAT, RADARSAT, and other satellites carry instruments specifically designed and operated for storm identification (Figure 6). Detection of hurricanes or tornados has indeed proven to be very effective in many different circumstances being considered an established methodology. However, only in the last few years, the advent of satellites capable of measuring the wind at the sea surface significantly, has improved forecast reliability. Indeed, the integration of these data with meteorological information and with a detailed tropical rainfall mapping has enabled a better determination of a storm's location, direction, structure, and strength, enabling the development of various predictive models.

Floods

Another important application of weather satellites is related to flood prevention. In this case, the information on rainfall prediction is essential to generate flood event forecasting. The reduction of flood deleterious effects can also be achieved using so-called flood impact prediction maps, which are sophisticated hydrological models developed using satellite-based measurements of landscape topography and surface wetness. For these particular cases, the role of satellite systems is also essential, which can provide

Figure 6. This three-dimensional image shows the shape of Typhoon Songda, which struck the Japanese homeland in 2004. This visualisation comes from data gathered by ESA's Envisat environmental satellite.

regular and frequent observation of the Earth's surface. To this end, LANDSAT and SPOT data would be extremely beneficial, while SAR instruments which are onboard ERS (Figure 7) and RADARSAT satellites can provide images during day and night and even in the presence of a thick cloud stratus, thus helping in keeping a real time flood extent mapping. Unfortunately, none of the previously mentioned satellites can provide real-time monitoring. For this reason, the knowledge of flood-related hydrogeological variables which could be helpful to prevent the flood evolution is also essential. Very high resolution sensors can be eventually employed to map damage assessment, direct response efforts, and aid reconstruction planning.

Desertification

The identification of desert areas requires advanced use of satellite technologies. In particular, drought monitoring requires information generated by different sensors to determine precipitation intensity, vegetative biomass, atmosphere humidity, and winds. A number of radiometers provide measurements of vegetation cover (Figure 8), including AVHRR/3, MODIS, MERIS, and the purpose-designed VEGETATION satellite.

Wildfires

Information derived through satellite missions such as NASA's MODIS is also essential in combating a wildfire hazard. The risk level can be, in fact, identified at the early stage of a forest fire

Figure 7. ESA's ERS-2 image of floods in northern Italy. This night image covers the Maritime Alps and the Western part of Piemonte, with the town of Torino clearly evident at the top right of the image. Some flooded areas are evident along the Po river.

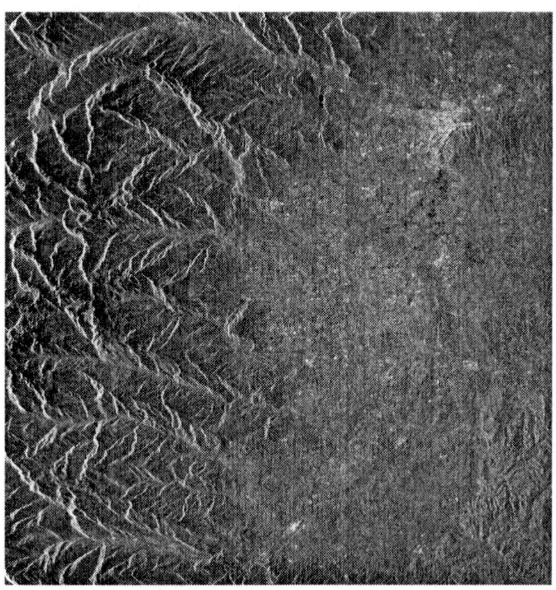

Figure 8. This set of examples illustrates the strong variations that can be detected in the vegetation cover, either at regional or global scales. The images present a sample of four images (VEGETATION, 2006) on Europe and Northern Africa which show vegetation density in January, March, July and September 1986.

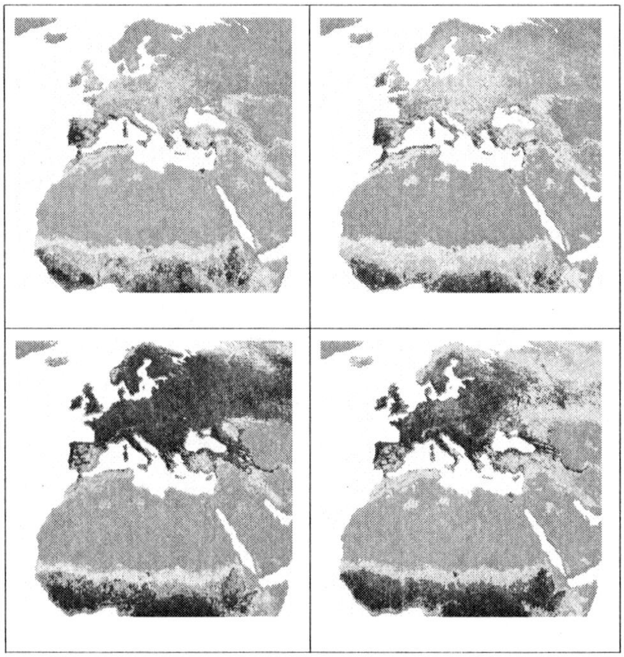

by combining meteorological forecasts, digital elevation maps, and other satellite imaging data which are essential for estimating the possible evolution of fires and for the assessment of a burned area.

Oil Spills and Pollution Dumping

An essential role is played by satellite technology and, in particular, by SAR systems, in the detection of oil spills and pollution dumping. Many countries have indeed established SAR-based surveillance systems which are capable in identifying oil slicks in a short time, providing information on the extent and helping in managing the most suitable response. Pollution-dumping dedicated missions are MOPITT (CSA) and SCIAMACHY (ESA).

Earthquakes

Satellite technology can support an earthquake event, assessing the vulnerability before in the predisaster event and providing the necessary data in the post-event phase. Crustal deformation can be measured by comparing SAR images of the interested before and after the earthquake. In-situ global position system (GPS) measurement can be used to improve the accuracy and the resolution of space-derived data. Unfortunately, the prediction of an earthquake is something still not possible, although some precursors have been identified in some ionosphere measurements (Jason, Pulinets, Silva Curiel, & Sweeting, 2002).

Volcanic Eruptions

The possibility to have accurate information on seismic or volcanic activity is instead achieved by using real time in-situ networks of GPSs receiving stations. Also, 3-D images reconstructed using panchromatic stereo imagery have proven to be helpful finding out the evidence of hazardous activities. Patterns of the thermal activity of a volcano can also be quantified making use of IR-based measurements such as the ones performed by IKONOS. Both visual and thermal satellite imagery can be very precious in identifying the locations and risk associated to volcanic ash, which could be a significant issue for aviation safety. In case of volcanic eruptions, the topographic changes and the deformations of the Earth's surface could be very accurately identified making use of interferometers, radars, or in-situ GPS stations.

Tsunamis

Tsunamis are generated by large-scale sudden movement of the sea floor, normally generated by earthquakes, landslides, volcanic eruptions, or man-made explosions. Even if it is difficult to prevent this event, it is essential to detect this phenomenon and to launch the correct warnings. Unfortunately, when a tsunami is generated, it is nearly imperceptible as the wave height is less than a meter. The most effective way to achieve real-time monitoring is to use a network of in-situ sensors which use satellite communications to send data to government warning centres.

COMMUNICATIONS SYSTEMS FOR DISASTERS AND OTHER EMERGENCY SCENARIOS

During disasters, telecommunications infrastructure failures occur through a variety of mechanisms (Townsend & Moss, 2005). Investigation of communications failures during large urban disasters in the past 15 years reveal three primary categories of causes:

- Physical destruction of network components
- Disruption in supporting network infrastructure
- Network congestion

The most common and well-documented cause of telecommunications failures in recent disasters has been the physical destruction of network infrastructure. Because of the time and funding needed to repair or replace systems, service disruptions caused by physical destruction also tend to be more severe and last longer than those caused by disconnection or congestion. The September 11 attacks caused collateral damage to an important telephone routing hub near the World Trade Center, disconnecting large portions of lower Manhattan from the telephone network. Newer telecommunications networks are designed to be more resilient to physical destruction. Wireless links, whose links are constructed out of intangible electromagnetic radiation, reduce some of the vulnerability of wired networks. Yet, as recent disasters have shown, they are vulnerable to physical destruction too. However, wireless networks have a high degree of variability in their vulnerability to physical destruction of nodes, and the loss of service that results. Broadcasting facilities are typically centralized at the metropolitan scale, making them extremely vulnerable. The destruction of One World Trade Center, where many television and radio broadcast antennas were located, disrupted the broadcast capabilities of numerous media outlets.

Electrical distribution systems are by far the most important supporting infrastructure for telecommunications networks. Electrical power is required to operate all modern telecommunications equipment, often in large amounts. Yet electric power distribution systems lack the "self-healing" capabilities of telecommunications networks, although future improvements are expected to give power networks greater capabilities in this area. Finally, failures in transportation disruptions can also impact the supply of fuel for electric power generation. After September 11, a key hub for transatlantic telecommunications—the Telehouse at 25 Broadway—which had already lost its main power supply, was knocked off-line due to failures in its backup generators caused by tainted diesel fuel. Ironically, one of the oldest technologies for telecommunications—amateur radio—remains the only communications infrastructure that has repeatedly demonstrated its ability to operate effectively when electrical power supplies fail.

According to all the previous influencing factors, providing broadband communication capacity during emergency or crisis times is always a necessary but difficult challenge (see Figure 4). Often, the volume of data to exchange in the disaster areas is important and critical (logistics information, medical data, etc.) and the communication terrestrial networks are not fully available. They are either damaged or saturated and even did not exist in remote areas.

The availability, quick deployment, and reliability of such emergency communication networks in disaster areas are strongly appreciated considering that lives and properties are often at stake. Satellites play a vital role in enabling critical activities, such as the collection and dissemination of disaster and emergency news, the issuance of warnings, the provision of back-up communications for the continuation of government and business activities, and the transmission of data from remotely located sensors, to continue.

However, the use of satellite to recover communications in disaster areas, for both rescue coordination and population support, is a more recent concern. The weakness and vulnerability of the terrestrial mobile and broadband networks, faced with disasters, have been revealed in major natural or man-made crisis. Modern society is more and more dependent on strong and efficient communication means.

Communication facilities include the specific communication links for transferring information between data acquisition satellites and ground stations in a generic space-based system in addition to data dissemination network essential for environmental disaster assessment. Information from satellites helps to identify areas at risk from disasters, enabling us to take action in advance to reduce the harm that disasters can cause. Maps created from satellite image processing are used to plan and support relief efforts. Up-to-date information is distributed quickly to local authorities and relief personnel on the ground.

Research Projects

Specific programmes and research projects supported by various spaces agencies (i.e., ESA, NASA) are aimed at incorporating the use of space technologies into operational disaster management programmes around the world. This is achieved by bringing together the existing users of space technology with those responsible for dealing with disaster management and space technology in developing countries. Related activities include training and pilot projects for the benefit of educating disaster managers and decision makers about the benefits of space technology.

Generally speaking, the potential applications of space technologies in the management of emergency situations have been identified a long time. National and international initiatives are numerous.

The European projects emergency management by satellite communications (EMERGSAT) (Rammos, Eldridge, Wu, Brownsword, Knight, Winder et al., 2000) and real-time emergency management via satellite (REMSAT) (Rammos, 2000) as well as the French project Prévention et Anticipation des Crues au moyen des Techniques Spatiales (PACTES) (Duncuing & Pierotti, 2001) are good examples of the use of satellite communication systems to anticipate and manage emergency situations.

The activity within the EMERGSAT intends to provide civil protection authorities with remote sensing images from existing satellite systems in near real time for risk management purposes; the project consists of a prototype development, technical verification, and validation of a satellite-based communication system for managing emergency situations involving the use of Earth observation techniques. It shall qualify this satellite communication platform as a cost effective and modular system for near-real time data dissemination in different topologies and different service level requirements.

The system shall allow a flexible set-up for serving a topology characterized by one fixed geographical location, for example, Earth observation satellite data acquisition station and a number of geographical areas in Europe and around the Mediterranean basin. Communications shall be enabled between the fixed location and portable stations to be placed in these areas, or mobiles moving in these areas, in particular ships.

The objective of REMSAT activity is to demonstrate the use of real time satellite communications, localization, Earth observation, and meteorology services, in emergency situations via a pilot demonstration involving end-users and making maximum use of existing technologies.

The REMSAT system provides timely information for the field operations staff to assist in the decision making process; it integrates several recent advances in satellite technology to improve the ability to provide current, spatially accurate, digital, and hard copy map products for use in briefings and distribution to field crews and support planning operations. This includes delivery of real-time position and status information on all resources such as aircraft, equipment, and personnel. In addition, the system can provide geostationary (GEO) and low Earth-orbit (LEO) communications services, digital mapping imagery, Earth observation (EO) products and services.

During a fire or emergency situation the REMSAT system breaks down into three primary points of contact and information dissemination:

- **Command center terminal (CCT):** Responsible for the current resource dispatch function as well as selecting, acquiring, preparing, and integrating digital map and EO data for distribution to the fire camp
- **Intermediate mobile terminal (IMT):** Consisting of a transportable unit able to provide the "missing" communication link between the fire camp and the Provincial fire headquarters
- **Hand held user terminal (HHUT):** Are provided to field crews and working equipment

Once these three levels of organization are activated and in full operation, the fire command is provided with the needed information to analyze and attack a large-scale fire.

The PACTES project initiated by centre national d'etudes spatiales (CNES) and the French ministry of research, aims at improving flood risk management over the following three main phases: (1) prevention: support and facilitate the analysis of flood risks and socioeconomic impacts (risk maps), and the preparation of contingency management plans; (2) forecasting and alert: improve the capability to predict and anticipate the flooding event; and (3) crisis management: allow better situation awareness, communication and sharing of information between the actors involved in crisis management (including on-site support personnel). In order to achieve its ambitious objectives, PACTES integrates state-of-the-art techniques and systems (integration of the overall processing chains, starting from the raw data provided by ground or space sensors, up to the final decision support tool); furthermore, the project takes advantage of integrating recent model developments in weather forecasting, rainfall, hydrology, and hydraulics, and in some cases of improving their robustness in order to make them usable in real-time and (as much as possible) by nonspecialist users. In this approach, space technology is thus used in the following main ways:

- Radar and optical Earth observation in which data are used to produce digital elevation maps as maps of soil conditions (i.e., soil moisture)
- Earth observation data are also an input to weather forecasting together with ground sensors

Other examples of international projects are:

Real time emergency management for forest fire services via satellite (REMFIRESAT) (Gonzalo, Martinez, & Martin, 2004) for using of satellite technologies in an emergency situation such as uncontrolled wildfires. The main satellite technologies to be exploited in the REMFIRESAT system are:

- **Satellite communications** providing global coverage, portable terminals, and wide bandwidth capabilities, the last having become very affordable in the last few years; furthermore, navigation systems based on satellite assets like the popular GPS can make an exact determination of the position of their resources.
- **Earth observation from space** demonstrated to be a very powerful tool for the management of disasters and other large event situations. The satellite systems will work in a synergetic manner with the ground facilities, reaching a more efficient utilization of available technology. The initial REMFIRESAT concept envisages handheld terminals, vehicle terminals, a set of command centers at user premises, and a mobile satellite communication centre to be located close to the fire.

Communication Recovery In Emergency Situation (CRIES) (Gayard, 2003a, 2003b; Gayard & Blanc, 2004) is aimed at proposing a satellite-based communication infrastructure for communication recovery and monitoring in emergency situations and natural disasters. The project includes three domains: mobile communication recovery for disaster area, broadband access in emergency situation, and scientific and logistic data retrieve and broadcast. The first domain is a system for short term recovery of mobile communication in emergency situations. Specially designed manned or unmanned aircraft carrying a communication payload could take off quickly after a disaster and circle above the area at high altitude. The communication payload is equipped with a base transceiver station (BTS) and a satellite terminal. It relays mobile phone signals via the satellite toward a central gateway connected to the mobile network. Such a system would be a major benefit to disaster relief, as they could quickly restore broken communications for several hours. In Europe, this system that includes a small fleet of aircraft, a dedicated satellite capacity, and a gateway station, would offer an inestimable tool for supporting both rescue teams and stricken populations. The second domain is a system for broadband communication capacity during emergency situations; a special designed satellite payload provides multimedia (Internet, e-mail, FTP...) communication capacity in disaster areas, Moreover, compact, lightweight, portable, and easy to use terminals are specifically developed for rescue teams. In Europe, a dedicated system, including its own satellite payload that will be a piggy back of commercial satellite, would offer an attractive tool for supporting fieldwork teams. The third domain is a system for the retrieval and broadcasting of data via a geostationary satellite. Remote and autonomous terminals will exchange data with a gateway station connected to the Internet. Each remote terminal could be accessed by a client/server mechanism from everywhere throughout the world.

Even if all the presented projects represent a big step forward in the disaster monitoring and recovery, most of the proposed architectures are not fully integratible with each other because they are based on different technologies and communication protocols, and thus "integration" and accessibility is the single most important feature and challenge in the next future.

FUTURE TRENDS

Civil populations are "addicted" to the comfort of phoning anybody from everywhere at anytime. Unfortunately, during disasters or emergency situations, mobile networks are either damaged, saturated, or totally absent. Furthermore, rescue teams and the stricken population are more and more dependent on fast, reliable ways of communicating, such as mobile phones. Mobile communications help to coordinate the rescue effort inside the disaster zone and to bridge the disaster zone to the rest of the world, providing news and logistic and psychological support. In some cases, roaming phones could help to locate victims.

Wireless solutions may solve the "last mile" problem, that is, the direct services delivery to customer's premises, offering high-bandwidth services without reliance on a fixed infrastructure. Furthermore, in many scenarios, especially in disaster situations, wireless represents the only viable delivery mechanism. A potential solution to the wireless delivery problem lies with aerial platforms, capable of carrying communications relay payloads and operating in a quasi-stationary position at altitudes up to 22km; moreover, any residual pointing error due to movement of the HAP is assumed to be compensated by appropriate station keeping mechanisms (Vishnevsky, Tereshchenko, & Lyakhov, 2003). The platforms may be airplanes or airships and may be manned or unmanned with autonomous operation coupled with remote control from the ground. Great interest lies with crafts designed to operate in the stratosphere at an altitude typically between 17 and 22km, which are referred to as high-altitude platforms (HAPs) (Pace, Aloi, De Rango, Natalizio, Molinaro, & Marano, 2004), (Grace, Daly, Tozer, Burr & Pearce, 2001).

HAPs can be rapidly deployed to supplement existing services in the event of a disaster (e.g., earthquake, flood), or as restoration following failure in a core network. In particular, such platforms could effectively integrate or substitute terrestrial satellite systems in different ways: as an example, they can be rapidly deployed to provide immediate coverage in disaster areas, or relocated, expanded, and upgraded with new payloads, reducing the obsolescence risk typical of traditional satellites (Avagnina, Dovis, Ghiglione, & Mulassano, 2001; Grace, Daly, Tozer, Burr, & Pearce, 2001). System flexibility is not only in the payload reconfigurability, but also in the possibility of changing platform demands, configuring the system according to needs.

Nowadays, HAPs are being actively developed in a number of programmes world-wide, and the surge of recent activity reflects both the lucrative demand for wireless services and advances in platform technology, such as in materials, solar cells, and energy storage.

Benefits of HAP Communications

HAP communications have a number of potential benefits, as summarized as follows:

- **Large-area coverage (compared with terrestrial systems):** The geometry of HAP deployment means that long-range links experience relatively little rain attenuation compared to terrestrial links over the same distance, due to a shorter slant path through the atmosphere. At the shorter millimetre-wave bands, this can yield significant link budget advantages within large cells.

- **Flexibility to respond to traffic demands:** HAPs are ideally suited to the provision of centralized adaptable resource allocation, that is, flexible and responsive frequency reuse patterns and cell sizes, unconstrained by the physical location of base-stations. Such almost real-time adaptation should provide greatly increased overall capacity compared with current fixed terrestrial schemes or satellite systems.
- **Low cost:** A small cluster of HAPs should be considerably cheaper to realize and launch than a geostationary satellite or a constellation of LEO satellites. A HAP network should also be cheaper to deploy than a terrestrial network with a large number of base-stations.
- **Incremental deployment:** Service may be provided initially with a single platform and the network expanded gradually as greater coverage or capacity is required. This is in contrast to a LEO satellite network, which requires a large number of satellites to achieve continuous coverage; a terrestrial network is also likely to require a significant number of base-stations before it may be regarded as fully functional.
- **Rapid deployment:** Given the availability of suitable platforms, it should be possible to design, implement, and deploy a new HAP-based service relatively quickly. Satellites, on the other hand, usually take several years from initial procurement through launch to on-station operation, with the payload often obsolete by the time it is launched. Similarly, deployment of terrestrial networks may involve time-consuming planning procedures and civil works. HAPs can thus enable rapid roll-out of services by providers keen to get in business before the competition. Furthermore, there is little reason why prepared HAPs should not be capable of being launched and placed on station within a matter of days or even hours. This will facilitate their use in emergency scenarios. Examples might include natural disasters, military missions, restoration where a terrestrial network experiences failure, and overload due to a large concentration of users, for example, at a major event.
- **Platform and payload upgrading:** HAPs may be on station for lengthy periods, with

Table 1. Broadband comparison with terrestrial-HAP-satellite systems

	Terrestrial	HAP	GEO Satellite
Station coverage (typical diameter)	<1km	up to 200km	Up to global
Cell size (diameter)	0·1–1km	1–10km	400km minimum
Total service area Maximum transmission rate per user	spot service 155Mbit/s	national/regional 25–155Mbit/s	quasi-global 155Mbit/s
System deployment	several base stations before use	flexible	flexible, but long lead time
Estimated cost of infrastructure	varies	$50million upwards?	>$200million
In-service date	2000	2005–2010?	1998

some proponents claiming 5 years or more. But they can be brought down relatively readily for maintenance or upgrading of the payload, and this is a positive feature allowing a high degree of "future-proofing."

- **Environmentally friendly:** HAPs rely upon sunlight for their power and do not require launch vehicles with their associated fuel implications. They represent environmentally friendly reusable craft, quite apart from the potential benefits of removing the need for large numbers of terrestrial masts and their associated infrastructure. Table 1 summarizes a comparison between terrestrial, HAP, and satellite delivery for broadband services.

Due to their low altitude, HAPs provide a better link budget with respect to satellites; however, their coverage area is limited to a diameter in the order of 200 km, and hence, they are conceived mainly to offer services on a regional basis. Therefore it is mandatory to clearly understand the potentialities resulting from a synergic integration of Earth, space, and stratospheric segments.

Three typical deployment scenarios are recognized to make clear the role of integrated infrastructures in emergency situations.

Satellite-Based System for Retrieval and Diffusion of Scientific Data

This first scenario defines a communications system capable of delivering broadband services during emergency situations. The data volume to exchange in disaster areas is often intense and critical (logistical, medical, environment, location data). In these areas, a communications network could be either unavailable for damage and congestion, or totally missing (rural areas).

Immediate availability, rapid deployment, and reliability are key factors for network infrastruc-

Figure 9. MANET-HAP, HAP-satellite and satellite-Internet communications

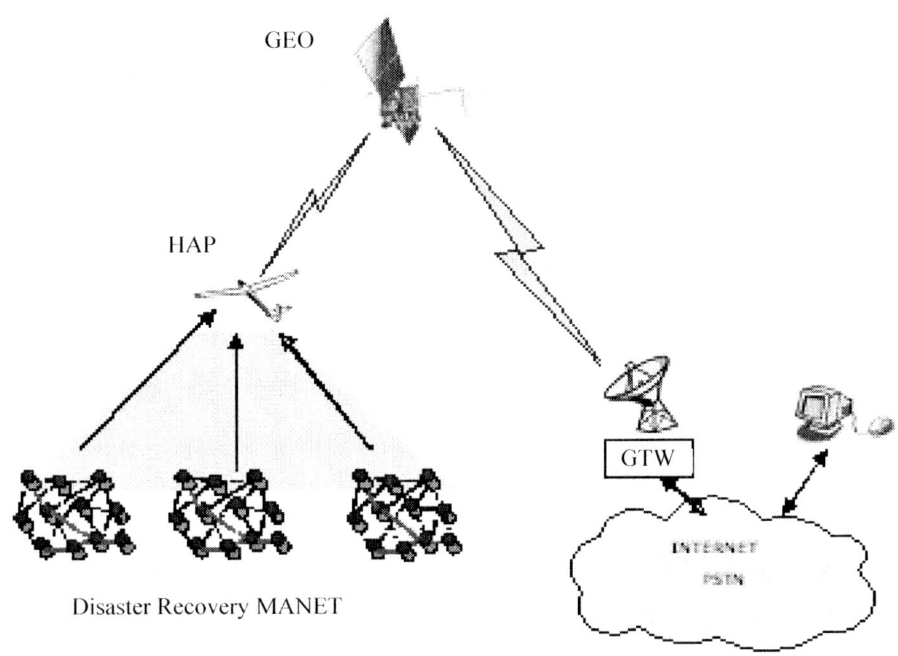

Figure 10. Wireless 3G - HAP/satellite, satellite-Internet communications

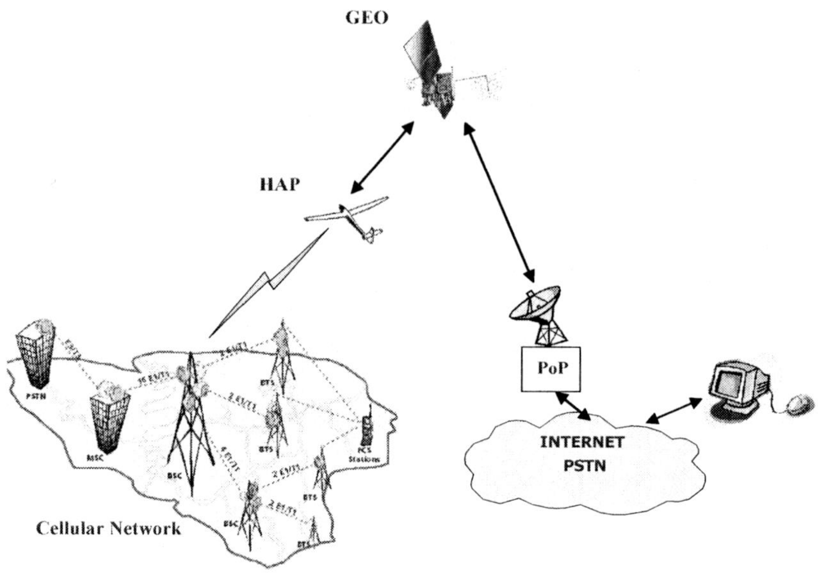

ture in emergency situations. A mobile ad-hoc network (MANET), for example, could be rapidly implemented in a disaster area. Rescue teams could be equipped with compact, portable, and easy to use terminals. A MANET could interconnect all operators in a peer-to-peer way. Now the problem is, how do operators get interconnected to the rest of the world?

Operators' terminals have stringent requirement in terms of power, size, and weight, so a direct connection to a satellite system is not so easy to do. The typical HAP coverage area is limited to a radius in the order of 200 km, considering a minimum elevation angle of 15° (Karapantazis & Pavlidou, 2005), the maximum altitude is about 22Km, and the HAP deployment is as fast as MANET. Therefore, HAPs are the best candidate to quickly offer services on a regional basis for low power terminals.

An integrated Satellite-HAP-MANET infrastructure (see Figure 9) should be able to deliver broadband services (Internet, e-mail, videocommunication, telemedicine, etc.) in critical areas, making them reachable more rapidly from the rest of the world. Operators' traffic could be forwarded toward the public internet using the HAP-Satellite segment as a transparent tunnel.

Recovery of Mobile Communication in Emergency Scenarios

Crises generate intense human need for communication to coordinate response activities, to convey news and information about affected groups and individuals, and as a panic reaction to crisis. Historically, major disasters are the most intense generators of telecommunications traffic, and the resulting surge of demand can clog even the most well-managed networks. In addition, for economic reasons, most communications networks are engineered for peak load at levels well beneath the demands placed on them during disasters.

HAPs can offer a wide range of services and, such services may be particularly valuable where existing ground infrastructure is missing or difficult. Specially designed HAP carrying a communication payload (see Figure 10) could take off quickly after a disaster and circle above the area at high altitude. Its communication payload is equipped with a base transceiver station (3G, GSM, etc.) and a satellite terminal. The BTS antennas will point toward the sinister area and will receive and transmit mobile phone signals. Such "airborne BTS" on HAP and the corresponding system would be a major benefit to disaster relief, as they could quickly restore broken communications for the first but critical hours. HAPs that are designed for short term recovery of mobile communication in emergency situations could also provide other provision of services. For example, services as TV/FM broadcasting over the disaster area to alarm population and provide early safety information or remote sensing of the disaster area to provide rescue organizations with data and picture (optical, infrared or RF picture, meteorological data, etc.) could be very attractive in emergency situations.

Integrated Satellite Multimedia for Monitoring Disaster Areas

High risk zones or areas damaged by disaster need to be deeply monitored. The use of the Internet to link electronic hardware is becoming more and more common through embedded IP software. Remote control equipment, scientific stations, telemetric beacons, and mobile medical imaging hardware will benefit from such a trend. The recent access to Ka frequency bands that offer large bandwidth and allow small aperture antenna, could provide a competitive way to build a satellite system dedicated to retrieve and broadcast data in IP datagram.

Earth observation satellites (Cianca, Prasad, De Sanctis, De Luise, Antonini, Teotino, & Ruggieri, 2005) are mainly located on low Earth orbit (LEO), usually less than 1000 km from the Earth's surface, and they are characterized by the need for downloading huge amounts of data, which are generated by their instrumentations and are stored onboard during the day. The trend for Earth observation satellites is toward new sensors with hundreds/ thousands of Mb/s data-generation capability. In the last decade, a large effort has been undertaken to improve the data-transfer capabilities of satellites. As an example, NASA has recently replenished its tracking and data relay satellite (TDRS) fleet with a trio of satellites called TDRS H, I, J, which enables Ka-band or Ku-band communications up to 300 Mb/s (up to 800 Mb/s with ground stations modifications).

At present, there are two ways for performing the data download:

- Direct data download to ground (mainly in X-band)
- Data forwarding through a GEO data-relay satellite

The limits on the data-transfer capabilities of current systems are mainly due to the short visibility period of LEO satellites with the ground station that should collect the data in the first case listed above, and to the transfer capability of the GEO data relay satellite in the second case. For instance, the GEO satellite ARTEMIS allows Ka-band communications only up to 100 Mb/s.

An integrated HAP-satellite scenario for data relay communications consists of two different links; an optical link between LEO and HAP and a link in the X-band or V-band between HAP and the ground station.

Data relay between HAP and ground could be done through a store and forward technique. This architecture has manifold advantages with respect to traditional data-delay architectures, specifically:

- The visibility between an Earth observation satellite and the HAP is longer than the case

Figure 11. Sensor networks data retrieval using HAP-satellite connections

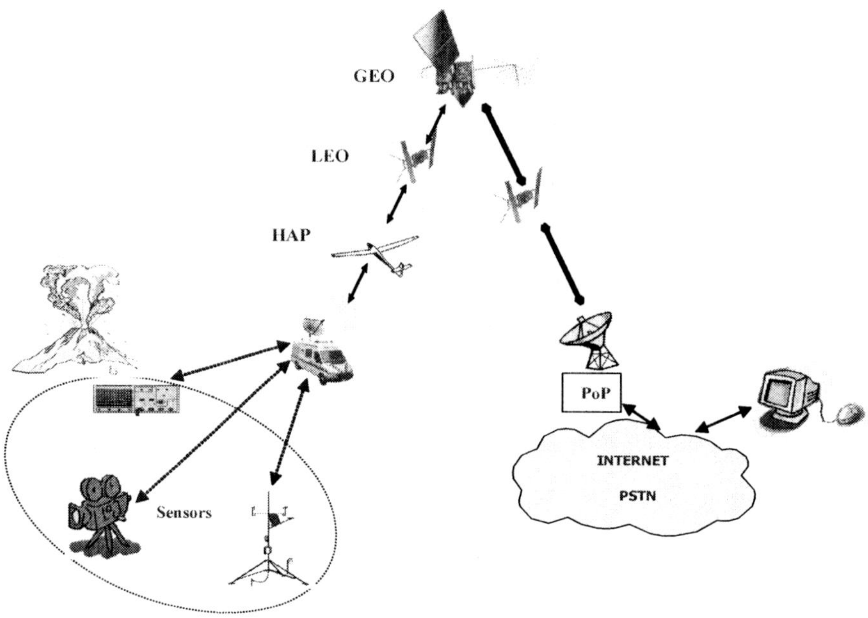

of a LEO satellite. For instance, the visibility time interval between an ENVISAT satellite and an HAP placed above the kiruna station is about 2.5 hours per day. A single HAP equipped with an optical payload could receive multiterabytes of data per day from a single LEO satellite. The issues of optical acquisition and tracking are not challenging (Miura & Suzuki, 2003).

- The distance between the LEO and the HAP is less than the distance between the LEO and the GEO, and less power is required by the link, thus relaxing the power requirements of the traditional RF payload.
- The payload of the HAP does not require space qualification and, hence the use of large solid-state memory (e.g., hundreds of terabytes in size) could be embarked upon.
- After the process of data regeneration, HAP can retransmit the data to a ground station without the typical problems of LEO satellites such as the visibility time. Data downloading could be done by using high-directive and high-gain X-band or V-band antennas at a high data rate (hundreds of Mb/s).
- A single HAP could serve several LEO satellites with multiple optical payload or in different time slots.

This scenario could be improved (see Figure 11) allowing the communication between satellite and terrestrial sensor and the public Internet using HAPs. Such type of purpose, where the timeliness is crucial, could be realized with the use of an integrated system satellite-HAP-sensors network (SHSN). Sensor networks have power limitations, so is almost impossible to deliver data directly from sensors to satellites. A HAP or (well-equipped) van may act as traffic collectors to deliver data traffic from sensors to satellites and may easy deployed in critical areas. Such system will reach remote stations and will provide bandwidth for exchanging technical, medical, and scientific data. All end terminals will exchange data with a gateway station connected to Internet (Point Of Presence). Each remote terminal will be accessed by a client/server mechanism from everywhere through the world thanks to Internet.

Cost Effective Disaster Monitoring Architecture

The cost for implementing monitoring schemes is strongly dependent on the space segment cost that is more expensive than the ground sensor or ad hoc networks used for sending and sharing data information coming from the disaster area. Because the whole architecture price depends on the number of satellites and HAPs within the network, some useful consideration and details about the cost of these network elements will be provide in this section.

The cost of using conventional satellites (>$100M each) to implement a network of Earth observation satellites tailored specifically to disaster prediction, detection, monitoring, and mitigation has proven prohibitive as the whole network would cost nearly $1000M to build and launch the LEO satellites.

However, recent advances in microelectronics have generated a new species of modern, highly sophisticated, computationally powerful, rapid-response microsatellites (and minisatellites) that have reduced the cost of a single satellite by more than an order of magnitude (see Table 2). These "faster, cheaper, better" microsatellites now make the implementation of such a disaster network both practicable and affordable. Moreover, the combination of various small satellite payloads and low-cost, responsive launch vehicles such as *Microcosm's Sprite* (Chakroborty, Wertz, & Conger, 2004) could provide improved weather prediction, real-time monitoring of a stricken area and nearly immediate restoration of communications services needed for relief efforts.

These modern microsatellites and minisatellites are designed and built within a different philosophy to that used for conventional satellites and offer:

- Low cost
- Rapid response
- Tailored mission and sensors
- High operational flexibility (reprogrammable)
- Low operational cost (high autonomy)
- Direct access (low cost terminals & Internet)
- Long operational lifetime in orbit (>10 years)

The typical manufacture time of these satellites is 9-12 months from "order-to-orbit." The typical cost of an optical Earth observation microsatel-

Table 2. Classification of satellites

Class	Cost	Mass
Large satellite	$ > 100 M	> 1000 kg
Small Satellite	$50 - 100 M	500 - 1000 kg
Minisatellite	$ 5 - 20 M	100 - 500 kg
Microsatellite	$ 2 - 3 M	10 - 100 kg
Nanosatellite	$ < 1 M	< 10 kg

lite is around US$2.5M, thus making a network of multiple small satellites for quick response observation of the Earth's changing environment entirely feasible.

The cost of building a complete disaster monitoring and mitigation network comprising seven optical Earth observation microsatellites is just $17.5M USD. Because the microsatellites are tiny (~50kg & 60 liters), it is possible to launch the complete network of 7 microsatellites on a single small launcher, minimizing the system launch cost. Each microsatellite will carry a cold gas station-keeping propulsion system sufficient for a 10-year lifetime.

Responsive space disaster monitoring satellites could be launched within hours for a cost of approximately $20 million per mission (including the cost of the satellite, launch, and operations). For example, Microcosm's Sprite (Berry, Conger, & Kulpa, 2001) low-cost, responsive launch vehicle is an integral part of a Responsive Space disaster monitoring program and requires minimal launch infrastructure. Sprite is capable of launching various small satellites weighing up to 810 lbs into low earth orbit within 8 hours.

On the other side, the cost of building the infrastructure using an HAPs network in more difficult to evaluate because this is a very new architecture and there are only a few prototypes all over the world. Estimates suggest that a HAPs infrastructure would be less than one tenth of a conventional satellite infrastructure (Tozer & Grace, 2001). Compared with satellites, an HAP will be able to serve 1000 times the number of users in a given area. HAPs also will not have a big launch cost.

INTERNATIONAL POLICY FOR DISASTER MANAGEMENT

The Internet and satellite communication services allow dynamic information sharing and exchange between partners in sustainable development within and outside the United Nations system, thus enhancing the benefits of complementary activities.

With active participation from international and national partners, the United Nations family is actively working toward internationally standardized interoperability for sharing and exchanging spatial data and information, often using open source software capacities. This has already significantly enhanced interagency cooperation, reduced duplication of efforts, and achieved tangible benefits within the United Nations family and for its stakeholders.

There are also a number of initiatives in place at the international level. The United Nations Office for the Coordination of Humanitarian Affairs and International Telecommunication Union have a working group on emergency telecommunication.

In order to protect their sovereignty, nations restrict the use of technologies that tend to ignore national boundaries such as remote sensing and satellite communications used also for providing high-resolution imaging of their territory. To avoid such impediments, it might be possible to devise international or bilateral protocols whereby a nation, in accepting disaster relief, would implicitly agree to limited acquisition of imagery necessary to facilitate relief operations.

Several nations (e.g., the United States and France) have both a space capability and a national organization charged with disaster response (The European economic community and the European Space Agency combined also have these responsibilities.) At this national level, development and utilization of the space capability may be more straightforward than at the international level. For this reason, procedures are recommended whereby such nations, when called upon to provide disaster relief, should, through bilateral arrangements with nations requesting disaster assistance, demonstrate the leadership of more highly developed nations, and pioneer the use of satellite technology for disaster management.

Numerous international agreements exist: the Tampere Convention (Oh, 2001) (drawn up in the 1990s and revised in 2001) addresses provision of telecommunication resources for disaster mitigation and relief operations; the ITU has put together a disaster communications handbook and is actively engaged in persuading its members to sign up to the Tampere Convention; moreover, the International Civil Aviation Organization is addressing standards for aeronautical emergency communications.

The Tampere Convention marks a milestone in the international effort in emergency telecommunications. The convention creates an international framework for the provision of telecommunication resources for disaster mitigation and relief between states and between a state and a nonstate entity. This international policy initiative, together with its national counterparts, when combined with modern technologies, could help in alleviating property loss and human suffering.

Under this framework, a state that perceives the need for disaster telecommunication assistance in its territory will request such assistance through the United Nations Emergency Relief coordinator, who is the operational coordinator under the convention. On the other hand, a providing state party is obliged to set down in writing the fees it expects to receive or have reimbursed. The fees, if any, will be based on an agreed model of payment and reimbursement, as well as on other factors, such as the nature of the disaster and the particular needs of developing countries. This framework does not preclude the existing or future arrangements between states and between a state and a nonstate entity in emergency telecommunication assistance.

The convention also recommends states to reduce or remove regulatory barriers that currently impede the use of telecommunications resources for disaster mitigation and relief operations. It further safeguards the privileges, immunities, and facilities accorded to persons providing disaster assistance by granting them immunity from arrest and detention and exempting them from taxation and duties.

Under the Tampere Convention, ITU works closely with the operational coordinator on several provisions of the convention. These include maintaining contact with focal points within states that are authorized to request, offer, accept, and terminate telecommunications assistance. States will also compile a telecommunication assistance information inventory listing, among others, competent authorities and points of contact. This inventory will be maintained and updated by the operational coordinator with the help of ITU.

After the adoption of the Tampere Convention, the European Space Agency invited broad industry support of the convention. The World Trade Organization, in its groundbreaking Telecom Trade Agreement, also enhances access by satellite communication operators into more than 50 countries, with potential emergency telecommunications applications.

In recent years, the space agencies decided to pool the satellite resources of different countries more effectively for the benefit of the international community, and for this reason they founded the International Charter on Space and Major Disasters. The aim of this charter, initiated by the French (CNES) and European (ESA) space agencies, is:

to supply during periods of crisis, to States or communities whose population, activity or property are exposed to an imminent risk, or are already victims, of natural or technological disasters, data providing a basis for critical information for the anticipation and management of potential crises.

CSA (Canada), ISRO (India), and CONAE (Argentina) are also party to the charter. Because the charter became operational on 1st November 2000, authorized civil defense organizations may enlist support from space by calling a telephone number, 24 hours a day, 365 days a year. Rescue and civil defense bodies of the country to which the participating agencies belong are registered authorized users. Civil protection authorities of other countries may also submit requests by contacting their sister organizations through existing cooperation mechanisms.

The charter has proven to be a highly effective and practical mechanism for delivering applications of Earth observation satellite data to those in society in most dire need. The charter has been activated over 55 times since its inception, and continues to support several events monthly in response to calls for assistance from countries all around the globe.

UNOSAT (Wiesmann, Wegmuller, Haeberlin, Retiere, Senegas, Strozzi, & Werner, 2004) and RESPOND (RESPOND, 2004) are further examples of initiatives that have recently been established between various space agencies and the United Nations.

UNOSAT is a consortium of UN agencies, remote sensing service companies, and space agencies, and is supported by a number of committee on Earth observation satellites (CEOS) members and their national governments. UNOSAT aims to encourage, facilitate, accelerate, and expand the use of accurate geographic information derived from EO-satellite imagery by professionals involved in achieving vulnerability reduction, crisis management, and recovery, as well as sustainable development at the local level.

RESPOND is an alliance of European and international organizations working with the humanitarian community to improve access to maps, satellite imagery, and geographic information. RESPOND is a 5-year staged programme providing access to global mapping, access to an archive of detailed base mapping, imagery, and thematic mapping, and access to rapid assessment maps for a major crisis. Satellite imagery collected during an emergency can later be used for postcrisis recovery and development. Images collected on different dates can be compared in order to monitor progress and plan further assistance. RESPOND is one of the services initiated under the global monitoring for environment and security (GMES) initiative of ESA and the European commission. Since 2004, the RESPOND consortium has been collaboratively producing products for the Sudan darfur crisis.

CONCLUSION

In this chapter, it has been illustrated how space technologies and satellite applications can mitigate the impact of natural and man-made disasters, providing the reader with an overview of the most important space technologies for both monitoring and telecommunications, and also showing the main issues in managing the disaster recovery stage. Even if in the last few years significant progresses have been made to reduce the deleterious effects of disasters, there are still a number of institutional and technical obstacles which should be overcome to reduce the human and economic costs of a catastrophic event by means of a pervasive use of satellite-based technologies.

Institutionally, there must be greater cooperation between satellite-operating agencies and entities related to disaster management. This cooperation, combined with modern technologies, is essential to increase the scale and the speed of the response in emergency situations. Technically, future efforts should be devoted at providing remote sensing data more rapidly and at higher spatial resolutions, thus satisfying the needs of an increasing number of disaster scenarios. Improved spatial and temporal resolutions, especially for storm tracing and wind measurements, is therefore required. Particular effort should be also aimed at increasing the compatibility of satellite derived information with the geographical information system (GIS) employed in disaster management programmes.

Moreover, future studies and applications of high altitude platforms (HAPs) shall provide, thanks to their rapid deployment, an added value to wireless communications in emergency situations because they offer reduced propagation delay and broadband covertures. Nevertheless, many issues and research aspects regarding the feasible integration of different architectures and technologies are still open and they require deeper investigation in the next years.

REFERENCES

ACE. (2006). Atmospheric chemistry experiment of the Canadian Space Agency. Retrieved February 22, 2007, from http://www.ace.uwaterloo.ca/

Avagnina, D., Dovis, F., Ghiglione, A., & Mulassano, P. (2002). Wireless network based on high-altitude platforms for the provision of integrated navigation/communication services. *IEEE Communication Magazine, 40*(2), 119-125.

Berry, J., Conger, R., & Kulpa, J. (2001, April 17-20). The sprite mini-lift vehicle: Affordable launcher for small spacecraft missions. In *Proceedings of the 19th AIAA International Communications Satellite Systems Conference*, Toulouse, France.

Chakroborty, S., Wertz, J. R., & Conger, R. (2004, October 4-8). Responsive access to space. The scorpius low-cost launch. In *Proceedings of the 55th International Astronautics Federation (IAF) Congress*, Vancouver, Canada.

Cianca, E., Prasad, R., De Sanctis, M., De Luise, A., Antonini, M., Teotino, D., & Ruggieri, M. (2005). Integrated satellite-HAP systems. *IEEE Communications Magazine, 43*(12), 33-39.

Duncuing, J., & Pierotti, P. (2001). PACTES: A project for improving the operational management of the flooding risk, by integration of Earth observation and advanced modeling techniques. In *Proceedings of the 52nd International Astronautical Federation (IAF) Congress*, Toulouse, France.

EM-DAT. (2006). The OFDA/CRED international disaster database—Université Catholique de Louvain—Brussels—Belgium. Retrieved February 22, 2007, from www.em-dat.net

EOH. (2005). Earth observation handbook. The committee on Earth observation satellites of the European Space Agency (ESA). Retrieved February 22, 2007, from http://www.eohandbook.com

Gayrard, J. (2003a, April 15-19). Satellite based systems for communication recovery in emergency situations. In *Proceedings of the 21st AIAA International Communications Satellite Systems Conference and Exhibit*, Yokohama, Japan.

Gayrard, J. (2003b, September 29-3). Roles of communication broadband satellites in public protection and disaster relief. In *Proceedings of the 54th IAC International Astronautical Congress*, Bremen.

Gayrard, J., & Blanc, F. (2004, May 9-12). Ka-band satellite system for public protection and disaster relief operations. In *Proceedings of the 22nd AIAA International Communications Satellite Systems Conference and Exhibit*, Monterey, CA.

Gonzalo, J., Martínez, S., & Martín, G. (2004, April 19-22). REMFIRESAT: Real time emergency management for forest fire services via satellite. In *Proceedings of the 2nd Symposium on Fire Economics, Planning and Policy*, Cordoba, Spain.

Grace, D., Daly, N., Tozer, T., Burr, A., & Pearce, D. (2001). Providing multimedia communications services from high altitude platforms. *International Journal on Satellite Communication, 19*, 559-580.

Holton, J.R. (Ed.). (1992). *An introduction to dynamic meteorology* (3rd ed.) International

Geophysics Series (vol. 48). San Diego, CA: Academic Press.

Jason, S., Pulinets, S., Silva Curiel, A., & Sweeting, M. (2002, August). Earthquake forecast science research with small satellite. In *Proceedings of the USU 13th Annual Small Satellites Conference* (SSC02-IX-5), Utah State University, Logan, UT.

Karapantazis, S., & Pavlidou, F. (2005). Broadband communications via high-altitude platforms: A survey. *IEEE Communication Surveys & Tutorials*, *7*(1), 1-35.

LandSat. (2006). Landsat Project—U.S. Department of the Interior. Retrieved February 22, 2007, from http://landsat.usgs.gov

Matar, M. A. (2005, June 9-11). Space-based facilities for environmental disaster monitoring, assessment and management. In *Proceedings of the 2nd International Conference on Recent Advances in Space Technologies (RAST)*, Istanbul, Turkey.

Miura, R., & Suzuki, M. (2003). Preliminary flight test program on telecom and broadcasting using high altitude platform stations. *International Journal on Wireless Personal Communications*, *24*(2), 341-61.

Oh, E. S. (2001, July 5-7). The international efforts in emergency telecommunications: An overview and analysis. In *Proceedings of the 6th Asia-Pacific Regional Conference of International Telecommunications Society*, Hong Kong, China.

Pace, P., Aloi, G., De Rango, F., Natalizio, E., Molinaro, A., & Marano, S. (2004, May 17-19). An integrated satellite-HAP-terrestrial system architecture: Resources allocation and traffic management issues. In *Proceedings of the 59th IEEE Vehicular Technology Conference—VTC Spring*, Milan, Italy.

Rammos, E., Eldridge, N., Wu, S., Brownsword, P., Knight, T., Winder, M. et al. (2000, April 10-12). REMSAT: A demonstration of the use of satellite techniques to help in the battle against forest fires. In *Proceedings of the International Symposium GEOMARK 2000*, France.

RESPOND. (2004). RESPOND alliance of European and international organizations. Retrieved February 22, 2007, from http://respond-int.org/Respond/

Scot, G. P. (2000, April 3-6). The contribution of VEGETATION/ SPOT 4 products to remote sensing applications for food security, early warning and environmental monitoring in the IGAD sub-region. In *Proceedings of the Vegetation 2000 Conference*, Lake Maggiore, Italy.

Townsend, A. M., & Moss, M. L. (2005). *Telecommunications infrastructure in disasters: Preparing cities for crisis communications* (Tech. Rep.) Robert F. Wagner Graduate School of Public Service, New York University.

Tozer, T., & Grace, D. (2001). High-altitude platforms for wireless communications. E*lectronics and Communication Engineering Journal*, *13*(3), 127-137.

VEGETATION. (2006). VEGETATION programme. Retrieved February 22, 2007, from http://vegetation.cnes.fr

Vishnevsky, V. M., Tereshchenko, B. N., & Lyakhov, A. I. (2003, October 19-22). Wireless communication networks using HAPs on the base of tethered balloons. In *Proceedings of the 6th International Symposium on Wireless Personal Multimedia Communications WPMC*, Yokosuka, Japan.

Wiesmann, A., Wegmuller, U., Haeberlin, Y., Retiere, A., Senegas, O., Strozzi, T., & Werner, C. (2004, September 20-24). SAR based products for the implementation of humanitarian aid and development assistance projects within the UNOSAT project. In *Proceedings of the IEEE International Geoscience and Remote Sensing Symposium IGARSS*, Anchorage, AK.

Chapter IV
Cospas-Sarsat Satellite System for Search and Rescue

James V. King
Communications Research Centre, Canada

ABSTRACT

This chapter outlines the development and evolution of the Cospas-Sarsat system, describes the principle of operation, presents the current status, and looks at the future of the system. Cospas-Sarsat, an international satellite system for search and rescue, started operating in 1982, and has been credited with saving many thousands of lives since then. More than a million aviators, mariners, and land users worldwide are equipped with Cospas-Sarsat distress beacons that could help save their lives in emergency situations anywhere in the world. A constellation of satellites is circling the globe monitoring for distress signals, while tracking stations on six continents receive the satellite signals, compute the location of the emergency, and quickly forward the distress alert information to the appropriate rescue authorities. This is a big improvement over the presatellite era, when distress signals from remote regions or far out at sea might not have been heard for many days or even weeks.

INTRODUCTION

In the 1960s, light aircraft and some marine vessels started carrying small, battery-operated radio transmitters operating at the international distress frequency of 121.5 MHz that could be activated in an emergency distress situation. Such transmitters, called emergency locator transmitters (ELTs) on aircraft as illustrated in Figure 1, and emergency position indicating radiobeacons (EPIRBs) on ships, emitted a low-power signal that could be picked up by a receiver in a nearby aircraft or by an air traffic control tower in the vicinity.

By the mid 1970s, more than 250,000 distress beacons were in service in various parts of the world, including Canada, the United States, and Europe. Lives of aviators and mariners were being saved thanks to these transmitters, but the service could be improved now that it was the "space age."

A NEW SATELLITE SYSTEM

To improve the detection of distress signals, particularly in remote areas, the concept of a satellite receiving system was proposed. One such study in 1971 entitled "Locating People in High Latitudes" (Stevinson & Baker, 1971) analyzed various satellite orbits that could be used, coverage areas, satellite waiting times, and so forth. By the mid 1970s, some trials were under way to use satellites for tracking animals and scientific buoys, and preliminary experiments were conducted in Canada, France, and the United States to use satellites for detecting and locating distress signals.

In Canada, the initial proof-of-concept tests were carried out using modified distress beacons, transmitting through an amateur radio satellite called OSCAR-6, to a prototype ground receiving station. A senior Canadian military search and rescue (SAR) official was convinced of the potential of such a system when he witnessed this rudimentary system compute the location of a distress beacon that had been deployed to a secret remote place in the wilderness, known only to him and the pilot who placed it there.

Agencies in those three countries then agreed to set up a joint experiment called search and rescue satellite-aided tracking (SARSAT), and initiate the development of space and ground segment prototype equipment (Winter, 1978).

About that same time, U.S. astronauts and Soviet cosmonauts met up in space when their Apollo and Soyuz capsules linked up high above the Earth, and following that success, there was mutual interest in cooperating in other joint space activities.

So in 1979, the former USSR (later Russia) joined the experiment and agreed to develop a compatible system called cosmicheskaya sistyema poiska avarinyich sudov (COSPAS), in Russian, meaning a space system for the search of vessels in distress, and the four founding countries, shown in the logo in Figure 2, established the Cospas-Sarsat system, as depicted in Figure 3.

Figure 1. 121.5 MHz ELTs are installed on many small aircraft (courtesy of CRC)

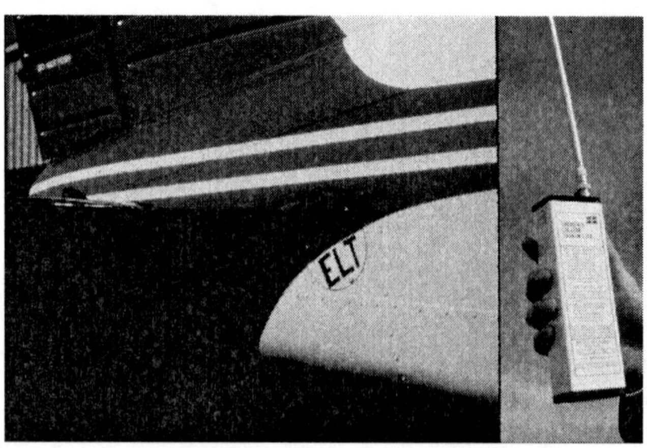

SYSTEM CONCEPT

The basic concept of the Cospas-Sarsat system (*Cospas-Sarsat, 1999*) is illustrated in Figure 4. The Cospas-Sarsat system comprises the space segment and the ground segment, to the point where the alert data leaves the MCC, as the RCCs and SAR response units are national or regional entities that utilize Cospas-Sarsat alert data to facilitate their operations. The distress beacons are owned and maintained by the users.

Distress beacons can be activated manually by the user or automatically in an emergency situation, as some can be triggered on impact or by immersion in water, depending on the type of beacon. The beacon antenna transmits the signal upward across much of the sky, so the user does not need to point the antenna or even know where the various satellites are in the sky. The user does not need to speak, or enter any data, as the distress message is self-contained in the beacon. Thus, the system works in any part of the world in any language.

Distribution of Alert and Location Data

121.5 MHz and 406 MHz distress signals from aviation ELTs, maritime EPIRBs, and personal locator beacons (PLBs) are detected by satellites in various orbits and relayed to ground stations that process the signals and compute the beacon

Figure 2. Cospas-Sarsat logo of the four founding countries (courtesy of Cospas-Sarsat)

Figure 3. Cospas-Sarsat satellites in polar orbit (courtesy of Cospas-Sarsat)

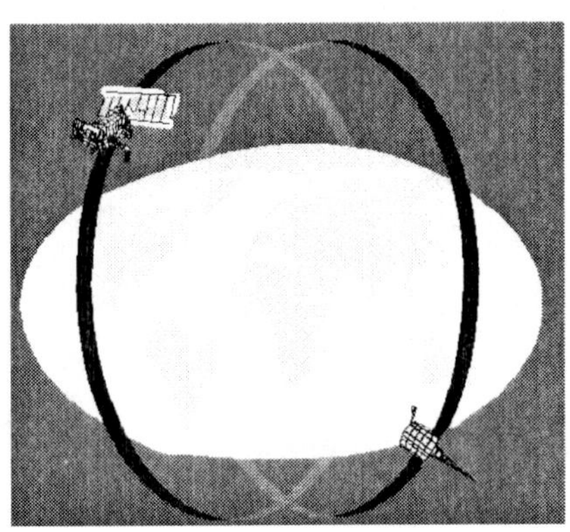

location. Distress alerts, along with their computed locations, are then automatically forwarded to mission control centres (MCCs) where they are sorted and routed to the appropriate rescue authorities at the rescue coordination centre (RCC) in the region of the distress.

Multiple satellites and ground stations normally detect the same beacon signal, providing additional data, and the MCCs screen this information and forward the most timely or most accurate to the RCCs. The additional follow-up data is still useful and provides good backup in the system. Distress data is in numerical form and presented in certain standardized formats, so it can be sent anywhere in the world, independently of local languages.

DEMONSTRATION AND EVALUATION PHASE

In the late 1970s and early 1980s, the design of the Cospas-Sarsat system was begun, radio frequencies were allocated, host satellites were arranged, satellite payloads were designed and built, and special ground receiving stations, called local user terminals (LUTs), were developed and installed.

The basic Cospas-Sarsat system utilizes a constellation of four low-Earth-orbit (LEO) satellites in near-polar orbit, as depicted in Figures 8 and 9, known as the LEOSAR system. With this type of orbit, a single satellite eventually scans the entire globe over a period of several hours.

When the first satellite was launched in 1982 the "experiment" was officially underway. A number of tests and demonstrations (Bogdanov, Ageev, Belousov, Buyanov, & Terekhin, 1984; Flatow, & Gal, & Hayes, 1984; Goudy, King, & Kissel, 1984; King, Hayes, & Jutras, 1983; Levesque, Hodgkins, & Drover, 1984; McGunigal, McKinnon, Brachet, & Zurabov, 1984; Renner & Kozminchuk 1984; Rogalski & Krupen, 1984; Werstiuk, 1983; Werstiuk, Ludwig, Trudell & Selivanov, 1984) were carried out around the world over the next few years to evaluate performance, optimize parameters, and demonstrate this new satellite system to SAR forces, as illustrated in Figure 5.

Figure 4. Basic concept of the Cospas-Sarsat system (courtesy of Cospas-Sarsat)

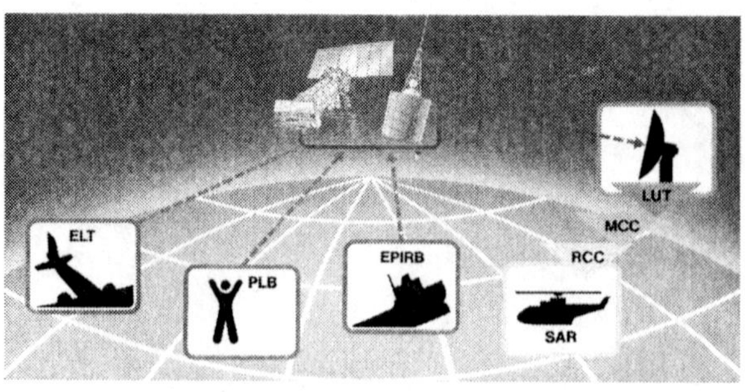

ELT	Emergency locator transmitter
EPIRB	Emergency position-indicating radio beacon
PLB	Personal locator beacon
LUT	Local user terminal
MCC	Mission control centre

Just days after the first satellite was activated in orbit to begin the testing phase, a real 121.5 MHz distress signal was detected and all three people aboard a small aircraft were successfully rescued soon after it had crashed in a mountainous region in northern Canada. There was extensive media coverage of this remarkable event (Tower, 1983) as it was the first use of this satellite system for detecting and locating a distress signal. A few weeks later, the first maritime rescue occurred in the Atlantic Ocean, off the east coast of the United States, and ever since then additional lives were routinely being saved with the help of the Cospas-Sarsat system.

In addition to providing distress alerting and locating services for the hundreds of thousands of existing owners of 121.5 MHz distress beacons, Cospas-Sarsat was also developing and testing a new, more sophisticated, distress beacon operating at 406 MHz. This type of beacon allowed the distress location to be pinpointed more accurately and to be recognized by its unique identification code. Search and rescue forces would then know in advance where they should go to search, and what they should be looking for, thus making for a more effective mission. 406 MHz beacons also transmit a low-power 121.5 MHz homing signal so SAR forces can pinpoint the beacon with existing homing equipment once they are in the vicinity.

OPERATIONAL SYSTEM

By the end of the 3-year demonstration and evaluation phase, several hundred persons had been rescued thanks to the experimental system, and the four founding countries issued a Cospas-Sarsat project report in 1985 (*Cospas-Sarsat, 1985*) declaring the system operational, and undertook to set up a more permanent, worldwide system. In 1987, the Cospas-Sarsat secretariat was established at the headquarters of the International Maritime Satellite Organization (Inmarsat) in London, and this secretariat later moved to Montreal in 2005. In 1988, the four countries signed a formal intergovernmental agreement (Cospas-Sarsat Document, 1988) ensuring the long-term continuity of the system in which three United Nations agencies were also involved:

- The International Maritime Organization (IMO) for worldwide shipping

Figure 5. Setting up beacon for testing with new satellite system (courtesy of CRC)

- The International Civil Aviation Organization (ICAO) for worldwide aviation
- The International Telecommunication Union (ITU) for radio frequency allocations

The system continued to expand as more countries joined and shared in its operation and use, more ground stations were built, and more distress beacons were installed on ships and aircraft, resulting in even more lives being saved by Cospas-Sarsat.

MANAGEMENT AND FUNDING

The Cospas-Sarsat global satellite system provides a valuable, humanitarian service, demonstrates international cooperation, and serves various user communities. The system is unique in the way that it is operated and funded, because its use is free of charge to the end user in distress. It is both a service to the public and a tool to help SAR forces do their jobs more efficiently and reduce their costs, although these cost savings are sometimes difficult to quantify.

The cost of implementing and operating the Cospas-Sarsat system is shared by various member governments, while the cost of buying and maintaining distress beacons is the responsibility of the users, but they pay no fee to access the system. Under the international agreement, the four founding countries provide the LEO space segment and they, as well as several other countries, own and operate ground receiving stations and mission control centres. The administrative costs of the secretariat are shared by all 40 member countries.

The program is managed by the Cospas-Sarsat council, comprising representatives of all member governments, and is supported by the secretariat and an international group of technical and operational experts, called the joint committee, which meets annually.

Because users in distress do not pay to access the Cospas-Sarsat system, no revenue is generated by the system. Hence, it was difficult to quantify the financial "return on investment" to implement and operate the system, but its success can be measured by the number of lives it helps save. A preliminary study in the United States in 1976 projected that the cost of a search and rescue satellite system compared to its benefits would be 1:6, showing that the system would be very beneficial.

Once the satellite system was in place and operating for some years, the United States did another cost benefit analysis (Mehta, 2006) for the year 2004. The cost for the United States to implement and operate their part of the system for that one year, carry out the searches and rescues with SAR aircraft and vessels, and amortize capital equipment totaled $26M. That year, 95 lives were saved in the United States thanks to the satellite system, and based on $3M per life (US DoT, 2002), the benefits were $285M. This resulted in a cost/benefit ratio of 1:11, which was even better than had been projected some 20 years earlier, clearly demonstrating the usefulness of the system. Furthermore, there are also a number of intangible benefits, including reduced search time by precious SAR resources and less risk for responders.

PRINCIPLE OF OPERATION

LEOSAR System Satellite Configuration

Figure 6 shows the path, or orbital plane, of a satellite circling the earth around the poles. The satellite travels in this plane while the earth rotates underneath it, enabling a single satellite to eventually view the entire Earth's surface. At most, it takes only one-half rotation of the Earth (i.e., 12 hours) for any location to pass under the satellite path.

Having more satellites reduces this "waiting time," so the nominal Cospas-Sarsat constellation is four LEO satellites, which provides a typical waiting time of less than one hour at midlatitudes. New satellites are launched every few years to replace older satellites, so occasionally there are five or six satellites in operation, but it would take many more LEO satellites to provide continuous global coverage, which is not feasible for this system.

Doppler Effect

Satellites at low altitude must move quickly over the Earth to stay in orbit. This movement causes a shift in the radio frequency, called the "Doppler effect," as illustrated in Figure 7, due to the relative motion between the satellite in space and the distress beacon on Earth. This principle is used to compute the locations of these devices, based on calculations around the time of closest approach (TCA). The Doppler effect always generates pairs of predicted locations, one on each side of the satellite subtrack, resulting in an ambiguity, because one is the real location and the other a mirror image. The ground stations analyze this pair and try to assess which is the most likely, based on the skewing of the Doppler curve due to the rotation of the Earth. Both possible locations

Figure 6. Orbital plane and footprint of one polar-orbiting satellite

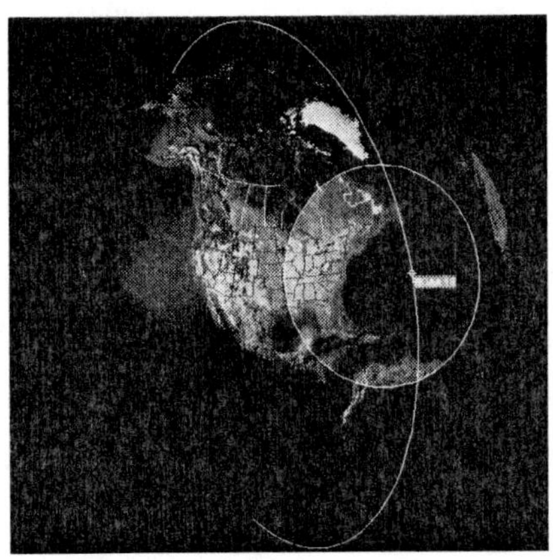

Figure 7. Doppler frequency shift vs. time

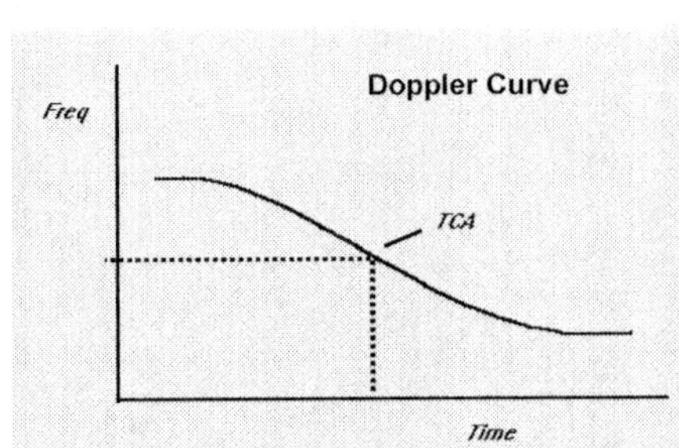

are always forwarded to the SAR authorities, with a calculated probability for each.

Space Segment

The nominal LEOSAR system comprises four satellites, two Cospas and two Sarsat, illustrated in Figure 8, equipped with SAR instrumentation that receives signals at 121.5 and 406 MHz, amplifies them, and transmits them back to Earth at 1544.5 MHz. In addition, the 406 MHz beacon signals are stored in the satellite's memory and replayed to all ground stations, thereby providing global coverage. Russia supplies two Cospas satellites placed in near-polar orbits at 1000 km altitude with an 83° inclination. The US National Oceanic and Atmospheric Administration (NOAA) supplies two multimission meteorological satellites, placed in sun-synchronous, near-polar orbits at 850 km altitude with a 98° inclination, carrying SAR repeaters supplied by Canada and SAR processors supplied by France.

Each LEO satellite makes a complete orbit of the earth around the poles in about 100 minutes, traveling at a velocity of 7 km per second. When viewed from the Earth, the satellite crosses the sky in about 10 to 15 minutes, depending on the maximum elevation angle of the particular pass. The satellite views a "swath" of the Earth more than 4000 km wide as it circles the globe, giving an instantaneous "footprint" or field of view about the size of a continent, as illustrated in Figure 6, and the four-satellite constellation has four such footprints moving over different parts of the Earth, as shown in Figure 9.

Distress Beacons

- **121.5 MHz beacons:** The original distress beacons developed in the 1960s were required to meet national specifications based on ICAO standards, but were not initially designed to work with a satellite system. Such beacons, as illustrated in Figure 10, transmit only fraction a Watt (about 0.05 to 0.1 Watt signal), having swept tone, amplitude modulation, which produces a warbling "wow, wow, wow" sound in a nearby aircraft receiver, similar to today's car alarms. This sound is similar for all such beacons, and the carrier frequency of the beacon is not very stable and is significantly affected by

Figure 8. Illustration of a NOAA Sarsat satellite (left) and a Cospas satellite (right) (courtesy of Cospas-Sarsat)

the ambient temperature, thus reducing performance through the satellites.

The 121.5 MHz Cospas-Sarsat system was designed and implemented to provide a better service for the thousands of beacons already in use, even though system performance would be constrained by their characteristics. Parameters such as system capacity, ambiguity resolution, and location accuracy would be limited, and no information could be provided about the operator's identity. Even with these limitations, the service for 121.5 MHz beacons was greatly enhanced with the inception of satellite detection and Doppler location techniques, and many more lives have been saved as a result. It is estimated that there are more than half a million 121.5 MHz beacons in use worldwide, primarily aboard small aircraft.

- **406 MHz beacons:** While Cospas-Sarsat was initially serving the 121.5 MHz beacon community, parallel development of a new type of beacon transmitting at 406 MHz commenced at the beginning of the Cospas-Sarsat program. Both frequency bands were built into all the satellites and ground stations, and new 406 MHz beacons were designed specifically for satellite detection and Doppler location by having:
  - High peak power output and low duty cycle
  - Improved radio frequency stability
  - A unique identification code in each beacon
  - An option to include the distress location in the signal
  - Digital transmissions that could be stored in a satellite's memory
  - Spectrum dedicated by the ITU solely for distress beacons

These parameters make the 406 MHz system far superior to the older 121.5 MHz system by providing the following features:

- Increased system capacity
- Better ambiguity resolution
- Improved location accuracy (10 times

Figure 9. Orbital planes and moving footprints of four LEOSAR satellites

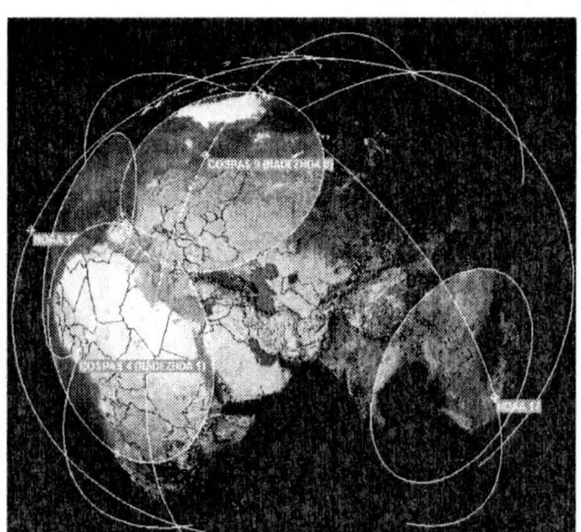

Figure 10. Typical 121.5 MHz ELTs used in aircraft (courtesy of CRC)

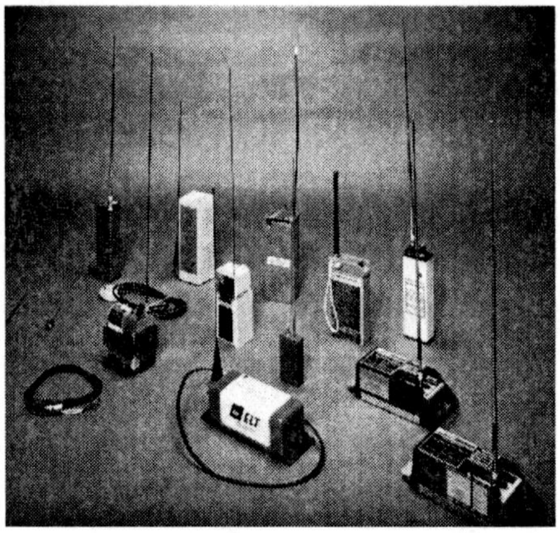

better, typically to within 2 km vs. 20 km for 121.5 MHz)

- o Identification of the user in distress
- o Optional location information
- o Global coverage
- o No interference from aircraft voice transmissions

406 MHz beacons, shown in Figure 11, transmit a 5-watt, half-second burst approximately every 50 seconds, and the carrier frequency is phase-modulated with a digital message. This low duty cycle of only 1% provides a multiple-access capability of more than 90 beacons operating simultaneously in view of a polar orbiting satellite, vs. only about 10 for 121.5 MHz beacons.

There is an option in 406 MHz beacons to include the beacon location as part of the distress signal, and this feature is now common in many of the newer 406 MHz beacons. In those beacons, after the beacon is activated there is space to encode the location, which can be derived from an external or internal satellite navigation receiver. If this data is not immediately available, for example, while the navigation unit is computing its first location, the distress beacon transmits a fixed default pattern, and then updates it when the location data does become available. It can update it again periodically if the beacon is moving, such as a beacon drifting at sea.

Some 406 MHz beacons, called personal locator beacons -PLBs, are designed to be carried by individuals and can be activated only manually. There are more than 40 manufacturers of 406 MHz beacons in 12 countries, and many more distributors around the world, selling over 100 different models type-approved by Cospas-Sarsat. The number of 406 MHz beacons in use increased dramatically, from virtually none in 1985 to half a million some 20 years later.

406 MHz Beacon Registration

Each 406 MHz beacon is encoded by the manufacturer or distributor with a "unique" digital message, making every 406 MHz beacon in the world identifiable, whereas 121.5 MHz beacons all transmit virtually the same analog signal, making them almost impossible to distinguish from one another.

The stored message includes such information as the country of beacon registration, the type of beacon, and a serial number or an identification number of the plane or ship that carries the beacon. The message also includes a mathematical error correction code to help ensure the distress message is received correctly through the satellites and ground stations.

To make the most use of this unique coding, all 406 MHz beacons should be registered in a database that can be accessed by SAR authorities whenever a distress alert is received from that beacon. That database would contain additional useful information, including a description of the plane or ship, the maximum number of passengers, emergency contact information, and so forth. Many countries control the allocation of unique serial numbers, operate a national database and periodically contact beacon owners to ensure the information is correct. Beacon owners should ensure they register their new beacons promptly, and inform the registry of any changes to the information. Several countries now have online beacon registration databases, allowing beacon owners to enter and update their information.

Beacon owners in countries that do not operate a national database could register their beacons in the online international beacon registration database operated by Cospas-Sarsat (IBRD, 2006), if appropriate arrangements have been made with that country.

Proper registration is very important to the beacon user, because that is the first place SAR authorities check for supplementary data when they receive a distress alert, so correct information is crucial for planning a SAR mission.

Ground Segment

- **Local user terminals:** Cospas-Sarsat LUTs track the satellites, receive the distress beacon signals via the satellites, compute the locations of the distress signals, and forward the alert data to Mission Control Centres. Most LUTs are fully automated and some are unmanned and installed in remote areas and can be operated remotely from the MCC. To track the LEO satellites, the LUT steers its antenna to follow the satellite across the sky, so each LUT needs to have the satellite orbit data on a regular basis.

Now there are more than 40 LEOLUTs operating on six continents (Cospas-Sarsat Documents,

Figure 11. Various types of 406 MHz distress beacons (courtesy of Cospas-Sarsat)

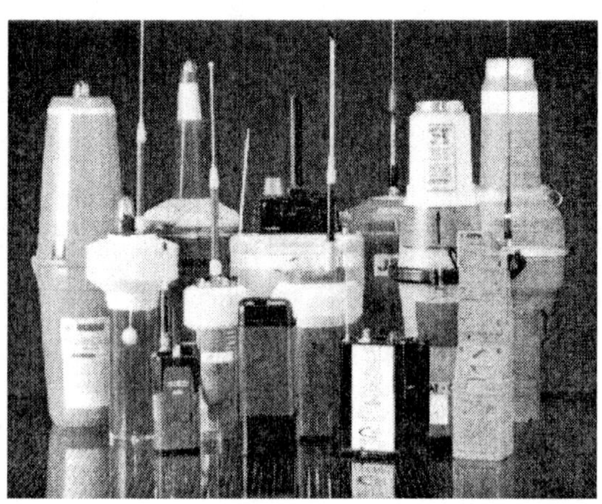

2005; Cospas-Sarsat, 2006a, Levesque, 2005), as shown in Figure 12.

- **Mission control centres:** MCCs have been set up in countries operating one or more LUTs, and because the LUTs are automated, it is generally at MCCs where people first see distress alert data from Cospas-Sarsat. The main functions of an MCC are to:
  - Collect, store, and sort the data from LUTs and other MCCs
  - Provide data exchange within the Cospas-Sarsat system
  - Distribute alert and location data to associated RCCs

All MCCs in the system are interconnected through appropriate networks for the distribution of system information and alert data.

ENHANCEMENT WITH GEOSAR SYSTEM

To help reduce the time waiting for a satellite to come into view, which is inherent in a limited LEO satellite system, Cospas-Sarsat started evaluating the use of geostationary-Earth-orbit (GEO) satellites (Keightley, 1987; Caron, 1991) as an enhancement to the 406 MHz system. Geostationary satellites, at an altitude of 36,000 km, remain at a fixed position in the sky, constantly viewing a huge area of the Earth. They could immediately relay 406 MHz beacon distress alerts, along with their identification codes, thereby eliminating the waiting time of the LEO system.

Initial experiments in the 1980s demonstrated the feasibility of such a concept, called GEOSAR, and a number of geostationary satellites were then equipped with 406 MHz repeaters. As depicted in Figure 13, two US weather satellites (GOES-East and -West), a European weather satellite (MSG), and an Indian satellite (Insat) now provide this service, and others are planned. To receive and process these satellite signals, new ground stations, known as GEOLUTs, were developed and built and there are now about 20 operating around the world.

This rapid relay of the distress signal, with the identification code identifying the user, was a big improvement, but the GEOSAR system could not automatically compute the distress location the way the LEO system does. There is virtually

Figure 12. Locations of Cospas-Sarsat LEOLUTs

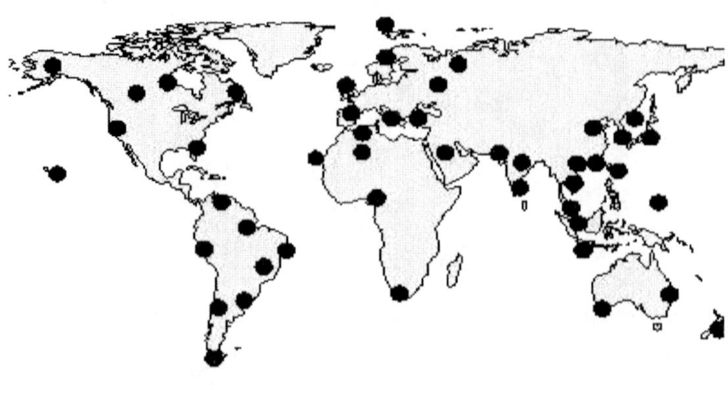

no relative motion between the geostationary satellite and the distress beacon, so no Doppler shift is available to locate the beacon. Beacons that include their location in the distress signal provide more complete data, but with or without such embedded location data, the rapid distress alert message, together with supplementary data obtained from a beacon registration database, make the GEOSAR system very helpful for search and rescue operations.

With the rapid distress alert, even without the location, authorities can begin checking the beacon registry for supplementary information, making telephone or radio calls to determine the possible whereabouts of the user, and start planning for SAR resources. If contact can be made with the owner, and it is found that the beacon had been inadvertently activated and there was in fact no distress, this case could be resolved quickly without deploying any resources. If no contact can be made and the distress transmission continues, the LEOSAR system would eventually detect the same beacon and provide the location for the SAR forces.

However, there are still some limitations because the beacon signal requires a direct line of sight to one of the satellites. There are some distress situations where this line of sight is impossible, such as in polar regions or when a plane crashes on the wrong side of a mountain or in a deep valley or when a maritime beacon is blocked by the ship superstructure.

ENHANCEMENT WITH MEOSAR SYSTEM

To further improve the 406 MHz system and complement the existing LEOSAR and GEOSAR systems, plans are being made to develop a MEOSAR system and fly 406 MHz repeaters on future navigation satellites, including GPS, GLONASS, and Galileo. These satellites are in medium-Earth orbit (MEO) at about 20,000 km, as illustrated in Figure 14. The MEOSAR element in GPS, being developed by NOAA, NASA, the US Air Force, and Canada, is dubbed the distress alerting satellite system (DASS). Russia is developing the GLONASS element, and Europe the Galileo element, which might also provide a satellite return link to the distress beacon to acknowledge receipt of the distress alert.

At that altitude, as shown in Figure 15, a MEOSAR satellite has a footprint much larger than a LEOSAR, and almost as large as GEOSAR, and it slowly moves around the world, providing long periods of coverage (King, 2006a). With multiple

Figure 13. LEO & GEO satellite constellations (courtesy of Cospas-Sarsat)

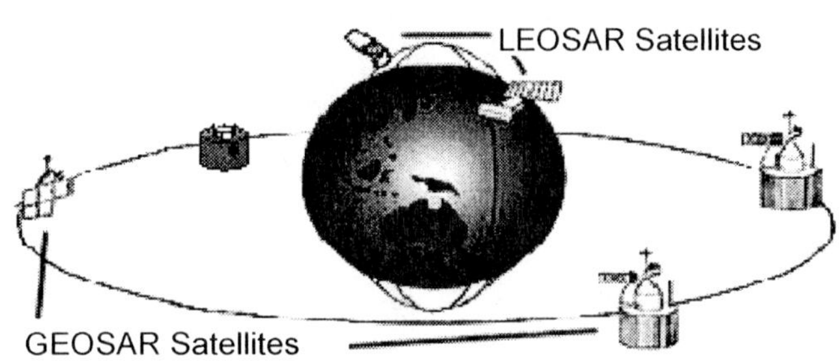

MEO satellites, the coverage is continuous everywhere in the world.

These constellations could each have about 20 to 30 SAR-equipped satellites that are slowly moving across the sky, thereby providing continuous, global coverage, including polar regions, with multiple viewing angles to the satellites, thus minimizing blockage by local terrain. This MEOSAR system could automatically detect and locate all 406 MHz beacons activated almost anywhere in the world with no time delay and monitor when each beacon was turned on and off.

Principle of Operation of MEOSAR System

The principle of the MEOSAR system is just the reverse of a satellite navigation system, and is still based on ranging or triangulation. The user in distress activates a transmitter, rather than a receiver, and its signal travels "backwards" via the navigation satellites (i.e., through the SAR payloads) to ground receiving stations called MEOLUTs. The exact same distress signal, relayed simultaneously via various navigation satellites in different posi-

Figure 14 A medium-Earth-orbit (MEO) satellite constellation is between LEO and GEO

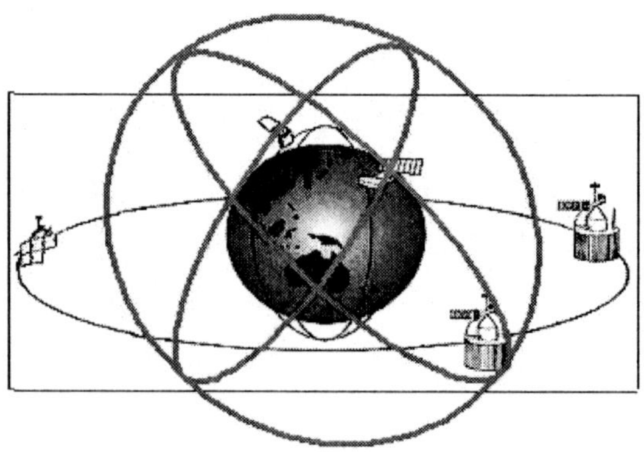

Figure 15. A MEO footprint is much larger than a LEO footprint, and moves much slower

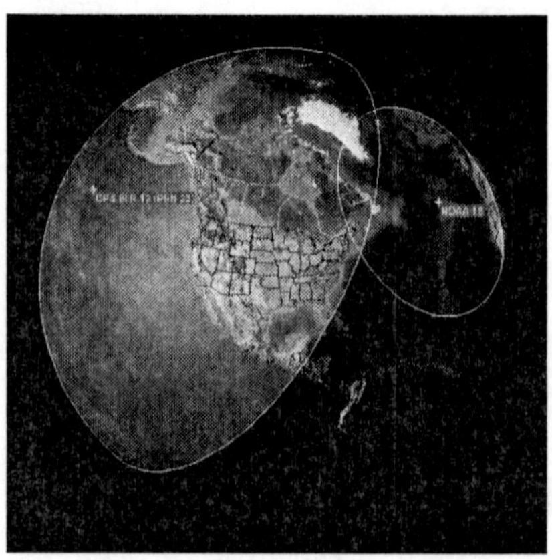

tions in the sky, as illustrated in Figure 16, arrives at the ground station at slightly different times, due to the different signal path lengths, much the way a yodel echoes off mountains.

These small time differences of arrival (TDOA) are accurately measured by the MEOLUT. Because the MEOLUT knows its own location, and that of each of the navigation satellites, as they broadcast that, it can quickly determine the location of the beacon. Because the MEOSAR satellites are all moving slowly in different directions across the sky, they also impart a small Doppler shift on the radio signal. This shift is slightly different through each satellite, so the MEOLUT also detects this frequency difference of arrival (FDOA), which it also uses to compute the beacon location.

These techniques allow a beacon to be detected and located on a single burst within a second, without having to wait for a LEO satellite to come into view, and then waiting for a Doppler curve to form over multiple bursts, taking several more minutes.

The MEOSAR system will offer many benefits, including:

- Continuous, global coverage
- Multiple signal paths so more reliable reception
- Almost instantaneous detection and location of beacons

Prototype MEOLUT ground stations, as illustrated in Figure 17, and satellite payloads are now being developed and tested in several countries (Affens, 2006; Cospas-Sarsat, 2006c, 2006d; King, 2006b) to demonstrate the MEOSAR proof of concept, and preliminary trials have confirmed the potential of the system. Over the next few years, evaluation testing and refinements will continue, and it is expected the operational MEOSAR system will be implemented over the next 5 to 10 years, thereby ensuring that 406 MHz satellite systems continue to provide the optimum distress alerting and locating service for worldwide operations.

During the initial period that MEOSAR satellites are gradually built and launched over a few

Figure 16. Beacon signals relayed simultaneously by multiple MEO satellites to a MEOLUT

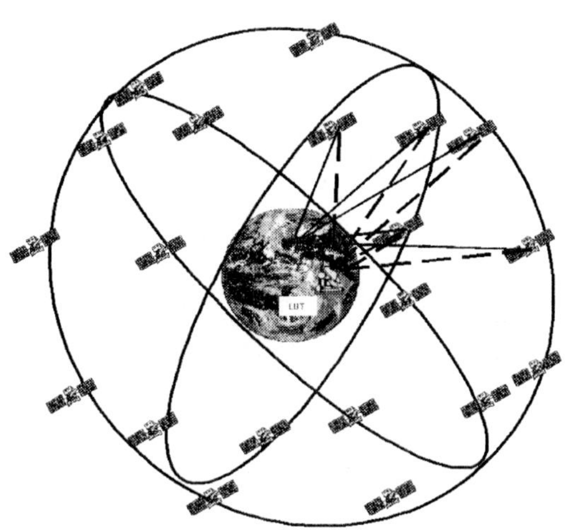

Figure 17. A typical tracking antenna at a prototype MEOLUT (courtesy of CRC)

years, there would not always be enough satellites in view to use triangulation to locate beacons. However, each individual satellite could be used to at least detect distress alerts by relaying beacon signals, much like a GEOSAR satellite, but with a slowly moving footprint. This could provide periodic GEO-like coverage in polar regions, which does not exist today. To achieve this capability in the early days, MEOLUTs would need only one or two tracking antennas initially, and more antennas could be added when more MEOSAR satellites are in orbit.

FUTURE

Most of the limitations of the original 121.5 MHz system, including many false alarms, inferior beacon performance, no user identification, poorer location accuracy, limited system capacity, and non global coverage, are overcome by the superior 406 MHz system. Little can be done to improve the 121.5 MHz system, so the international SAR community requested Cospas-Sarsat to phase out the 121.5 MHz satellite alerting system. The termination of the 121.5 MHz satellite service will occur in February 2009, after more than 25 years of continuous service. A low-power 121.5 MHz signal will continue to be transmitted by 406 MHz beacons for final homing, but this will not be relayed and processed by the satellite system.

All new satellite developments, including GEOSAR and MEOSAR, serve only 406 MHz beacons. The existing LEOSAR and GEOSAR 406 MHz systems will continue operating for many more years, and the new MEOSAR system is expected to start coming online around the same time as the 121.5 MHz service is being phased out.

CONCLUSION

Since its inception in 1982, the Cospas-Sarsat system has provided assistance to SAR forces, as depicted in Figure 18, in rescuing more than 20,000 persons worldwide, and this number continues to grow by more than a thousand a year.

The use of older 121.5 MHz beacons will be phased out in the future and the more sophisticated 406 MHz beacons will become commonplace, particularly with the faster alerting service provided by the 406 MHz GEOSAR and future MEOSAR systems, and the inclusion of navigation chips in many 406 MHz beacons. This international satellite system for search and rescue, as illustrated in Figure 19, will continue providing service for many years to come, serving the many thousands of users who are mandated to carry distress beacons and many more who choose to carry them.

Additional information about Cospas-Sarsat can be obtained from administrations in participating countries and from the Web site (Cospas-Sarsat, 2006b).

Figure 18. Typical SAR resources at work (courtesy of Canadian Defence Department)

Figure 19. Cospas-Sarsat, the international satellite system for search and rescue (courtesy of NOAA)

REFERENCES

Affens, D. W. (2006, April). NASA Goddard Space Flight Center DASS Ground Station. In *Proceedings of the CS/EWG-2 Meeting*, Montreal, Canada.

Bogdanov, V., Ageev, V., Belousov, A., Buyanov, A., & Terekhin, V. (1984, April). Ground segment of Cospas System. In *Proceedings of the International Symposium on Satellite Aided Search and Rescue* (pp. 217-222). Toulouse, France.

Caron, M. (1991, June). *Application of Geolocation Techniques using Satellites in Geostationary Orbit*, Communications Research Centre Canada-CRC Report No. 1435, Ottawa, Canada.

Cospas-Sarsat (2006a, February). Information bulletin (No. 18). Retrieved February 22, 2007, from www.cospas-sarsat.org

Cospas-Sarsat (2006b). Retrieved February 22, 2007, from www.cospas-sarsat.org

Cospas-Sarsat (2006c, June). Status of the European MEOLUT prototype. In *Proceedings of the GJU/ESA at CS/JC-20 Meeting,* Montreal, Canada.

Cospas-Sarsat (2006d, June). SAR/GLONASS system status. In *Proceedings of the CS/JC-20 Meeting*, Montreal, Canada.

Cospas-Sarsat Document (1988, July). *International Cospas-Sarsat Programme Agreement*. July 1, 1988, signed in Paris, France.

Cospas-Sarsat Document System Data (No. 31). (2005, December) Retrieved February 22, 2007, from www.cospas-sarsat.org

Cospas-Sarsat (1999, October). Introduction to the Cospas-Sarsat System, C/S G.003, Issue 5 – Revision 1, October 1999 www.cospas-sarsat.org

Cospas-Sarsat Project Report (1985, September). Document C/S R.001. Retrieved February 22, 2007, from www.cospas-sarsat.org

Flatow, F. S., Gal, C., & Hayes, E. J. (1984, April). 406 MHz ELTs/EPIRBs. In *Proceedings of the International Symposium on Satellite Aided Search and Rescue* (pp. 195-208). Toulouse, France.

Goudy, P. H., King, J., & Kissel, F. (1984, April). Sarsat 406 MHz System technical evaluation. In *Proceedings of the International Symposium on Satellite Aided Search and Rescue* (pp. 429-440). Toulouse, France.

IBRD. (2006). International 406 MHz Beacon Registration Database—IBRD Web site. Retrieved February 22, 2007, from www.406registration.com

Keightley, R. J. (1987, May). Simulation of the Canadian GOES Signal Processor. *IEEE Transactions on Aerospace and Electronic Systems, AES-23*(3), 361-370.

King, J.V. (2006a, April). Coverage of MEOSAR compared to LEOSAR: Preliminary study. In *Proceedings of the CS/EWG-2 Meeting*, Montreal, Canada.

King, J. V. (2006b, June). Status of Canada's prototype MEOLUT development. In *Proceedings of the CS/JC-20 Meeting*, Montreal, Canada.

King, J. V., Hayes, E. J., & Jutras, G. J. (1983, September). Results of Cospas-Sarsat system performance testing (Paper No. 83074). In *Proceedings of the 1983 International Electrical, Electronics Conference and Exhibition- IEEC&E '83* (pp. 180-183). IEEE Catalogue No. 83CH1955-4, Toronto, Canada

Levesque, D., Hodgkins, K., & Drover, P. (1984, April). The development of an institutional structure for the future global satellite system for search and rescue. In *Proceedings of the International Symposium on Satellite Aided Search and Rescue* (pp. 597-620). Toulouse, France.

Levesque, De. (2005, October). Status of the Cospas-Sarsat System. In *Proceedings of SARSCENE 2005*, Charlottetown, Canada.

McGunigal, T. E., McKinnon, D. J., Brachet, G., & Zurabov, Y. (1984, April). Cospas/Sarsat- An overview. In *Proceedings of the International Symposium on Satellite Aided Search and Rescue*, Toulouse, France.

Mehta, A. (2006, May 12). Cost benefit analysis for SARSAT program. In *Proceedings of the Beacon Manufacturers Workshop*, Newport Beach, CA, USA.

Renner, R. C., & Kozminchuk, B. W. (1984, April). 121.5/243 MHz ELT identification. In *Proceedings of the International Symposium on Satellite Aided Search and Rescue* (pp. 361-384). Toulouse, France.

Rogalski, V., & Krupen, V. (1984, April). The Cospas System tests technical results. In *Proceedings of the International Symposium on Satellite Aided Search and Rescue* (pp. 441-449). Toulouse, France.

Stevinson, H. T., & Baker, D. A. (1971). Locating people in high latitudes (Laboratory Tech. Rep. No. LTR-FR-32 and –32A). National Aeronautical Establishment. Ottawa, Canada: National Research Council of Canada.

Tower, C. (1983, April). Rescue from outer space. *Reader's Digest*. 49-54.

US DoT (2002, January 29). Treatment of value of life and injuries. U.S. Department of Transportation's Inspector General Memorandum.

Werstiuk, H. L. (1983, September). Cospas-Sarsat system overview (Paper No. 83071). In *Proceedings of the 1983 International Electrical, Electronics Conference and Exhibition -IEEC&E '83* (pp.168-171). IEEE Catalogue No. 83CH1955-4, Toronto, Canada.

Werstiuk, H. L., Ludwig, D., Trudell, B. J., & Selivanov, A. S. (1984, April). The Cospas-Sarsat system. In *Proceedings of the International Symposium on Satellite Aided Search and Rescue*, Toulouse (pp. 95-110). France.

Winter, A. E. (1978). Assessment of potential ground stations. Communications Research Centre Memorandum, Ottawa, Canada.

Section II
Satellite Internet and Navigational Technologies

Chapter V
Global Navigation and Satellite Systems and Services

Justo Alcázar Díaz
UPM, Spain

Tirso Velasco
Universidad de Valencia, Spain

ABSTRACT

This chapter introduces the concept of satellite navigation in the context of space infrastructures and technologies that can contribute to improvement of life on Earth. It includes a review of the motivations for developing a satellite navigation system, and the applications and services these systems have in daily life. Furthermore, currently existing global navigation satellite systems (GPS and GLONASS) and other GNSS systems under development (GALILEO) are described from different perspectives, from the technical and architectural aspects to the ways chosen to finance their development and operations. To round up this chapter, an analysis of the expected trends in GNSS systems is presented and potential scenarios for future evolution of global satellite navigation are discussed.

INTRODUCTION

Satellite-based navigation is, together with telecommunications and Earth observation, one of the three main legs of the so called space applications. The objective of this chapter is to provide a review of how global navigation satellite systems and services (GNSS) can influence the improvement of life on Earth.

GNSS systems are currently offering, or will offer in the near future, a set of services that are contributing to the gradual transformation of society and bringing benefits to users worldwide. Nowadays, the GPS systems and receptors are familiar and can be acquired at acceptable prices by anyone, achieving positioning precisions of a few meters. The applications fields are very wide in areas as different as transport, oil exploration, insurance, banking, environment, geology, science, and so forth. The market outlook is promising, as demand for satellite navigation services and derived products around the world is growing at a rapid rate and is expected to improve in the next years, with the generalization of its use in cars and its integration in the last generation of personal mobile communications.

As explained later along the chapter, the operation of a GNSS system is based on a time-difference-of-arrival concept. A receiver receives a signal from a satellite, containing the information about the position of the satellite and its clock. This receiver is then able to determine the distance between both by measuring the travel time of the signal. With the information of at least three satellites, the receiver will know its position with an accuracy ranging between centimetres and few meters, depending on different factors. A fourth satellite will provide the time reference.

GNSS services are based on huge infrastructures that ask for big efforts and investments for its development and maintenance. This infrastructure includes a float of satellites, carrying high precision clocks, and a complex system of high capacity ground stations (basically big antennas and processing centres) worldwide distributed. Few governments or organisms are able to set up such a system. As a result, currently, there are only two global operative systems (GPS and GLONASS, although this last one has never been fully developed), both with a military origin. This scenario is expected to change considerably in the coming years, with the ongoing development of the first civilian system (GALILEO), the improvement of the current existing systems (third generation of GPS and full deployment of GLONASS), and the availability of regional augmentation systems.

This changing scenario, with an increasing number of available systems (both globally and regional) and the consequent gain in performances (positioning and time accuracy, continuity, integrity) will revolutionize the navigation applications and services at many different levels. Concepts like interoperability of the systems, security issues, or integration with other growing communications and information technologies will draw a new landscape in the following decade for the applications and services of GNSS systems.

In this context, this chapter intends to provide a full understanding of the following main issues related to global navigation satellite systems:

- Basics and concepts related to global positioning and satellite navigation systems
- Description of the current and near future infrastructure for the already existing GNSS systems (GPS and GLONASS) and for the system currently under development (GALILEO)
- Description of the main application and services of the GNSS systems
- Analysis of the business perspective for satellite navigation services
- Trends and analysis of scenarios for the expected evolution of GNSS systems in the near future

BACKGROUND ON GLOBAL NAVIGATION SATELLITE SYSTEMS (GNSS)

Definition of Global Navigation Satellite Systems

GNSS refers collectively to the worldwide positioning, navigation, and timing (PNT) determination capabilities available from satellite constellations. A GNSS system can be defined considering the meaning of the four letters of its acronym (see Figure 1).

Currently, there are only two systems fulfilling the four characteristics in the above definition: the American global positioning system (GPS) and the Russian global navigation satellite system (GLONASS), although the second one has not

Figure 1. Definition of a global navigation satellite system

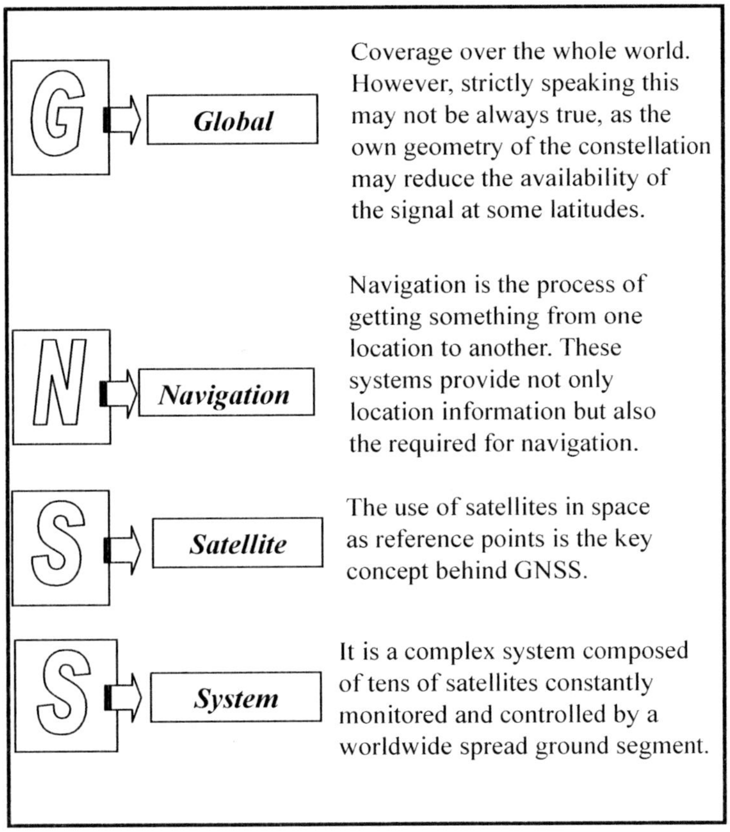

been fully deployed yet. A third one (European GALILEO) is under development.

Together with the GNSS systems, other satellite-based navigation systems exist, commonly referred to as augmentation systems, which complement and enhance, on a regional basis, the performances and services provided by the GPS constellation. As it will be further detailed later on during the chapter, the main satellite based augmentation systems (SBAS) are the American wide area augmentation system (WAAS) and local area augmentation system (LAAS), the European euro geostationary navigation overlay service (EGNOS), Japanese Quazi-Zenith satellite system (QZSS), and the Indian GPS and GEO augmented navigation (GAGAN).

Other independent satellite-based systems, such as Argos or SARSAT, can also provide localisation services for different applications, but they do not provide global navigation services.

Brief Historical Overview of Satellite Navigation Systems

Like other big technological steps of the last decades, global navigation satellite systems have their origin in military applications, in the context of the Cold War. The first GNSS is the American global positioning system (GPS), initiated by the Department of Defence (DoD) of the USA in the sixties. The first conceptual design was finished by 1972, and the first "Block I" satellite was launched in 1978. In parallel, the Soviet Union was developing its own system, GLONASS, which put into orbit a first Satellite in 1982. At European level, the first studies of the so called NAVSTAR (a forerunner to GALILEO) were started in 1980.

The exclusively military orientation of GPS was kept until 1983, when initial civil applications of the GPS signal started to be explored but still maintaining the military control of the USA's Department of Defence over the system. During the 1980s, the combined efforts of both military and the civil contributions from research institutions and private companies helped to shape the system as it is now. Examples of these contributions are the development of initial GPS data processing software in the University of Bern (Switzerland) or the inclusion of a GPS receiver in the Chrysler "car of the future" in the mid 1980s. In 1988, DARPA presented the first GPS handheld receivers.

The development of GPS quickly completed its milestones: the first satellite of Block II was launched in 1987, and initial operation capability, with 21 satellites, was reached in 1993. In parallel, new technologies and algorithms allowed a spectacular increase of performances and new services in the 1990s. It was also during that decade when the development of the currently existing satellite-based augmentation systems (WAAS, EGNOS, MSAS) was initiated, for improved regional coverage. The economical problems of Russia, after the fall of the Soviet Union, have not allowed a similar evolution and the full development of GLONASS. However, the end of the Cold War has facilitated the potential cooperation of both systems (GPS and GLONASS). In 1997, the first integrated GPS / GLONASS surveying system was developed. It is not until the late nineties when the dual purpose (civil and military) of GPS is finally formalized.

The first years of the 21st century have seen the birth of the European System GALILEO, approved by the European commission in 2003 and the first GALILEO-GPS interoperability agreement, signed in June 2004. In parallel, the first satellite-based augmentation systems (WAAS, EGNOS, MSAS) have started their operations.

In a global perspective, there are different potential motivations why a country or organisation is willing to set up a complex system such as a GNSS. Both GPS and GLONASS origins are clearly driven by the military views of the United States and Soviet Union governments in the context of the Cold War. Together with the change in the political context, different motivations are on the table for the development of GALILEO by Europe or the completion of GLONASS. European authorities recognise the independence from the United States and Russia as the main reason for developing GALILEO. Other motivations are the participation in a growing market, the expected improvement of services and performances with a new independent system, the development of technical knowledge and creation of technical employment in Europe, or even the security needs of European countries. The development of alternative GNSS systems is at the same time one of the main drivers for the actualisation plans of GPS by the United States government.

Applications and Services

GPS is an invaluable service for safe and efficient movement, measurement, and tracking of people, vehicles, and other objects anywhere from the Earth's surface to geosynchronous orbit, as well as providing timing and synchronization for global communications, electronic transactions of all types, and power-distribution networks.

As a dual-purpose system, GPS uses are two-folded and civil and military applications can be distinguished. A similar process is being followed by the Russian system GLONASS that is also shifting from a purely military system to a dual one. Civil applications can also be divided into the pure commercial uses for the global market and the scientific applications of a worldwide ubiquitous satellite network. GPS is used by national mapping agencies worldwide as a basic component of geographic information systems and for natural hazards mitigation (earthquakes, volcanoes, sea-level rise), climate monitoring, severe storm predictions, and in characterizing space weather (ionosphere). Applications involving GPS on-board Low Earth orbiters are revolutionizing a great variety of scientific disciplines such as weather forecasting and gravity field determination from space.

Military Applications

GPS provides an unparalleled force-enhancement tool for the United State's - and its allies - military missions. It was conceived as a war-fighting system, and is expected to remain, at least partially, under military control in the future. GPS aids in all aspects of military combat operations thanks to its common-datum, common-grid, and common-time capabilities, with contributions to many aspects. According to the report of the Defence Science Board Task Force (2005), the most relevant applications are listed here:

- Air operations for manned and unmanned platforms, allowing point-to-point air navigation without ground-based navigation aids
- Naval operations, enabling seamless global maritime navigation on the open ocean, harbours, and inland
- Land operations, allowing for safer and more efficient operations
- Space operations, enabling continuous and precise orbit determination of military spacecraft
- Precise weapons delivery and targeting
- Special operations in day/night all-time all-weather conditions
- Communications, by the provision of precise timing and frequency synchronisations for encrypted data transmissions
- Other operations, such as logistic and supply, mine fields charting and clearing, battle space awareness, or search and rescue operations, are also performed with increased efficiency with the use of GPS services

Civil Applications

Applications of GNSS are present in many different sectors of human activity from the most obvious, such as air navigation or rescue operation, to others, including science, electricity distribution, or leisure applications. The constant improvement of GPS performances and services and the increased availability of navigation signals are contributing to the increase of the applications of GNSS, not only for the big industry or infrastructure, but also for the common user. Currently, small GPS receivers are wide spread in cars, boats, and so forth. In the coming scenario with three different infrastructures (GPS, GALILEO, and GLONASS) this range of applications is expected to increase even more.

Detailed information about current civil applications and their foreseen evolution in the next years is provided in the epigraph for satellite navigation "enabled" markets.

Scientific Applications

Many branches of science have benefited from GNSS services. The availability of a large constellation of satellites, together with a wide world ground station network, providing precise positioning, and timing is a powerful tool used in many different scientific applications, such as:

- **Weather and atmospheric science:** Multifrequency signals of the GNSS satellites are affected by atmospheric conditions. The different effects in these signals are used to extract information of the troposphere and ionosphere physics and to improve atmosphere and weather models.

- Global timing and geo-reference systems use the feedback provided by GNSS systems to improve the behaviour of their models.
- Environmental studies: For example, the observation of tides and currents, with the deployment of buoys provided with GNSS receivers, and tracking of animal movements or bird flocks migration paths, which help us to understand wildlife patterns.
- Earthquake warning systems based on satellite navigation are being developed in seismic and volcanoes areas. Geologists use GNSS receivers in order to precisely determine the tectonic plates' movements, tidal effects, or local phenomena.
- Earth stations such as radiotelescopes or satellite ground tracking systems, requiring high precision tracking, use GNSS receivers to help determine local and atmospheric conditions and to improve their behavioural models.

HOW DOES A SATELLITE NAVIGATION SYSTEM WORK?

Principles of Operation of Global Navigation Satellite Systems

Basically, navigation satellite systems implement a time-difference-of-arrival concept. This is done using precise satellite position and on-board atomic clocks to generate navigation messages that are continuously broadcast from each of the satellites in the GNSS constellation. These messages, containing information about the position and clock of the satellite sending the message, can be received and processed by users anywhere in the world (within the coverage of the system). The receiver, upon reception of those messages, determines the distance between both, by measuring the travel time of the signal. The user needs the information of at least three satellites in order to determine its position with enough accuracy (ranging between centimetres and a few meters, depending on different factors). The signal from a fourth satellite would provide the accurate (in the order of nanoseconds) reference of time.

Geometrically, the idea of how distance measurements from three satellites can give us the position is as follows:

- With the measure of the distance (range1) from the user to the first satellite, the possible locations of the user are narrowed down to the surface of a sphere that is centred on the satellite and that has a radius equal to range1.
- Measuring then the distance (range2) from the second satellite, implies that the user has to be not only on the first sphere but also on a sphere of radius range2 from the second satellite. In other words, the user has to be somewhere in the circle where both spheres intersect.
- Measuring the distance (range3) from a third satellite, the potential positions are narrowed down even further, to the two points where the sphere of radius range3 cuts to the circle that is the intersection of the two first spheres.

In summary, by ranging from three satellites, the position is narrowed down to just two points in space. To decide which one is the true location, a fourth measurement could be done. However, usually one of the two points is a ridiculous answer (either too far from Earth or moving at an impossible velocity) and can be rejected without measurement. The fourth measurement is used for other purposes, as checking the other three and providing an accurate time reference. This accurate time reference is needed because the clocks of the receiver are imperfect (not affordable to use atomic clocks) and a difference of the timing of just one thousand of a second would translate into some hundred miles of error. With an ideal receiver clock, all the ranges would intersect at a single point (the user's position). Then, the idea to get a perfect timing is to make an extra satellite measurement. If three perfect measurements can locate a point in 3-dimensional space, then we would need four imperfect measurements to get an equivalent result.

The principles of operation of GNSS systems can be summarized as follows:

- GNSS operation is based on trilateration of satellite signals.
- GNSS measures distance (ranges) using radio signal travel times.
- Each satellite in the constellation is continually sending its location and the precise time of its transmission.
- The signal from each satellite is "modulated" by a code. Usually, each satellite has a different code in such a way that the receiver is able to distinguish the satellite. These codes will also be generated by the receiver.
- GNSS user equipment receives the signal from each satellite and records its position and the signal arrival time.
- The difference between the code received from the satellite and the one generated by the user corresponds to the time difference (signal propagation delay). This time can be easily translated into a distance between the user and the receiver.
- The GNSS receiver computes its position and time from the calculated delays and distances. Different corrections are applied to these calculations, in order to take into account the propagation phenomena through the atmosphere. GNSS usually operates

Figure 2. GNSS operation

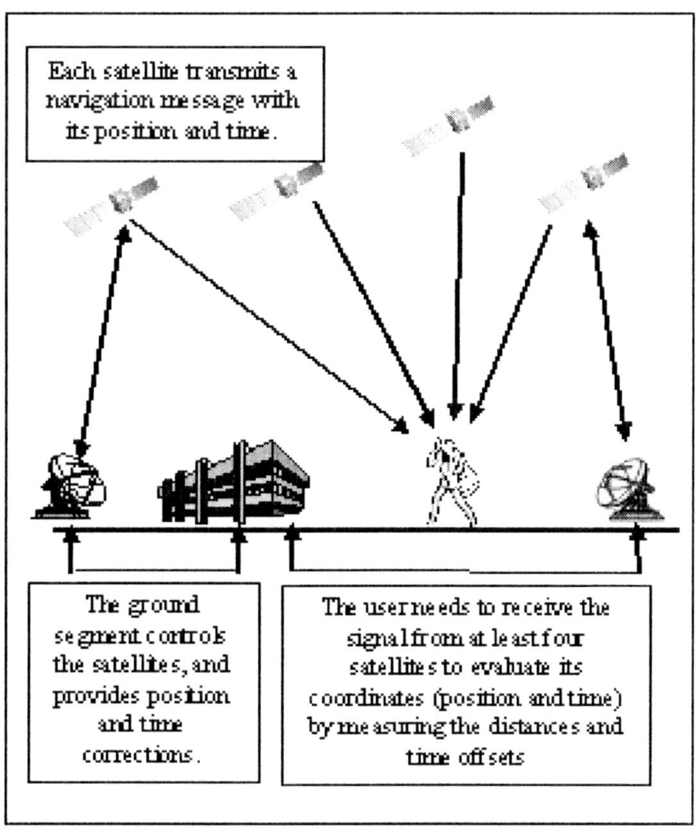

within an Earth-centred-Earth-fixed (ECEF) reference frame, so position determinations are independent of the local topography, which must be accommodated by geodetic models within the GNSS receivers.

Although the principles of operation seem quite simple, they have associated some nontrivial technical issues. The most important in order to guarantee the accuracy of the system is to maintain a precise reference for the satellites orbital location and system time. This is done by a global network of ground stations that keeps constant track of these parameters and continuously uploads the necessary corrections. The uploads include orbit position projections for each satellite in the constellation, based on sophisticated models and effective for several weeks, as well as corrections to on-board satellite clocks. A centralized control centre or master station is in charge of evaluating these corrections and performing the global monitoring and control of the whole mission.

System time is maintained globally in the control centre, synchronized with the UTC or other similar reference, and aboard each satellite by on-board atomic clocks accurate to within a few nanoseconds to the system time and individually stable to a few parts in 10^{13} or better. Usually, this means using cesium or rubidium atomic frequency standards. Although several standards are provided in each satellite for redundancy purposes, only one standard is operational aboard each satellite at any given time.

A third issue concerns the maintenance of the constellation of satellites, so each user is able to see at least four satellites, separated sufficiently and geometrically oriented in three-dimensional space so that processing will define a precise signal intersection. This basic requirement will influence the topology used for the constellation and the number of satellites that has to be kept in operation. The technical complexity and costs associated with the procurement, launch, and maintenance of the spacecrafts (typical lifetime of in the order of 10-15 years) make a challenge to continuously keep the satellites in their position with the required performances.

Components and Segments of Satellite Navigation Systems

The infrastructure of a GNSS system can be basically divided into three so-called segments: the space segment, the ground segment, and the user segment.

- The space segment in composed of the satellites which transmit the navigation signal, also called signal in space (SiS), used by the user receiver to calculate its position, time, and velocity.
- The ground segment or ground control segment includes all the means used by the owner or operator to keep the system working appropriately, tracking the satellites and transmitting the corrected orbital parameters. This includes a control centre or master station, and ground stations to control and monitor the satellite constellation and provide the satellites and user with the information needed to generate an adequate SiS.
- The user segment is composed of the user receivers which get the signals from the satellite, perform the calculations using different algorithms, and provide the user with the required service.

In addition to these three segments, another two components, which are not strictly part of the GNSS, have to be taken into account, the launch segment and the external entities. The launch segment includes all the means (basically launch services and early operations) in order to place the satellites into their defined orbits from where they will provide the signals. The term external entities refers in a broad sense to any entity that interacts with the GNSS system. Some examples are the entities providing time reference information (i.e., UTC) and precise Earth-based position reference (i.e., GTRF system), other GNSS systems (i.e., GALILEO and GPS will be coordinated to obtain their time offset), other systems providing regional integrity information, the scientific community and so forth. A GNSS is not conceived as an isolated system, but open to different international public and private institutions, including other GNSS systems to improve the interoperability among them.

Finally, the augmentation systems can also be considered as part of the GNSS. Those are systems that use space or ground-based infrastructure to enhance the navigation signals with greater performances or value added services in a determined local or regional area. Depending on the infrastructure used for the augmentation, the augmentation systems can be classified as ground-based (GBAS) or satellite-based (SBAS).

As an example of the components in a real GNSS, Figure 3 shows the implementation of these components in GALILEO.

Table 1 summarizes the inputs, functions, and products of the different segments as implemented in GPS.

Performance and Sources of Errors for Satellite Navigation Systems

The performances experienced by the user of a GNSS are driven by different parameters. The most obvious are position and timing accuracy, but there are others, related to the quality of the signal provision, which are even more appreciated by the user. Table 2 defines the seven relevant parameters related to the GNSS performances (Lewis, Kennedy, Ghashghai, & Bitko, 2005).

Figure 3. GALILEO architecture and segments composition

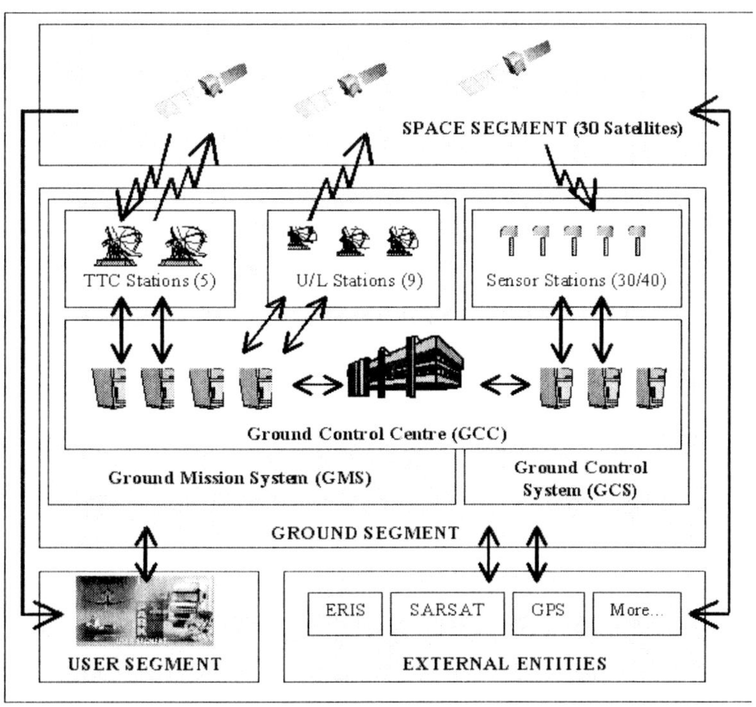

Table 1. Inputs and products of the three main GPS segments

Segment	Input	Function	Product
Space	Navigation message	Generate and transmit: code & carrier phase navigation message	P-Code; C/A Code; L1 and L2 carrier; Navigation message
Ground	P-code observations; Time (UTC)	Produce GPS time; Produce satellite ephemeris; Constellation control and management	Navigation message
Users	Code observations; Carrier phase observations	Navigation solution; Surveying solution	Position; Velocity; Time

Different sources of error can impact the performance of the user, causing loss of position and timing accuracy, or decreasing availability of the system. These can be nonexhaustively classified as:

- **Bad geometries:** The orbits of the satellite and the relative position with respect to the user can influence the quality of the signals available. Although the systems are designed to have an adequate number of satellites in the user's view, not all the configurations have the same performance, and not all the satellites are available at any time. Land irregularities, such as obstacles or urban canyons, are also included in this group.
- **Technical limitations:** Put a constrain on the final performance of the service. These include the performance of the technical equipment of every segment: space (spacecraft payload, on board antennas, attitude control, etc.), ground segment (ground antennas, ground communications, satellite control algorithms, etc.) and user segment (user receiver capabilities). The most critical issues are related to the timing (clocks and methods used for synchronisation on board the satellites and on ground) and the difficulties of the provision, maintenance, and replacement of the satellites.
- **Physical propagation phenomena:** The propagation of the signal between the satellite and the segment through the atmosphere is subjected to different effects (tropospheric and ionospheric delays, multipath, etc.), which limits the potential accuracy of the signal. Modeling and subsequent correction of the signal are keys issues.

Other Issues for Global Navigation Satellite Systems

Security Issues

GNSS systems should not only be considered as innovative solutions to PNT requirements, but also as key elements of national space policies (for the United States in the case of GPS, for Russia with GLONASS, or for Europe in the case

Table 2. Main parameter related to GNSS performance

PARAMETER	DEFINITION
Position Accuracy	Statistical value of error between true position and estimated position; measured as a distance at a stated confidence level
Availability	Percentage of time that a position accuracy meets specified accuracy performance level
Continuity gap	Maximum continuous length of time that the specified position accuracy is not met without advance notification
Integrity	Ability to determine whether the system is providing reliable navigation information
Time-to-Alarm	Length of time required to provide notification to the user interface that service is unavailable
Timing Accuracy	Statistical value of error between UTC and estimated time; according to specifications, at 95% confidence level, timing accuracy shall be not greater than 20ns for static user and 35 ns for dynamic user
Guarantee	Concept of ensuring the services for applications in which a disruption of service would have significant safety or economic effects.

of GALILEO). The potential dual (civil and military) nature of these systems gives them a range of defence-related applications. It is then critical to consider the security implications of GNSS systems and take practical steps to mitigate the risks derived of a potential abuse and misuse. In this sense, two main concerns could be identified: the need to protect access to the signal for users with safety-critical needs because of potential threats, and the security issue resulting from the potential use of the highly accurate navigation signal by hostile users, including the need to be able to deny access to such users. The jamming or intentional interferences to the signal leading to a jeopardised performance can be prevented using antijamming techniques. The prevention of the use of the signal by hostile users is managed through the concept of selective availability (SA).

The selective availability refers to the capability of the owner or provider of the system to provide a lower performance service to the user, or even to interrupt the service in a region. This is done through the use of encrypted codes, while the user is provided only with "corrupted" codes. This ability has been used in GPS by the DoD, for national security, in order to assure a better service for governmental or military applications. In this sense, during many years the civil users have been deliberately given codes with slight corruptions. The removal of the SA from the GPS civil signal was in fact done on May 2000 at least in part as a response to plans for GALILEO.

Other threats for the secure use of the GNSS signals are: intentional signal disruption, signal spoofing, receiver data spoofing, augmentation data spoofing, software modifications, and receiver hardware modifications.

Regulatory Issues

Legal issues related to GNSS systems will also have to adapt to the new scenario of GNSS systems.

Currently, various international laws are applicable to the GNSS services and systems, such as those coming from space law for the space segment (i.e., Outer Space Treaty, Liability Convention), air law for air navigation (i.e., Chicago Convention), international communications law (ITU frequency assignments), and so forth. Besides, there is a specific text, issued by the International Civil Aviation Organisation, the ICAO charter on GNSS. However, these existing conventions do not cover all GNSS specific liabilities.

The current scenario is driven by the existence of only one service provider, GPS. The United States government does not assume specific liabilities derived from the provision of GPS signals to other parties. However, the United States has signed different bilateral agreements with other countries or organisations (European Union, Russia, Japan, China, or Israel). These agreements are in part a result of the new situation with the development of new systems such as GALILEO or Regional Augmentation Systems.

The new scenario—with the development of new systems, the shift from military-oriented systems to dual systems, or the nongovernmental management of some systems—is expected to change the regulatory frame of GNSS. In this context a new international convention would be necessary to establish a legal frame in which the GNSS services are provided, and addressing all issues regarding satellite navigation: rights and obligations of service providers, liability chains, and so forth. Liability, understood as the legal obligation and responsibility to provide and guarantee the signal, is a key concept for the new regulatory frame, especially for GALILEO, which aims to provide value-added services.

Dual Use: Military vs. Civil

Generally, complex space-based systems like telecommunication satellites and earth observation systems are dual-use, in the sense that they have both civil/commercial and military applications. In this way, satellite navigation is currently provided by two systems under military control (GPS and GLONASS), both developed initially for military use but shifted later to dual use. GALILEO, on the other side, was conceived as a civil system and it has the opportunity to complement other GNSS systems in the dual-use domain. The combined use of GNSS can increase significantly the positioning performance for all civil and governmental end-users, hence dual users.

As mentioned above, this duality of uses is generally common to most systems. In this sense, the United States is currently re-evaluating the dual-use nature of GPS, and shifting it more toward civil applications, especially in a frame where new systems like GALILEO will be available. The provision of civil services with increased guarantees or the cooperation with other systems is an issue that can be of particular concern for the United State's national security interests. On the other hand, GALILEO, although conceived as a civil system, will also be used for the security interests of the European union countries, through its public regulated service (PRS). These duality issues will be more complex in a hypothetical future scenario with GPS and GALILEO, with interoperability agreements and agreements with third nations.

GLOBAL NAVIGATION SATELLITE SYSTEMS (GNSS) INFRASTRUCTURE

The global positioning system (GPS) is the reference for satellite navigation systems, as the Russian GLONASS has never reached fully operability and presents worse performances, and the European GALILEO is still under development. Table 3 shows the main design characteristics of the three systems.

This section reviews the current existing infrastructure for GNSS and the expected developments and improvements of that infrastructure during the following years. In particular, it includes a detailed study of the technical characteristics, types of service, and expected evolution of the three GNSS systems in the previous table. Furthermore, other existing or in development satellite navigation systems, such as the augmentation ones, are reviewed.

Global Positioning System (GPS)

Overview of GPS

GPS is a space-based PNT system developed by the American's DoD. It emerged in the late 1960s and early 1970s as a merger of synergistic Navy and Air Force programs for timing and space-based navigation, respectively. Currently, it is managed by the United States government through an interagency process that seeks to fuse civilian and military interests and with the objective of supporting the establishment of GPS as the standard for GNSS systems at the United States and international level. The United States Air Force (by delegation from the DoD) finances and operates a system of nominally 24 GPS satellites (distributed in six orbital planes) and a control segment with associated ground monitoring stations located around the world.

GPS was the first, and still remains the only, global, three-dimensional radio navigation and timing system providing continuous operational service nowadays. With a military origin, it has evolved in the last years to a dual-use system. Its signal represents a commodity service, provided as a public good by the United States government and freely available to all without direct cost or other encumbrance. Its military signals are encrypted for exclusivity of access by the United States and allied military forces.

It is remarkable that GPS has been an active program for over 30 years. The launch of developmental satellites began in 1978, while the first operational satellites were launched in 1989. The initial operational capability started in 1993 and the full operational capability was reached in 1995.

Table 3. Design characteristics for various GNSS systems

PARAMETER	GPS	GLONASS	GALILEO
Nr of Satellites (nominal)	24	21	30
Orbital Planes	6	3	3
Altitude (km)	20200	19100	23616
Orbital period (hr)	11h56min	11h15min	14h21min
Orbit Inclination (deg)	55	64.8	56
Frequencies (MHz)	1575.42 1227.60 1176.45	1602-1615.50 1246-1256.50 3rd freq (TBD)	1188-1215 1260-1300 5010-5030
Time Reference	UTC	Moscow Time	UTC
Geodetic Reference	WGS84	SGS85 / ITRF	ITRF

Figure 4. GPS satellite. Image credit: NASA

GPS Technical Characteristics

From its beginning, the GPS architecture was designed to minimize the vulnerability of military navigation and timing by moving the vital electronic processing and transmitting equipment into space to make them extremely difficult to reach. The GPS constellation, as of mid2006, is composed of 29 operating satellites to ensure 24 at any moment, of the following types: 16 Block II/IIA satellites; 12 block IIR satellites; and one block IIR-M. Those satellites are at semisynchronous attitude (12 hour period), distributed in six planes, each of them inclined 55 degrees.

GPS broadcasts two signals in the so-called L1 and L2 bands at 1575.42MHz and 1227.60MHz. Civilians using low-cost receivers only have direct access to the L1 signal, using the coarse acquisition code (C/A Code). This means that such receivers are unable to correct for delays to the signal as it passes through the ionosphere, which is now the dominant cause of errors. Precise positioning receivers can access the ranging code, so-called P-Code, now encrypted as the Y-code under a policy of antispoofing, on both the L1 and L2 signals, which enable correction of ionospheric errors. Two additional civil signals are being introduced in the modernised versions of GPS.

GPS is operated within an Earth-centred-Earth-fixed reference frame, which makes the position determinations independent of local topography. This reference must be accommodated by geodetic models within the GPS receiver, being the baseline geodetic reference for GPS the world geodetic system 1984 (WGS-84).

In what concerns the ground segment, corrections are uploaded to each satellite at least once a day by the worldwide operational control segment with the master control station at Schriever Air Force in Colorado. These uploads include orbit position projections for each satellite, based on sophisticated models developed on ground from the measurements continuously performed on the constellation, and corrections to the satellite clocks.

An overview of the main technical characteristics of GPS has been presented previously in Table 3.

Levels of Service and Accuracy

Depending on the user, GPS distinguishes two levels of service:

- **Standard positioning service (SPS) for civilian users:** The C/A code allows only direct L1 measurements. Using this service, the typical accuracy is often less than 10 meters, although the officially stated standard for worst case horizontal and vertical positions is less or equal to 22 and 77 meters, respectively, at the 95% confidence level (only based on the signals in space).
- **Precise positioning service (PPS) for military/governmental users:** This service provides enhanced accuracy and reliability because the Y-code is also used to obtain the direct measurement of pseudorange on both the L1 and L2 signals.

Depending on the achievable accuracy, applications are classified as follows:

- **Single point positioning (SPP):** This technique delivers the SPS performance and for which GPS was originally designed
- **Differential GPS (DGPS):** This technique overcomes some of the limitations of GPS by applying corrections to the basic pseudorange measurements. Those corrections are obtained by making measurements at known points (reference stations). The accuracy achievable in this case ranges from a few meters down to a few tenths of meters, depending on the quality of the receiver and the technique used.
- **GNSS surveying:** This technique also works differentially but can achieve centimetre accuracy using a special measurement technique. Receivers used in surveying and geodesy measure the phase of the underlying carrier wave signal (the so-called carrier phase). Because civilian users can only access to the SPS, receivers for this technique should employ sophisticated signal processing techniques for measuring the phase of the L2 signal. This makes the civilian surveying receivers much more expensive than those used for SPP and DGPS.

At its current level of performance, GPS is providing, on average, better than 5-meter horizontal accuracy, better than 10-meter vertical accuracy, and absolute time within 0.1 microsecond of UTC. With differential GPS techniques, local accuracies of 1 meter and better are routine. Indeed, global GPS civil service performance commitment has been met continuously since December 1993 and the current performances for the civil service far exceeds the current specification from the SPS performance standard, as shown in the following table (U.S. Coast Guard, 2001)

Expected Evolution of GPS

Although GSP is the only fully operational GNSS, and has being successfully providing its services for several years, there are several issues and limitations that need to be improved in future generations of the system. In general, these issues can be grouped in three major areas: the need to adapt to the evolving requirements from the military and civilian users (improvements in accuracy, availability, and reliability); the need to provide a worldwide efficient service to a global market; and the need to face the competency of the incoming European system, GALILEO. A detailed description of these issues can be found in Defence Science Board Task Force (2005) and Lewis et al. (2005).

At this time, GPS is updating its IIR generation; the first satellite of this modernised version (GPS IIR-M) was launched in September 2005. The

Table 4. Actual performances of GPS SPS, compared with the specification (U.S. Coast Guard, 2001)

PARAMETER	SPECIFICATION	ACTUAL
PDOP Availability	6m or less, 98% of time	99.98798%
Horizontal Service Availability	95% Threshold of 36m, 99% time	2.74m
Vertical Service Availability	95% Threshold of 77m, 99% time	3.89m
User Range Error	6m or less, constellation average	1.22m (in July 2005)

IIR generation spacecraft features a modernized antenna panel providing increased signal power to receivers on ground, two new military signals with improved accuracy, enhanced encryption and antijamming capabilities for the military, and a second civil signal that provides users with an open access signal on a different frequency. The following block of GPS II, the IIF, will launch a first satellite in 2007. Finally, the third generation of GPS, block III, is being designed, with the objective of having the first satellite in orbit by 2013. The diagram in Figure 5 depicts the evolution of the different blocks and the main features of each block.

The deployment of these generations is usually very slow, as the full constellation has to be replaced to include the new capabilities. Therefore, the modernizations always have a long and gradual introduction. It is required that each new satellite or navigation payload includes the capabilities from the previous versions, in order to guarantee backwards compatibility. The estimated time for the full deployment of a new generation is about 8 years, and usually satellite from different generations coexist.

GPS IIR-M and GPS IIF

Main assets of these new generations currently under development are:

- **New civil frequencies:** The deployment of these two modernisations of the second block of GPS will introduce second and third civil signals (L2 and L5). These will significantly increase the performances of the current L1 C/A signal. The L5 signal will be protected through the ITU. This means that the signal will be relied upon for safety of life applications, such as civil aviation and emergency service operations.
- **Ground control segment:** A new master control station will be developed to introduce the modernized GPS capabilities, including the improvements in signal integrity and errors estimation. An alternative master control station will be added, with full mission capability.
- The network of sensor stations will be increased from the current 6 stations to a total of 17, some located in non United States territory.

Figure 5. GPS generations and main introductions

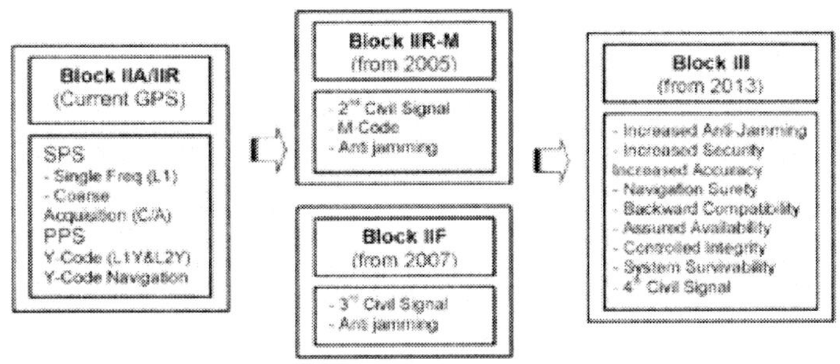

- The full deployment of these blocks will not be complete until 2012 for GPS IIR-M and 2015 for GPS IIF.

GPS III

The United States government and the private sector have completed the preliminary studies for this new generation of GPS, with the completion of the system requirements in 2005. The main goals for GPS III are:

- To increase the system accuracy
- To improve signal availability and integrity
- To assure and improve the level of unaugmented integrity
- To maintain the backward compatibility with existing receivers
- To guarantee a smooth transition from block II to block III

GPS III will incorporate the modernised L1 civil signal (L1C), which includes new modulation techniques to allow increased robustness and potential accuracy for the users. This signal has been proposed as a common baseline L1 open service signal for GPS and GALILEO.

GLONASS

Overview of GLONASS

GLONASS is a space-based PNT system developed by the USSR as the Soviet answer to GPS. It was started in 1982 and although it was declared operational in 1990, the system has never been fully operative and it is mainly used as a complement to GPS observations. GLONASS is developed and managed by the Russian Space Agency and the Russian Ministry of Defence.

The current convergence of market forces, technology, and governmental policy is accelerating the uptake of GNSS in Russia, which were on hold after the fall of the Soviet Union and the Cold War. Despite the fact that the GLONASS program has appeared to have stopped during the last years, and has never been fully operative, PNT services seem to be back as one of the top priorities of the Russian Federation and special emphasis is being paid to GLONASS as the key element for the development of these services. The original objectives, focused in national security, have evolved into a dual-use system, and GLONASS is seen now by Russian authorities as a critical piece to ensure national security and economical growth. A considerable amount of resources is being dedicated to the modernization of the system, in order to achieve a full operative system and improved overall performances by the end of this decade.

The GLONASS signal is currently open for civil use worldwide without a direct fee being charged for its use. In addition, there is an open access to the GLONASS civil signal structure for user equipment manufacture, applications development, and value-added services. It is expected that the GLONASS system will be fully compatible and interoperable with GPS and GALILEO, and will reach a total number of receivers between 15 and 20 million in 2020.

GLONASS Technical Characteristics

The technical specifications of GLONASS are very similar to the ones of GPS (see Table 3). The main difference in the design of both systems may be that GLONASS makes use of frequency division multiple access (FDMA). This means that each satellite broadcasts its own particular frequency with the same code, while GPS uses code division multiple access (CDMA), which means the same frequencies for all satellites but a different code for each one.

The constellation of GLONASS is designed to be composed of 24 satellites (21 nominal plus 3 redundant), distributed in three orbital planes, with 8 satellites in each of them, and an inclination of 64.8°. In what concerns the orbits, the height is 19,100 km and the revolution time 11 h 15 m. However, as of February 2006, there were only 16 satellites in orbit (13 operational, 2 in commissioning phase, and 1 in maintenance).

A particular feature of the design of GLONASS is an orbit inclination (64.8 degrees) significantly higher than those for GPS and GALILEO, in order to cover the Russian higher latitudes. In what concerns the ground segment, GLONASS includes only stations in the Russian Federation. These stations provide monitoring of the constellation status, correction of orbital parameters, and navigation data uploading. Further details on this system can be found at CSIC (2002).

Evolution of GLONASS

In parallel to the updating of GPS and the development of GALILEO, the Russian Federation has commenced a program to update the configuration of GLONASS and revitalise the system, including the launch in 2005 of three new satellites. The objective is that GLONASS becomes fully operational, with 24 Satellites, in 2009, with a performance comparable to the one of GPS and GALILEO (Revnivykh, 2006). In 2006, 13 satellites were operative, and two more in the commissioning phase.

Modernization of GLONASS encompasses the following actions:

- A second civil frequency at L2 frequency band (GLONASS-M) was added in 2003 for higher accuracy. Three GLONASS-M satellites are planned to be launched in 2006, with an improved 7-year design lifetime and broadcasting in the L1 and L2 bands.
- From 2007 to 2008, it is planned to launch GLONASS-K satellites that include a third civil signal at L3 frequency band, for improved performance and higher reliability and accuracy, especially for safety-of-life applications.
- Provision of a search and rescue service
- Satellite modernization, with improved clock stability and dynamic model
- Increase of the ground stations network
- Geodesy system refinement and adoption of the ITRF reference model. Global differential ephemeris and time corrections in the third civil signal—sub meter real time accuracy for mobile users.
- The full constellation is planned to be broadcasting three sets of civil signals by 2012.

In parallel with the modernization, negotiations are taking place to increase the international cooperation with GALILEO and GPS representatives, in order to facilitate future interoperability. Agreements are also being discussed with other countries potentially interested in the system, such as India. GLONASS, as GPS, will remain as a dual-use system, but the civil and commercial applications are now driving its modernisation.

GALILEO

Overview of GALILEO

European interest for satellite navigation comes from the first studies of the European Space Agency, NAVSAT, started in 1980. The joining of the efforts from both ESA and EC opened the path to the development of GALILEO, which was proposed to the EU in February 1999 and finally approved in March 2003 after long negotiations, with a budget of 3900M€. In its White Book for a Transport Policy, the European Commission pointed out the satellite navigation as a critical technology to push up the transport infrastructures in Europe.

The two main differences with respect to GPS are the civil orientation and the guarantee and integrity of the signal. This guarantee is translated in the continuity of the availability of the signal while the integrity assures the quality of the received signal. In spite of GALILEO being conceived as a civil system, it seems clear that it will be also a strategic infrastructure and most likely will be used in parallel for security and defence purposes.

From a European perspective, the motivations to set up a programme such as GALILEO were to develop a GNSS system for Europe, that allows independence from GPS and control of a global navigation system. Together with these political interests, the expected economical and industrial returns from the satellite navigation market in the next years is too interesting to let it pass by, without taking part in the benefits. In addition, GALILEO should be seen as a strategic program in the EU for the development and harmonization of the European industry in the front edge of the technology.

Figure 6. GALILEO constellation. Image credit: ESA

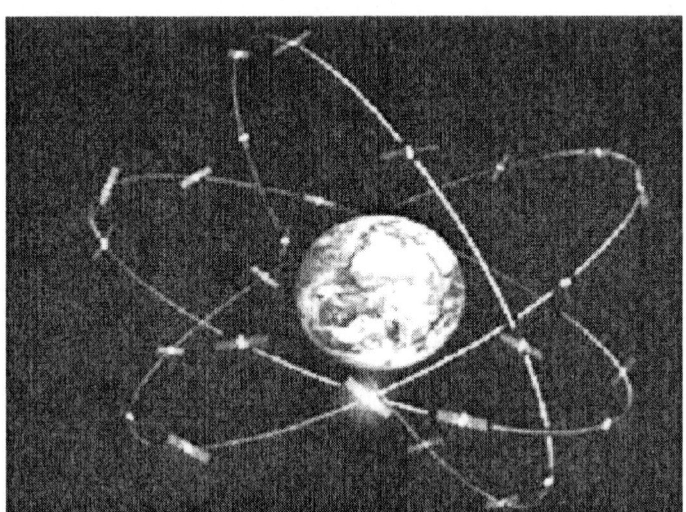

GALILEO Technical Characteristics

Like the other GNSS already reviewed, GALILEO's architecture is basically composed of three segments: the space segment, the ground segment, and the user segment. The space segment will consist of 30 medium size satellites in a medium orbit (23,616 km high) distributed to reach global coverage. Three navigation bands are used: E5, E6, and L1. The ground segment foresees a global network of stations. It is functionally divided in the ground mission system (GMS), with the task of controlling the whole mission and generating the navigation message, and the ground control system, with the mission of controlling the satellite constellation. Physically, both are placed in the GALILEO control centre (GCC), which is replicated in two different locations for redundancy. Apart from this location in the GCC, there are three different types of remote stations: the uplink stations (part of the GMS), for message transmission to the satellites; the TT&C stations (part of GCS) to control the constellation; and the GALILEO sensor stations (GSS), for measuring the satellite messages. This basic architecture was shown in Figure 3 and an overview of the technical specifications is included in Table 3.

The frequencies used by GALILEO satellites are within the 1.1 to 1.6 GHz band, a range of frequencies particularly well suited for mobile navigation and communication services. Each GALILEO satellite is supposed to broadcast 10 different navigation signals and a distinction is made between signals containing navigation data (so called data channels) and signals without data (pilot channels). In Figure 5, a complete overview of the signals is provided, and data channels and pilot channels are depicted in orthogonal planes, because they are shifted by 90 degrees in phase to allow their separation in the receivers.

GALILEO, as GPS, also makes use of CDMA, which means that all the satellites in the constellation transmit at the same frequency, and a code is added to the signal to distinguish which satellites the signals are coming from. One of the reasons there are so many signals in GALILEO is because different alternative codes with different characteristics were necessary in order to satisfy the different types of users (an indoor, static user may like long codes, while outdoor, fast-moving users would prefer short codes). Another reason is to allow the receiver to estimate the ionospheric delay error, because by combining measurements to the same satellite at two different frequencies, it is possible to cancel out the ionospheric error.

Figure 7. GALILEO navigation signals

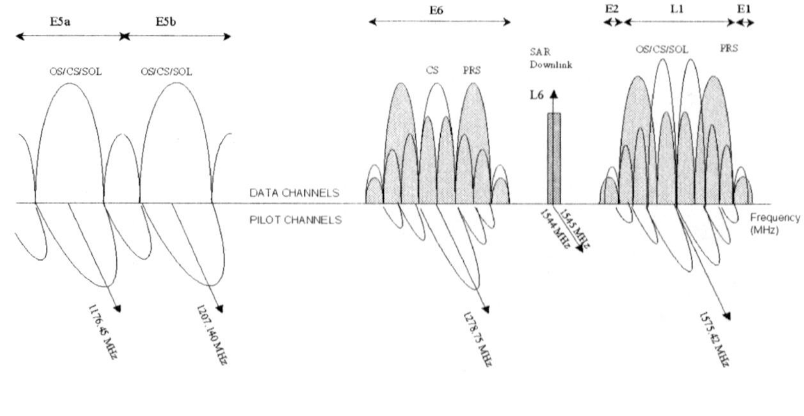

This is the reason why GALILEO services are generally realized using pairs of signals.

Levels of Service and Accuracy

As further detailed later on in the epigraph of GALILEO/GPS business models, four main levels of service are intended to be provided by GALILEO. In Figure 8, an integrated overview of the signal frequencies used by each service is provided.

- **Open access service (OS):** Simple positioning with the following specifications—L1 single frequency (15m Horizontal Positioning, 35m Vertical Positioning), and dual frequency (4m H, 8m V). For this service, the signals L1, E5a, and E5b, either data or pilot, are used. Several combinations are also possible, such as a dual frequency service based on using L1 and E5a (for best ionoshperic error cancellation) or single frequency services (at L1, E5a, E5b, or E5a and E5b together), and even triple frequency services using all the signals together (L1, E5a, and E5b), for very precise centimetric applications.
- **Commercial service (CS):** In addition to the L1 navigation band, the E5, E6 may be also used.
- **Safety of life (SoL):** E5 and L1 signals to get an integrity performance of 20 m for the vertical alert limit and 40 m for the horizontal alert limit. This service is based on the measurements obtained from the open signal and uses the integrity data carried in special messages designated for this purpose within the open signals. The SoL service is like a data channel within the open signals.
- **Public regulated service (PRS):** This uses two signals in the 1575.42 MHz and the 1278.75 MHz bands. The signals are encrypted allowing the implementation of an effective access control scheme. Increased robustness and integrity.

In addition to those four main services, GALILEO also provides a Search and Rescue (SaR) service for the transmission and localisation of distress beacons in combination with the COSPAR/SARSAT SaR system.

The distinctive shape of the spectrum of the signals is due to the special modulation adopted for GALILEO in order to avoid interference with other GNSS within the same band. Indeed, this is the case for GPS at L1. The modulation adopted is the so called BOC(1,1) or binary offset carrier of rate (1,1). This kind of modulation allows GPS and GALILEO signals to share the same frequency while avoiding mutual interference. The compatibility and interoperability of GALILEO and GPS has been one of the main issues since the early days of the GALILEO project. After long discussions about this key topic, an agreement was signed between the United States and the EU to guarantee the compatibility and to allow the common use of both systems and the improvement of services.

Table 5 summarises some of the most relevant requirements for GALILEO performances once it is fully deployed, and for the different services provided. While the accuracy requirements are similar to those for GPS, as mentioned before, one of the differences with current GPS services is the requirements for signal availability, integrity, and continuity, which is translated into the requirements stated in the Table 5.

Implementation of the GALILEO Programme

The implementation of GALILEO is characterised by a phased approach. Indeed, in order to minimise the risks and to compensate for the short experience of Europe in navigation systems, the strategy adopted for GALILEO has been to validate the system with a reduced configuration before full deployment. The following phases can be distinguished:

Figure 8. GALILEO signal frequencies

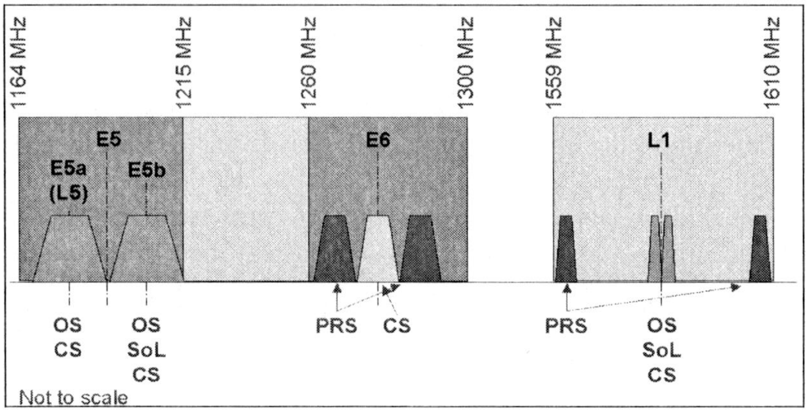

Table 5. GALILEO services performances requirements

	OPEN SERVICE	COMMERCIAL SERVICES	SoL SERVICES
Positioning Accuracy	15 m H + 35 m V (single frequency) 4 m H + 8 m V (dual frequency)		4 m H + 8 m V (dual frequency)
Timing Accuracy	30 nsec		30 nsec
Availability	99.5%	99.5%	99.5%
Integrity	None	None	Required
Integrity Risk			3.5 $*10^{-7}$/150 Seconds
Continuity Risk			10^{-5} / 15 seconds
Access Control	Free open access	Controlled access of ranging code and navigation data message	Controlled access of navigation data message

- **Definition phase (2002-05):** Where the basic services, the system specification, and the global architecture have been defined; within this phase it is possible to include the development of the GPS augmentation system, EGNOS, conceived as a first step before GALILEO.
- **In orbit validation phase (2005-08):** With the scope of validating and verifying GALILEO before the deployment of the overall system; this reduced architecture will include one of the two redundant GCCs, a reduced number of stations, and the minimum number of spacecraft to perform the validation (4 satellites).
- **Full deployment phase (2008-2010)**: To complete the deployment of GALILEO until full operational capability (FOC)
- **Long term operations phase:** To provide the defined services starting at FOC

The implementation is in principle intended to be funded through a public-private partnership (PPP) by both the European institutions and industry. This approach is further described in the following sections. In what concerns the implementation, the biggest uncertainties are now driven by the economic and funding issues, as technical issues seem to be well under control. On one hand, the private-public funding will mean the assumption by the industry of about two thirds of the total investment, a level of contribution that is currently far from become a reality. The capacity of the market to absorb these investments is still uncertain. On the other hand, the cost overruns have started at a very early stage of the project. Before the in-orbit validation phase is started, the EU and ESA have announced a deficit of 360M€, which will have to be assumed by both organisations.

Despite the big impulse given in Europe to GALILEO and the possibilities in the growing market of satellite navigation, there are a number of problems and uncertainties in the development of the programme. Apart from the financial issues, it is generally recognized that the current joint

Figure 9. GALILEO planning and time schedule

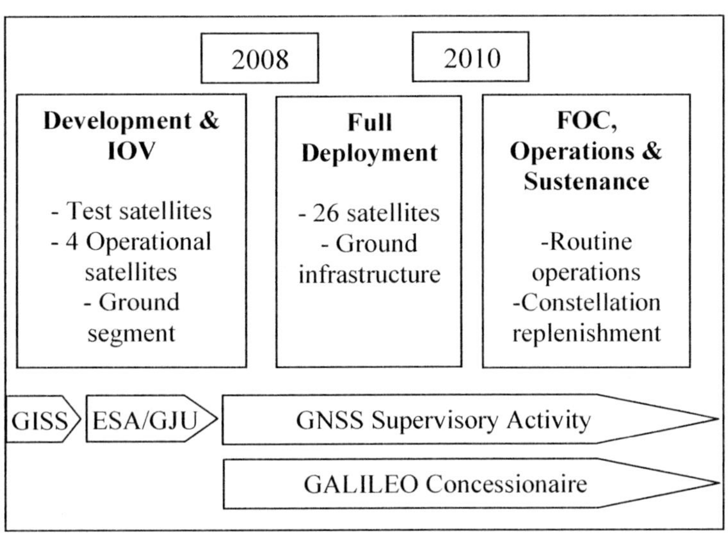

structure for GALILEO (European Space Agency / European Union plus private sector) is by far too complex. Negotiations between the European Commission, the Agency, the member countries of both institutions (including the organisations created ad hoc, the GJU, and the GSA), and the private sector are not easy to handle. Decisions are taking several months before being agreed upon among all the participants and account for most of the delays in the normal development of the activities.

Augmentation Systems

Augmentation systems use the GPS signal as an input to provide an improved service, usually for a specific region of interest. These augmentation systems can be ground- (GBAS) or space-based (SBAS), depending on the infrastructure used.

The first terrestrial augmentation to GPS provided meter-level accuracy for oil exploration and precision scientific applications. These involved base stations that had to be located to within 30-50 km of the receiver. However, over the course of the last 20 years, this capability has migrated to services provided by terrestrial beacon systems and to satellite broadcast.

Several satellite augmentation systems are nowadays available or under development. These have the advantage of a greater coverage, and usually make use of satellites in the geostationary orbit (in a fixed position in the equator). These systems can increase the accuracy, integrity, availability, and continuity of GPS signals, and allow the control of a national positioning system. In other words, they represent an augmentation to the core service provided by the GPS satellites, and none will operate on its own without the presence of the basic GPS signals. In most cases, the augmentations serve as checks on the quality of the basic signals, but together, they will soon represent an absolutely critical component of transportation economy and public safety.

Figure 10 shows the geographical zone of influence of each of the most significant satellite-based augmentation systems: WAAS, EGNOS, and MSAS.

Wide Area Augmentation Systems (WAAS)

In the United States, the wide area augmentation system (WAAS) is a multibillion dollar program developed by the Federal Aviation Administration (FAA) and the Department of Transportation (DoT) to increase the accuracy, reliability, and availability of GPS-based services for aviation users by transmitting special augmentation signals

Figure 10. Geographical distribution of SBAS. Image credit: ESA

over satellite communication links. Currently, GPS alone does not meet the FAA's navigation requirements for accuracy, integrity, and availability. WAAS corrects for signal errors caused by ionospheric disturbances, timing, and satellite orbit errors.

WAAS comprises about 25 ground reference stations across the United States to monitor GPS satellite data. Two master stations, located on each US coast, collect data from the stations and create a correction message. This message is then uplinked by three plinked stations and broadcasted through one or two geostationary satellites. In parallel to WAAS, the FAA has developed the local area augmentation System (LAAS), based on ground augmentation for aviation applications.

EGNOS

Europe through the European Commission (EC) and ESA has developed the European geostationary navigation overlay service (EGNOS) as a first step in its navigation policy before the development of GALILEO. EGNOS will augment the two GNSS operative systems (GPS and GLONASS) and make them suitable for safety critical applications such as aircraft or ship navigation.

EGNOS consists of three geostationary satellites and a network of ground stations, transmitting a signal that contains information on the reliability and accuracy of the available navigation signals. It allows users in Europe to increase the capabilities of these signals. EGNOS V1 (advanced operational capability) was technically qualified in June 2005. Initial operations started in July 2005. Operational stabilisation was expected by the beginning of 2006.

MSAS

In Japan, the MTSAT satellite-based augmentation system (MSAS) will improve the accuracy, integrity, continuity, and availability of GPS signals in the Japan's aerial region by relaying augmentation information to users via the MTSAT geostationary satellites. The system also consists of a network of ground monitor stations and the master control stations, in Japan, and monitor and ranging stations outside Japan.

Other Augmentation Systems

Other countries are planning to develop augmentation aystems. This is the case of China with the Beidou system, with 4 satellites in geostationary orbits (Chinese Defence Today, 2004), and India, with Gagan.

Other Initiatives Based on GPS

QZSS (Quasi-Zenith Satellite System)

The QZSS project is a Japanese government-private sector corporation program aiming for new satellite business using the quasi-zenit satellite system. The private sector is responsible for mobile communications, mobile broadcasting, and satellite systems. The government is responsible for technology development and demonstration, including GPS complement and augmentation with QZSS (Tsujino, 2005). It is a satellite navigation and communications system, which uses and complements GPS signal for the Asia-Pacific region, and is more sophisticated than the described augmentation systems.

QZSS is composed of three satellites 120 deg apart in inclined and slightly elliptic orbits in the different orbital planes that pass over the same ground track. The plan is to launch three satellite broadcasting GPS-like signals, to increase the number of satellites available at high elevation angles (close to zenith) over Japan. The satellites are in geosynchronous but not geostationary orbits This configuration allows increasing the availability of navigation, broadcasting, and communication in downtown/dense urban areas, and mountainous areas. The service will not be limited to Japan, but will also cover East Asia and the Oceania region (Petrovski, 2003).

The system is planned to be compatible with GPS and GALILEO. The master control station is located in Japan, and different time management and monitor stations are deployed in the Asia-Pacific area. The first satellite launch is foreseen in 2008.

GPS Multinational Networks

Some countries or organisations have developed multinational networks to enhance the signal from GPS. This is the case of:

- **Nationwide differential GPS system (NDGPS):** An international nonproprietary standard for surface transportation operated in about 50 countries
- **Continuously operating reference stations (CORS) network:** A cooperative endeavour involving more than 130 public and private organisations with close to 1000 sites, which include the NDGPS and WAAS sites
- **International GPS service (IGS):** A network of over 350 stations monitoring GPS signal, and providing increased accuracy; it is coordinated by NASA/JPL.

Other Satellite Navigation Systems

Apart from the GNSS, other satellite-based systems have been developed for different applications over the last two decades. This is the case of Argos and SARSAT. However, the trend is to replace these systems with new commercial communication systems or by the use of the GNSS capabilities.

Argos

The Argos system is a joint French and United States satellite-based system dedicated to environmental studies. The system acquires, processes, and broadcasts data received from fixed or mobile automatic transmitted. Some applications of Argos are:

- **Marine applications:** Ship transportation, buoy location, iceberg drift measurements, pollution data monitoring, weather data collection, or fleet tracking
- **Terrestrial applications:** Snow cover studies, hydrological, and flood monitoring
- **Aerial applications:** Migration studies, balloon monitoring

Probably the largest use is for wildlife tracking, to follow the migrations and distribution of wild animals, which carry transmitters. Position accuracies of 100 to 200 m are available, also providing telemetry transmission. The Argos payload is carried onboard NOAA satellites, in low polar orbits, developed for Earth observation applications.

COSPAS-SARSAT

This is a satellite-based system designed to support global search and rescue (SaR) services. It was started in 1982 with the cooperation of the United States, USSR, France, and Canada, while many other countries joined afterwards. The system uses small transponders in aircraft or boats that transmit automatic distress signals in case of crash or accident.

The space segment is onboard two Russian navigation satellites (COSPAS), and two NOAA environmental satellites, in low polar orbits. These satellites receive the transponder distress signal and transmit it to the ground stations. These ground stations, called local user terminals (LUT), receive the signal transmitted by the satellites, and calculate the coordinates of the distress call and pass them to the mission control centre (MCC).

GALILEO plans to develop a search and rescue service, compatible with COSPAS/SARSAT, which would significantly increase the localisation performances.

BUSINESS PERSPECTIVE OF SATELLITE NAVIGATION SERVICES

GNSS Industry Sector and Value Chain

The GNSS industry sector can be considered as a subgroup of the position, navigation, and timing (PNT) sector, which is simultaneously a subsection of the geospatial information & communication technology Geo-ICT within the information & communication technology ICT market.

Ever since the launch of the first satellite navigation systems, companies worldwide have been developing products and services to stimulate and serve a civilian market for positioning technology. Continued innovation in technology has led to huge improvements in the price and performance of user equipment. By 2006, several thousand companies were already involved in satellite navigation device production and service provision. This growth of the size of the market also contributed to the development of the structure, and what started as an industry that supplied stand-alone navigation units, has evolved to combine both navigation and communication technologies.

Importance of GNSS to Industry

As it was described in the previous sections, the implementation of GNSS systems represents huge investments in infrastructure and technology, which could be justified in terms of United States, Russian, or European strategic policy. The importance for the civilian industry has rapidly grown since the nineties, currently being one of the main drivers for the update of the GNSS infrastructures. The following issues have to be considered when analyzing the impact of GNSS systems in industry.

- GNSS is an enabling technology in the adoption of information & communication technology (ICT) within mobile and remote workforces.
- Knowledge of local position and time is often critical to the efficient operation and management of mobile equipment, assets, and resources.
- Ability to precisely measure position and motion of equipment allows improved / automated operation.
- It is the use of GNSS as an enabling technology which has driven its growth.
- GNSS is an important tool for "traditional" users of positioning information (air and marine navigation, geodesy and surveying). These users were among the first to adopt GPS and are expected to embrace additional GNSS systems, such as GALILEO.

GNSS Value Chain

The GNSS value chain can be represented using the comprehensive turnover model for navigation products, developed by the GJU (2005), based on the total potential addressable market for each separate application and information on current pricing and future trends. In Figure 11, this model is depicted and the different segments of the GNSS market are identified.

In particular, for the "net turnover," defined as the turnover directly associated with positioning systems hardware, the overall prediction for annual net product turnover until 2020 is shown in Figure 12, rising from € 5 billion in 2004 to €30 billion by 2020. In what concerns the "gross turnover," or the turnover associated with the entire navigation system in which the positioning system is placed, the prediction is shown in Figure 13, evolving from €23 billion in 2004 to €178 billion by 2020.

With this model and figures, GJU concludes that the market characteristics are evolving in the sense that while in 2004 an important percentage is dedicated to the segments of components and receivers, it is expected that this shifts toward a more mature and integrated systems market by 2020.

A third type of turnover that can be defined is the "service turnover," or the turnover associated with GNSS-based services. While by 2005 this service turnover was in the order of € 7 billion and represented the 23% of the total market value, by 2020 it is expected it will grow to €98 billion and around 35% of the total market value.

Satellite Navigation Market Drivers

The following market drivers have been identified for the GNSS market.

1. **Technological trends:**

- **Chipset miniaturisation:** The progress made in silicon technology allows for size reduction, power consumption reduction, and integration of additional functionalities in the receivers. Falling chipset prices increase volume and market scope. Miniaturisation and receiver advances improve prices vs. performance and favour its use in portable devices. In parallel, advances in geographical information systems and digital mapping provide the necessary data to support new Location Based Services, gathering position and commercial information. The combination of geographical information systems with satellite navigation is at the root of most of the new applications already on the market.
- Mobile communication technologies push the markets for positioning and timing services. The integration of satellite navigation modules within mobile communication handsets is most likely one of the biggest evolu-

Figure 11. GNSS value chain (Adapted from "Business in Satellite Navigation," GJU, 2005)

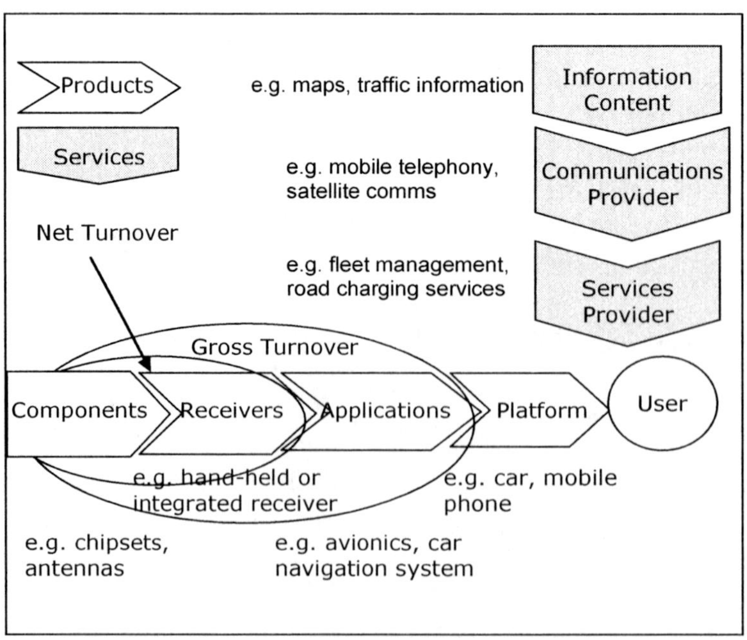

Figure 12. Annual net turnover for satellite navigation products (data from GJU, 2005)

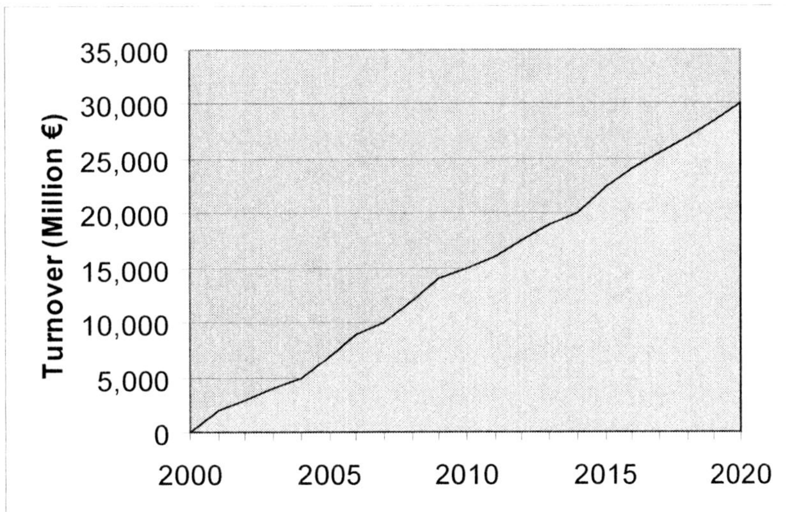

Figure 13. Annual gross turnover for satellite navigation products (data from GJU, 2005)

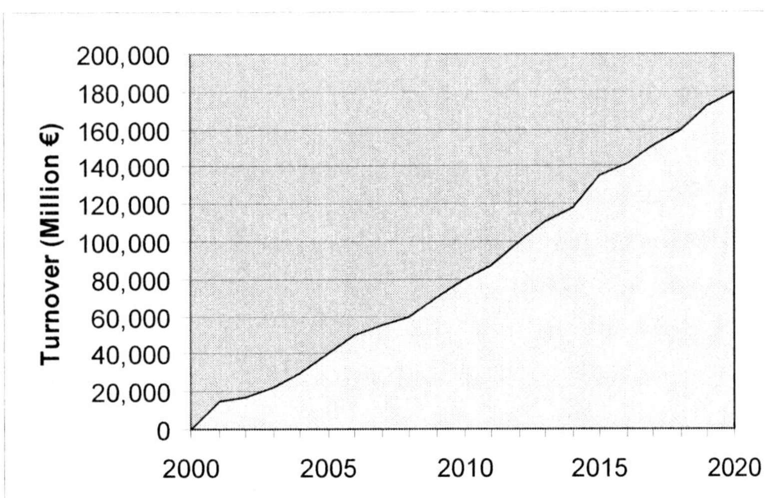

tions over the next decade for the mass-market user. This convergence is further enhanced by the development of mobile communication networks (GSM, UMTS, etc.) that enable real time data to move between system components. Another aspect to consider is that hybridisation of communication and satellite navigation signals will enhance indoor positioning.

2. **Political and regulatory environment:** Regulation both at national and international levels will indirectly drive the use of satellite navigation systems. In particular, the following political and regulatory environments could have a relevant influence:

    - Regulation concerning new or more efficient and safer transport networks (road, rail, maritime, aviation)
    - New measures to enhance public and consumer protection
    - Measures in support of people with disabilities, for regional developments or for humanitarian aid in poor countries, and so forth

3. **Economic and industrial developments:**

    - Overall economic activity and production patterns
    - High product replacement rates in consumer electronics industry

4. **Social factors:**

    - Changing work patterns and increased mobility heightening demand for efficient travel in both personal and public modes.
    - Increased demand for consumer electronics
    - Concerns about transport and personal safety, drive better safety performance and monitoring needs.

Satellite Navigation "Enabled" Markets

The variety of scenarios where GNSS services are proving their capabilities is very wide. GNSS applications have been used to alleviate some of daily problems (increased traffic congestion and urban planning/infrastructure constraints) and have led to advances in telematics, asset management, and emergency response, among others. In addition, mass market applications are already common, and examples of the growing penetration of theses services in the civilian markets are numerous:

- For car navigation, more than 5 million units are already installed in Japan with more than 1 million being produced annually.
- For telematics, GM OnStar has more than 4 million subscribers; ATX (Mercedes, BMW) more than 700,000.
- For cellular phones, CDMA phone networks depend on GPS time for network synchronization (more than 100 million CDMA phones in North America and more than 285 million users worldwide rely on GPS timing for their networks).
- For GPS enabled cell phones for E-911 as well as other location-based applications, more than 50 million per year are sold.

Some civil applications currently established or expected for the next future, are identified hereafter, focusing in the potential markets for the GNSS applications:

- **Location based services (LBS):** The LBS which are backed by the mobile phone market are expected to be the main driver for the commercial development of GNSS user applications (expected revenues of 2.1 billions € in Europe by 2009 and 10 billions € by 2009). The incorporation of the navigation receiver in mobile phones and computers, adding the capacity to compute location and precise timing, is currently a fact. The navigation capabilities provide to the user information for tracking and destination location, criteria for journey optimisation, emergency assistance services, and so forth. These services can be incorporated in markets as different as gaming, social services, family protection, workforce management, tourism, or social networking.

- **Road transportation:** GNSS receivers are commonly installed in cars as a key tool for providing new services to people on the move such as road user charging, real time traffic information, emergency calls, route guidance, or advanced driving assistance. Transportation companies will broadly use navigation information for the efferent management of their fleets. The key point for the generalisation of these applications is their integration with other industries (such as automotive, computing, or wireless terrestrial communications) and services (traffic, road assistance). In the United States, for example, by combining navigation tracking and wireless communications, the new telematic systems will offer postcollision notification integrated into the emergency networks. It is expected that by 2020 there will be over 330 million cars with GNSS-based navigation systems onboard.
- **Aviation:** civil aircrafts and air traffic controllers have been largely using positioning and timing services for their regular operation during different phases, take-off, on-route, and landing, under all weather conditions and increasing operation safety.
- **Satellite systems:** GPS is also a commonly used tool for guidance and control of satellite, during all phases of launch and operation and with the capacity to track orbits up to geostationary orbits.
- **Maritime:** Satellite navigation is becoming a fundamental tool for improving the safety of maritime navigation in open sea and inland waters. Apart from the direct applications in vessels guidance and navigation, there are many other activities where satellite navigation signals can be used. In fishing, for example, it helps to locate traps and nets; it is used for fleet management and cargo monitoring, commercial harbour operations, and so forth. A key issue for GALILEO will be its contribution to the international Search and Rescue (SaR) service, enhancing the worldwide performance of the current COSPAS-SARSAT system. The GALILEO SaR service will decrease the time to alert from hours to minutes, and the determination of the stress beacon from a few kilometers to meters.
- **Rail transportation:** Satellite-based navigation helps railway operators in safety and traffic management applications.
- **Oil and gas:** The oil and gas industry is extensively using satellite navigation services for on-shore and off-shore exploration and exploitation activities. Many applications rely on high accuracy positioning and timing services, including geophysical exploitation, geotechnical evaluation, rig and platform services, underwater inspection, pipe lying, and so forth.
- **Agriculture:** GNSS services, in combination with satellite imagery or short range tagging, are used in different precision agriculture applications, including the measurement of crop yield during harvesting, management of soil sampling, variable rate fertiliser spreading, tracking of animal movements, and so forth. GPS farming systems provide precise guidance for field operations, or collection of map data on tillage, applications, planting, weeds, insect and disease infestations, cultivation, and irrigation.

- **Mapping and GIS:** Using GPS for GIS data collection and data maintenance is essential for timely decision-making and the wise use of resources. Any organization or agency that requires accurate location information can benefit from the efficiency and productivity provided by GPS technology.
- Civil Communications infrastructures use GPS as the primary distribution mechanism for time and frequency synchronization.
- Electrical Power Distribution Networks also use the very accurate timing functions of satellite navigation.
- **Electrical commerce and finance:** Many banking and financial firms employ GPS timing for synchronization of their encrypted computer networks.
- **Emergency services:** Police, fire, and ambulance providers are increasingly using GPS as means of managing their fleets of emergency vehicles.
- Search and rescue (SaR) systems currently uses satellite navigation systems for the reception and distribution of emergency beacons and response, and localization of the originator of these signals. GALILEO foresees the inclusion of a specific SaR service with dedicated frequencies and increased performance (high integrity), which is expected to dramatically increase the efficiency of the international SaR services.
- Many "social" applications have also being proposed for a great variety of aspects: services to help people with disabilities, insurance companies, car theft, border control, police and fire-fighters, and so forth.
- **Leisure applications:** This sector can bring many unexpected uses of GNSS services. GPS has been used, for example, by golfers to help measure shot distances. Treasure hunting and geo-catching are becoming popular sports in many countries.

Funding of GNSS: Public vs. Public-Private Partnership

The huge economical effort of setting up complex systems as a GNSS makes the way of financing a critical issue. GPS and GLONASS, conceived as military systems, were exclusively developed using public military funding. However, the situation has changed, and the possibility of making revenues with the provision of these systems has raised the possibility of a different way of funding, with the intervention of the private sector.

The way GALILEO is coming together in its business plan is quite different from the GPS and GLONASS approaches. GPS is an Air Force program, with funding coming from the U.S. Department of Defence. The needs of the military therefore come first in the program specifications, and the needs of the civil market, although also addressed in GPS, are not in general a defining priority. For GLONASS, a equivalent approach is followed by Russia. GALILEO, on the other hand, is primarily market-oriented and dedicated to the civil user community, and funding might be obtained from other sources.

Public-Private Partnership (PPP)

The strategy that might be used to finance the GALILEO program is a public-private partnership (PPP) introducing the private sector in the ownership and management of the system. This means that there would be public funding, mainly through the European Space Agency (ESA) and the European commission (EC), and equity funding coming from private sector shareholders. A similar PPP approach is also under construction for some other satellite-based navigation systems, such as the QZSS augmentation system in Japan.

The idea behind the PPP approach is to transfer risk and responsibility to the private sector at the same time that it optimizes the benefits of the public sector. Some of these benefits would be, on the one hand, the introduction of the private sector efficiencies, ensuring overall value for money and improved project design, and on the other hand, the spread of the public expenditure over a longer period. At the same time that the PPP approach reflects the fact that GALILEO combines public service and commercial aspects, it confirms private sector commitment to the project and contributes to keep costs under control because much of the risks of cost over-run would fall on the private sector.

In general, the PPP structure proposed for GALILEO could represent a possible approach for the development of the space sector in order to:

- Encourage competitive market position of industry in accessible engineering market
- Promote research and development and new space applications with spin-off potential into the commercial sector
- Create synergies between public and private services and needs
- Put pressure for return on investment to better focus on user and market needs

Within the PPP schema, there was the possibility to choose between two approaches: the corporate joint venture in which public and private sectors would be coinvestors, and the concession structure in which there is a clear separation between the functions and responsibilities of the private and public sector. This second approach, the concession structure, was adopted for the GALILEO project: the private sector would be asked to finance, operate, maintain, replenish, and exploit the system in exchange for the right to develop market revenues for a specified period of time. Under this approach, the joint undertaking (GJU), created by the European commission and ESA to manage the concession award process, would be responsible for the Development Phase of the GALILEO project (fully publicly funded). The GJU would subsequently award a concession for the deployment and operation of GALILEO to a private sector consortium through a competitive tender. This second phase of deployment is expected to be funded partly by the public sector (one third) and partly (two thirds) by the concessionaire.

After years of political and economical discussions, the GJU decided to award the concession to operate GALILEO to a consortium, a result of the merging of the different European companies biding. The concession contract is still as of 2006 under discussion, mainly in what concerns cost overruns, completion, revenues (market), performances, design (interface between IOV and concession program), overall risk coverage (spare, contingency, insurance), deployment program, compensation on termination, and replenishment. Once the concession contract is completed, the EU and ESA would control GALILEO under the GALILEO supervisor authority (GSA).

Sources of Revenue within the PPP/ Concession Scheme

According to projections from the European Commission, the expected overall cost of GALILEO is around 7 billion € (1.5 billion € for the development, 2.1 billion € for the deployment, and around 200 million € /year for the exploitation, with the concession duration 20 years). In parallel, the overall projection of revenues is 8.5 billion €, but it is not clear at all whether that figure of revenues would ever be reached.

Revenue for the operating consortium in GALILEO is expected to be generated from two main sources:

- **Royalties on chipset sales from manufacturers:** This can be a source of revenue while maintaining free open access service. As GALILEO signals are encoded, the chipsets must contain decoding software and the operating consortium may hold copyright in software. It is envisaged that up to € 0.50 per chipset could be a reasonable level of the royalty to be paid by manufacturers.

- **Service revenues from service providers:** The services based on a fee (high accuracy, contractual quality of service, etc.) may generate additional revenues to the operating consortium.

With this schema, it was projected to have by 2020 over € 500 million in revenues for the operating consortium of GALILEO. From this amount € 300 million would come from the royalty revenue and the rest from the service revenue. It has to be noted that early revenues would come mainly from royalties, while service revenues would develop as the market evolves. In what concerns the different applications, the main contributors to those revenues would be the personal communications and LBS, followed by the aviation sector.

In addition to those two main sources of funding, Price Waterhouse, in the GALILEO inception study, proposed some other additional revenue streams that may be open for the Concessionaire. For example, the so called "assisted GNSS," based on the combination of navigation signal from a GNSS and a communication signal via a cellular network, such as GSM / UMTS. Using assisted GNSS, the user would benefit from a reduction in the time required for positioning determination following activation or reactivation of the terminal and improved availability of the positioning in urban areas, for all of which he could be willing to pay a premium. Other potential source of revenue could be providing the service of user authentication (determining whether someone or something is, in fact, who or what it is declared to be) and signal authentication (guarantee that the signal received is in fact from GALILEO rather than another source, and providing protection against "spoofing").

Risks within the PPP

The involvement of the private sector in the PPP schema might open some issues. On one hand, the private investors need to obtain revenues from the provided services, in a market still in development and uncertain, taking into account that both the GPS and GALILEO signals will be free of charge. In Alcazar and Velasco (2004), one of the main risks identified for the development of GALILEO is finding sufficient equity. In this sense, probably the main challenge is not to find the funding itself but instead to find it on time, in order to avoid delays in the deployment. On the other hand, some security issues can be opened by the fact that a key strategic infrastructure is managed by private companies.

In relation to the involvement of the private sector in the PPP schema, it is the issue of whether the industry and users are going to be willing to pay for additional services (extra signal integrity and coverage). If companies feel they are subsidising navigational services for other parties and they are not receiving the service they need, most likely they will be reluctant to participate in the funding and will try to develop alternative systems. In a similar way, final users may not be willing to pay for these extra services, taking into consideration the availability of free signals. However, according to some initial studies in the area, it is foreseen that emergency services, airlines, and airport operators would be among the group of end-users willing to pay due to the benefits they could get.

GALILEO Business Models

GPS currently provides two different services: civil (SPS) and military (PPS). The first one is free of charge. This is possible due to the fact that GPS is exclusively funded by the United States government, who is interesting in developing and managing a system for its national security interests.

GALILEO, on the contrary, as explained before, foresees the provision of two kind of civil services: one free (open sService, OS), and one fee service (commercial service, CS). The last one is expected to provide revenues to the private companies investing in the development of GALILEO. This service will provide added value information (mainly integrity) and liabilities.

The GALILEO joint undertaking groups the previously defined applications for GNSS systems in three main groups, depending on the characteristics and type of service the users would need to receive. In the next figure, a nonextensive list of applications within each of the three groups is shown:

For the group of "safety of life," the driver for the service to the user is the "integrity." In other words, applications within this group expect to receive an error free service because of the nature and sensitivity of the applications, where any error could have critical consequences. Within the second group, "mass market" applications, the driver for the service to the user is the "low cost" of the service provided. Finally, for the "professional" applications, the main driver would be the "precision" of the service.

Based on the previous three types of applications identified and their drivers, different business models have been defined for GALILEO. In particular, the following five sets of services were already mentioned when describing the system:

- **Open access service (OS):** Oriented to the "mass market" and free of charge; simple positioning and interoperable with other GNSS systems.
- **Commercial service (CS):** This is a fee service with high accuracy and integrity of the signal, for the "professional use." It is based on the open service standard but provides added value services with respect to the open service as the higher accuracy, data broadcast, and authentication. It is also a guaranteed service, and has the possibility of additional commercial encrypted data. The access is though external service providers.
- **Safety of life services (SoL):** For emergencies and rescue systems; integrity and authentication of the signal; guaranteed service with certification-liability.
- **Public regulated service (PRS):** For the public and strategic use of the different governments; it is encrypted which means that the access is restricted to authorised users and that it has a service denial capability; uses separate signals which provides improved service robustness. In addition, it is also characterized by the integrity or quick alarm in case of malfunction, and had continuous availability, even in times of crisis.
- **Search and rescue (SaR) service:** For the transmission and localisation of distress beacons in combination with COSPAR/SARSAT SaR System; the main features are near real-time detection, precise, and return link (acknowledgements) feasible.

In addition to this sets of five main services, GALILEO will also provide other services in combination with external systems, as for example, the support to the international time reference bodies and the support to the external regional integrity services. In combination with additional information provided to the users on a local basis, GALILEO will offer other services, such as:

- Local precision and high-precision services, with local differential correction signals
- Local assisted services, which provides two-way communications services to assist the user for PNT calculations
- Augmented-availability services, with supplementary transmissions overlaying the GALILEO signals

Business models developed for GALILEO are based on an optimum use of the five services in terms of revenue generation. In particular, it is expected that the OS, CS, and SoL will raise most of the revenue, with integrity and a certain level of service guarantee expected to trigger most of the business development. The following issues have to be taken into account:

- There will be, most likely, no real business associated with services in PRS and SaR, as the signal users will mainly be the signal owner.
- New services based on the SoL signal can be provided in aircrafts, trucks, rail transport, and so forth. A concern to the development of the SoL market lies with the cost of the service.
- There are different opinions as to the outlook of the open service. Some experts claim that very little market growth is foreseen for OS, while some others disagree. The fact that the OS can be used free of charge does not mean that a market for it is not created. The value-added services could be extremely important. Also, there might be innovative processes that can add features to the open signal (maintaining free access) such as authenticating the signal and providing some level of guarantee.
- Commercial services will grow around numerous applications. Experts list several, such as distribution, transport, telematics, and fleet management, as well as road tolling, in accordance with new EU regulations. Time stamping for financial services is another promising development area.

Figure 14. Applications and features of the three types of services for GALILEO (Source: GJU)

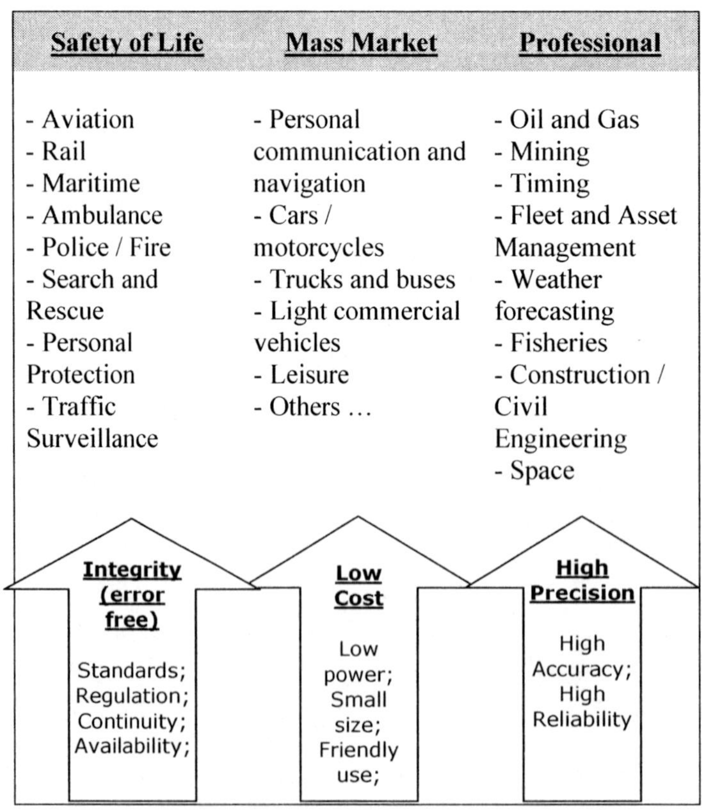

TRENDS AND ANALYSIS OF FUTURE SCENARIOS

Trends

The following years will likely witness a huge development of GNSS infrastructure, with the development of the next generations of GPS (Blocks IIR-M, IIF, and III), the European system GALILEO, and the completeness of GLONASS. Not only will there be three complete systems available, but also the interoperability possibilities will allow the improvement of the individual performances of each one, yielding increased accuracy, integrity, availability, and guarantee of the global services.

In particular, the following trends can be identified in what concerns the improvements of the systems for the near future:

- **Space segment:**
  - **Constellation configurations:** More satellites, additional signals, longer lifecycles
  - **Signals enhancements:** Higher power, more robust codes, antijam, antispoofing capabilities, flex-power
  - Integrity alerting
  - Signal authentication
  - Higher data rates will enable search and rescue, and other value-added services
  - **Greater system performances:** Accuracy, availability, or continuity
- **Ground segment:**
  - Satellite tracking and maintenance
  - Civil signal monitoring
  - Increased worldwide networks
- **User segment and applications:**
  - Leading role in the definition of future satellite navigation systems
  - Augmentations—Dissemination and use of PNT data
  - Receiver design—Ultrasensitive, software-defined radios
- **Technology drivers:**
  - Intersatellite link and intersatellite ranging functions for the GNSS constellation(s)
  - New functions onboard (orbit determination, integrity channel, communication links, etc.)
  - New cooperative concepts among different GNSS systems
  - Navigation related communication services
  - Systems interoperability

As a general trend, not only can an increased integration of GNSS systems be expected (in particular of the next generation of GPS and GALILEO), but also an increasing integration of application and services regulation.

Potential Scenario for 2015

Although the expected schedules for the currently planned GNSS systems (see Figure 15) foresee the availability of the three systems by the end of this decade, it is very uncertain whether these schedules will be fulfilled, and the scenario that GPS, GALILEO, and GLONASS will be fully operative should be considered around 2015. Within this timeframe, there would be three fully operational global GNSS systems, comprising 60 to 90 satellites in the sky, plus an increasing number of regional systems. This will lead to improved availability and more robustness, but a limited impact on accuracy. Integration with other technologies will mean that indoor positioning will be mature and robust. While currently only the GPS L1 C/A and GLONASS L1 signals are available for civilian users, by 2015, up to nine signals are expected, together with integrity information distributed by the three systems. This evolution is shown in the next table, where current and future signals for the three GNSS are presented.

This increased infrastructure both in space and on ground, together with the possibility of intersystem operability, will drastically change the scenario of GNSS, and open it to the mass markets. This should lead to low-cost receivers measuring carrier phase from combined constellations.

In parallel, the role of regional ground augmentations might be redefined, given the inherent robustness of the satellite-based only solutions:

- Sustenance of the systems will become even more of an issue. Pressure to reduce operational costs will result in system optimization taking advantage of the advances in technology and a larger number of satellites in the sky.
- Integrity service will be maintained. Standards will evolve to cover multiconstellations and services will increasingly become multimodal.
- Accuracy-based services will move from meter to centimetre level.

In Figure 15, the expected developments of GNSS systems in the following years is shown. Some issues can be highlighted from this schedule. In the first place, it is shown in the GPS bar the cost in time needed to update the constellation: completion of GPS Blocks IIR-M and IIF takes about 8 years from the launch of the first satellite of the new generation. This is in contrast with the "optimistic" approach of both GALILEO and GLONASS, which assume times from 3 to 5 years for the full deployment of their systems. For GPS III, a full replacement time of 5 years from 2013 to 2018 is also expected. The experience of past developments shows that the efforts needed for the deployment of these infrastructures usually delay the schedules more than expected.

A second aspect shown in this figure is that GPS has a longer schedule than the other two. This could mean that accumulated experience by Unites States' authorities is reflected in a more detailed and extended schedule, at least for the deployment of the modernised II Blocks. Schedules from GALILEO and GLONASS should be considered as desired approaches and, on one hand, the low experience of Europe in navigation systems, and, on the other hand, the recurrent economical difficulties of Russia will be reflected in a delay on their schedules.

Table 6. Expected GNSS services and signals by 2015 (Note: in bold, signals not existing today)

Service	GPS	GLONASS	GALILEO
Basic Positioning	Standard Positioning (SPS) L1 C/A, **L2C, L5**	Standard Precision (SP) L1, **L2, L3**	Open Service (OS) **L1, E5a, E5b**
Integrity / Safety	-- **L5**	-- **Integrity Message**	Safety of Life (SoL) **L1, E5a, E5b**
Commercial / Value Added	--	--	Commercial Service (CS) **E6**
Security / Military	Precise Positioning (PPS) L1 P(Y), L2 P(Y), **L1 M, L2 M**	High Precision (HP) L1, L2, **?**	Public Regulated (PRS) **L1, E6**

Figure 15. Overview of GPS, GALILEO and GLONASS schedule for next decade

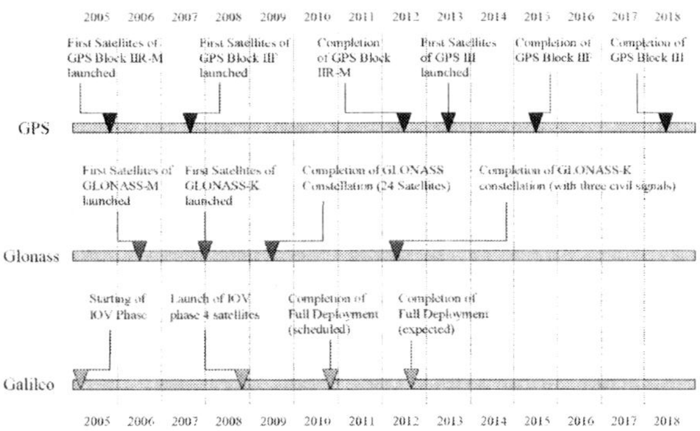

Long-term evolution of GALILEO, after its completion, has also been considered. A potential GALILEO block II should eventually include the following features, which are in line to those intended for GPS block III: additional signals, use of bands with greater frequencies (C-band), spot beams, burst transmission, inter satellite links, use of a dedicated payload on GALILEO satellites for supporting system operations, secondary payloads, and next generation clocks.

CONCLUSION

Navigation, positioning, and timing (PNT) are today vital capabilities for modern economies, and a powerful tool for the national security of developed countries. GNSS are rapidly evolving into a key part of the global infrastructure, not only limited to the navigation services, but integrated with other information technologies. For these systems in general, and for GPS in particular, there is a clear tendency to change from military-oriented systems to dual-use infrastructures (although the de-facto control of GPS still remains in the U.S. Department of Defence). We are attending to an increased civilian use in the mass commercial markets, with new application and services introduced continuously. Most of the applications of GNSS were never predicted by the designers of GPS, who most likely only had in mind positioning for military purposes.

GPS and GLONASS combined have already demonstrated the benefits of extra satellites. In addition, GALILEO is still to be developed. It will be Europe's own global navigation system, providing high accuracy and guaranteed global positioning service under civil control. In the near future, a user will be able to acquire its position and time with the same receiver from any of the satellites in any combination (GPS, GLONASS, GALILEO). For a future scenario with two (or more) satellite navigation systems, the generation of a wide market is mandatory, and for this, there are two key issues: the integration in the personal mobile systems of the third generation, and the increase of the coverage in urban environments, which is currently only at 50% with GPS and is expected to reach 95% with the combination of GPS and GALILEO.

During the chapter, the differences and similarities among the three main GNSS systems have been discussed, not only in what concerns technical characteristics and architecture, but also in what concerns the business model chosen for each of them. As explained, both GPS and GLONASS are developed using governmental funding only. On the other side, the business model for GALILEO is completely different. It intends to use a public private partnership (PPP) where the physical system (satellites, ground stations, etc.) would remain a public asset but a "concession holder" would be responsible for the day-to-day operation. The concession holder can recover costs and generate profit through a commercial service, while also delivering agreed service levels for the other public-oriented services.

The main economic challenge for the PPP system of GALILEO is whether or not private industry and final users will be willing to pay for it, even if the assumptions made in the business plan for GALILEO are correct and it is possible to raise the required level of funding. In addition, there are serious uncertainties in what concerns the complex set-up for the management of the program, that might impact in the schedule and even make the PPP scheme feasible. In a worst case scenario, if industry has the perception that they do not need the extra features offered by GALILEO, companies may refuse to pay for this service and would look for alternative systems, relying on the public sector to assume the financial risks. Apart from the funding problems for the PPP that this situation could generate, it should be taken into consideration that this could also impact the schedule, delaying the deployment of the system. Any delay in the deployment, either by financial reasons or for other reasons (political, technological, etc.) may have a severe impact in the market opportunities for GALILEO as other GNSS systems may take a considerable advantage or alternative technologies might occupy the market originally targeted by GALILEO. These issues, now reflected in the current delays in GALILEO, are increasing the uncertainty of the programme.

Other important issues for the development of GNSS systems in the next years refer to security and safety. On one hand, GNSS systems are sensitive infrastructures and can become a target for hostile elements because of the economic ramifications its temporary shutdown or disruption would have due to the increasingly important role of navigation systems in the proper functioning of society. On the other hand, governments have to be sure that these services are not used to the benefit of those potential hostile forces. It is then necessary to ensure not only adequate physical protection of vital infrastructure (control centres, communication networks, etc.), but also protection against spoofing and other forms of misuse or interference with the signal in space.

During the next 10-15 years, we will witness a significant growing in the available GNSS services and infrastructure, passing from only one global system with 25-30 satellites to three systems with up to 90 satellites in view. In parallel, interoperability and introduction of new signals will increase the performances of the GNSS applications and services. Although the economic perspectives are still uncertain, a significant increase in the GNSS markets is expected, and their integration in the information and communications technologies.

REFERENCES

Alcazar Diaz, J., & Velasco, T. (2004, October). *The GALILEO programme: Risks and challenges.* Paper presented at the 55th International Astronautical Congress at Vancouver. Ref. IAC-04-M.4.08

Chinese Defence Today. (2004). *BEIDOU-1 satellite navigation system.* Retrieved February 22, 2007, from http://www.sinodefence.com/strategic/spacecraft/beidou1.asp

CSIC—Coordination Scientific Information Center, Russian Federation, Ministry of Defense. (2002). *GLONASS Interface and Control Document Version 5.0.* Retrieved February

22, 2007, from http://www.glonass-center.ru/ICD02_e.pdf

DSBTF—Defence Science Board Task Force. (2005, October). *The future of the global positioning system.* Report of the Office of the Under Secretary of Defense for Acquisition, Technology and Logistics. Washington D.C.: Department of Defense.

GJU—GALILEO Joint Undertaking. (2005). *Business in satellite navigation: An overview of market developments and emerging applications* (2nd ed.). Brussels: GALILEO Joint Undertaking and the European Commission.

Lewis, R., Kennedy, M., Ghashghai, E., & Bitko, G. (2005). *Building a multinational global navigation satellite system: An initial look.* RAND Corporation.

Petrovski, I. (2003, June). QZSS—Japan's new integrated communication and positioning service for mobile users. *GPS World, 14*(6), 24-29.

Revnivykh, S. (2006). GLONASS: Policy, status and modernization plan. In *Proceedings of the Munich Satellite Navigation Summit*, Munich. Retrieved from http://www.munich-satellite-navigation-summit.org/Summit2006/Beitraege/Ses_1_4_Revnivykh.pdf

Tsujino, T. (2005, July). Effectiveness of the Quasi-Zenith satellite system. *Ubiquitous Positioning, Science & Technology Trends, Quarterly Review* (16). Japan: National Institute of Science and Technology Policy.

US Coast Guard Navigation Center. (2001). *GPS standard positioning service (SPS) performance standard final.* Retrieved February 22, 2007, from www.navcen.uscg.gov/gps/geninfo/2001SPSPerformanceStandardFINAL.pdf

APPENDIX: GLOSSARY OF ACRONYMS

- **C/A:** Coarse Acquisition
- **CDMA:** Code Division Multiple Access
- **CORS:** Continuously Operating Reference Stations
- **CS:** GALILEO Commercial Service
- **DARPA:** USA's Defense Advanced Research Projects Agency
- **DGPS:** Differential GPS
- **DoD:** USA's Department of Defense
- **DoS:** USA's Department of State
- **DoT:** USA's Department of Transportation
- **EC:** European Commission
- **ECEF:** Earth-Center-Earth-Fixed
- **EGNOS:** European Geostationary Navigation Overlay System
- **ERIS:** External Regional Integrity System
- **ESA:** European Space Agency
- **EU:** European Union
- **FAA:** Federal Aviation Administration
- **FDMA:** Frequency Division Multiple Access
- **FOC:** Full Operational Capability Phase
- **GAGAN:** GPS and GEO Augmented Navigation
- **GBAS:** Ground Based Augmentation Systems
- **GCC:** Ground Control Centre
- **GCS:** Ground Control System
- **Geo-ICT:** Geospatial Information & Communication Technology
- **GIS:** Geographic Information Systems
- **GISS:** GALILEO Interim Support Structure
- **GJU:** GALILEO Joint Undertaking
- **GLONASS:** GLObal NAvigation Satellite System
- **GMS:** Ground Mission System

- **GNSS:** Global Navigation Satellite System
- **GPS:** Global Positioning System
- **GSA:** GALILEO Supervisory Authority
- **GSM:** Global System for Mobile communications
- **GTRF:** GALILEO Terrestrial Reference Frame
- **ICAO:** International Civil Aviation Organization
- **ICT:** Information and Communication Technology
- **IGS:** International GPS Service
- **IOV:** In Orbit Validation Phase
- **ITRF:** International Terrestrial Reference Frame
- **ITU:** International Telecommunications Union
- **JPL:** Jet Propulsion Laboratory
- **LAAS:** Local Area Augmentation System
- **LBS:** Location Based Services
- **LUT:** Local User Terminal
- **MCC:** Mission Control Centre
- **MSAS:** MTSAT Satellite-based Augmentation System
- **MTSAT:** Multifunction Transport Satellite
- **NASA:** USA's National Aeronautics and Space Administration
- **NDGPS:** Nationwide Differential GPS
- **NOAA:** USA's National Oceanic and Atmospheric Administration
- **OS:** GALILEO Open Service
- **PDOP:** Position Dilution of Precision
- **PNT:** Positioning, Navigation, and Timing
- **PPP:** Public Private Partnership
- **PPS:** Precise Positioning Service
- **PRS:** Public Regulated Service
- **QZSS:** Quasi-Zenith Satellite System
- **SA:** Selective Availability
- **SaR:** Search and Rescue
- **SBAS:** Satellite Based Augmentation Systems
- **SiS:** Signal in Space
- **SoL:** GALILEO Safety of Life Service
- **SPP:** Single Point Positioning
- **SPS:** Standard Positioning Service
- **TBD:** To Be Defined
- **TTC:** Telemetry, Tracking, and Control
- **UMTS:** Universal Mobile Telephone System
- **USA:** United States of America
- **USSR:** United Soviet Socialist Republics
- **UTC:** Universal Coordinated Time
- **WAAS:** Wide Area Augmentation System
- **WGS:** World Geodetic System

Chapter VI
The Satellite Internet:
The Convergence of Communication and Data Networks

Agnieszka Chodorek
Kielce University of Technology

Robert R. Chodorek
AGH University of Science and Technology

ABSTRACT

The aim of this chapter is to show the satellite Internet as a new quality, which was created thanks to the convergence of satellite communication and data networks. The chapter describes the development of satellite communication and satellite data networks, presents methods of Internet access via satellite, and discusses the opportunities and challenges of building effective commercial services based on satellite Internet. The main advantages of the satellite Internet are high bandwidth, very good availability (in practice, anywhere in the world), and natural IP multicasting. Although getting broadband Internet access by satellite is considered very expensive, independence from the local infrastructure results in the satellite Internet being a good solution for both business communications (a corporate network or its fragments) and remote area communications (rural communications and services to isolated communities).

INTRODUCTION

One of the most important applications of satellites is their usage as a part of telecommunication infrastructure. Satellites in communication systems work as space relay stations, which contain several transponders. Each transponder receives, amplifies (or regenerates), and rebroadcasts the transmitted signal (analogous transmission) or transmitted block of information (digital transmission). To avoid signal or block damages, caused by interactions between simultaneous uplink (to the satellite) or downlink (from the satellite) transmissions:

- Uplinks and downlinks operate at different frequencies (typically, lower for downlink and upper for uplink).
- Downward beams, which operate at the same frequency, are pointed at different geographical areas.

Transponders can be transparent or they can have additional functionality. Transparent transponders are the most widely used and serve as radio-relay repeaters. Nontransparent (so-called onboard processing or OBP) transponders can function as hubs, switches, or routers (in the context of data transmission) or exchanges (in the context of telephony).

According to their orbit, communication satellites can be divided into satellites in geostationary orbit (GSO) and satellites in nongeostationary orbit (nonGSO) with lower altitude. The most known solutions for nonGSO satellites are satellites in medium-Earth orbit (MEO) and in low-Earth orbit (LEO) (ITU, 2002).

The application of satellites to communications includes telephony (both digital trunks – long-distance links, and personal communication systems – mobile satellite telephony), radio and television, and widely understood data transmission. The official International Telecommunications Union (ITU) definitions divide satellite services into several types, including (ITU, 2002):

- **Fixed-satellite services (FSS):** A radiocommunication service between given positions on the Earth's surface (specified fixed points or any fixed point within specified areas)
- **Intersatellite service (ISS):** A radiocommunication service between two or more satellites directly interconnected without an intermediate earth station
- **Mobile-satellite services (MSS):** A radiocommunication service between mobile earth stations
- **Broadcasting-satellite services (BSS):** A radiocommunication service, in which signals transmitted by satellites are intended for direct reception by the general public (using very small receiving antennas, smaller than needed for FSS)

Some other, usually application-oriented services, mentioned by ITU, are: radionavigation-satellite service, meteorological-satellite service, and so forth.

In practice, we observe the convergence between the above listed services. In FSS, MSS, and BSS services, one or more satellites are used. Typical MSS-based service, such as mobile satellite telephony, uses FSS links to communicate with the exchange located on the Earth's surface. Typical BSS service—satellite television broadcasting—uses feeder links from an earth station to a space station, and so forth.

Due to the large sizes of antennas and the large cost of earth stations, early satellite systems were not able to render services for individual users. In the 1980s, the introduction of satellite technologies based on small and low-cost earth stations, like DBS and VSAT, was the turning point in the applicability of satellite technology to a public consumer. Satellite communication services became readily available to millions of people and satellite technology became a part of everyday life for many people.

The current DBS and VSAT technologies are a base for several services, among which the most important is broadband Internet access. Thanks to these technologies, the broadband Internet is available anywhere—in almost all locations on Earth—and for millions of people. Thus, the key problem is a convergence of the Internet and space technologies. This issue, addressed in this chapter, will be important for IT professionals who will build the next generation communication satellite infrastructures for s-commerce.

The chapter consists of six sections. In the second section, the development of satellite communication and satellite data networks is presented. The third section gives a brief overview of the Internet technology. The fourth section is devoted to a convergence of the Internet and space technologies, including commercial services based on the satellite Internet. The fifth section outlines future trends and, finally, the sixth section concludes the chapter.

DEVELOPMENT OF SATELLITE COMMUNICATION AND SATELLITE DATA NETWORKS

This section is devoted to the basics of the satellite Internet. The first part of this section briefly describes satellite communication, from the first experiments to current systems. The second part shows the past and the present technologies of satellite data transmission.

Development of Satellite Communication

The very first experiments with satellite communication were carried out with the usage of the natural satellite of the Earth, the Moon. The Moon has been used as a radio signal reflector. Very short, decimetric radio waves (at ultra high frequency, UHF), sent from the Earth, pass the atmosphere, rebound from the surface of the Moon, and return to the Earth. Such experiments were carried out many times, mainly by amateur radio operators. Moon relay experiments, carried out by the United States Navy during the 1950s, resulted in the building in 1959 of a fully operational lunar relay system, which linked Hawaii with Washington, DC.

Nevertheless, only when artificial satellites appeared toward the end of 1950s, was it possible to build a more flexible and efficient communication systems. The father of modern satellite communication is considered to be Sir Arthur Charles Clarke, whose paper (Clarke, 1945) has addressed the principles of satellite communication based on GSO satellites. However, in early days of space technology, precise positioning of geostationary satellites was very difficult. Thus, the first satellites were nonGSO.

The first telephone and television transmissions which used artificial satellites were carried out using passive satellites. Echo-1 (1960), the first communication satellite, was built as a 30.5 meter (100 foot) diameter balloon with a metallized surface. Like in the moon relay experiments, Echo-1 reflected the UHF radio waves. Also in 1960, the first communication satellite equipped with an active device—a transponder—was launched. Five years later, the first commercial geostationary communication satellite, Early Bird (Intelsat-I), working in the so-called C-band (6 GHz for uplink and 4 GHz for downlink), was placed in service. It provided communications between Europe and North America and enabled up to 240 simultaneous telephone calls or one TV program. For the sake of comparison, Intelsat IX satellites (2000) enable up to 160,000 simultaneous calls and the Intelsat K-TV (1999) satellite enables 210 TV programs.

The half century of satellite communications was a period of constant development. Satellite transmission has evolved from analog to digital. Parallel to satellites, earth stations also underwent constant evolution. First, earth stations had to be equipped with large antennas (e.g., a standard A type of station initially required 30-m antennas, standard B: 15-to-20-m (Elbert, 2001)) to meet signal-to-noise requirements. The progress of science and technology enabled significant reduction of antenna diameter (e.g., Standard A type of station uses now 15-to-18-m antennas, standard B: 11-to-13-m (ITU, 2002)) as well as a reduction of earth station costs. On the other hand, business and entertainment created the demand for satellite systems based on very small earth stations. As an example, TV content providers (network broadcasters) need a fast and efficient method of distributing TV programs to cable TV headends. Thanks to satellite-based delivery system, content distribution can be carried out simultaneously to multiple locations in large geographical areas. A relatively low cost of earth stations with small (3 to 5 m in diameter) antennas makes satellite distribution more competitive than terrestrial radio links.

In the beginning of the 1980s, usage of the Ku-band range of frequencies (14 GHz for uplink and 11-12 GHz for downlink), different from C-band, led to further reduction of antennas' diameter. As a result, very small aperture terminals (VSAT), which use 0.6 to 2.5 m antennas for bidirectional data and voice transmission, and direct broadcast satellite (DBS), which use 0.6 to 1.8 m antennas for receiving TV signal, have appeared. Analogous DBS was later replaced by "satellite versions" of digital video broadcast (DVB) and Advanced Television Systems Committee (ATSC) standards.

By the end of the 1990s, very promising MSS projects of global mobile personal communication services (GMPCS) had started. Satellite telephony was intended to fill a gap in the market of mobile telephony. However, the considerable risk to terrestrial cellular telephony companies caused them to join forces against their "mutual enemy." They introduced "practically impossible" services (such as roaming for prepaid users), and significantly reduced the cost of telephone calls. As a result, the competition between satellite and terrestrial cellular telephony companies led to the fast bankruptcy of Iridium, the first commercial GMPCS operator.

Development of Satellite Data Networks

Data transmission has become an important element of communication since the 1970s. It resulted from a rapid development of computer systems and a necessity to exchange data between distant endpoints (e.g., located at two branches of a company, situated at a distance of thousands of kilometers). Simultaneously, development of mini- and microcomputers meant that even the smallest branch of a company could be equipped with such a system, and communications with head offices significantly improved the efficiency of companies.

Also in the 1970s, a computer network called ARPANET, which marked the beginnings of the Internet, was developed. ARPANET, initially intended for research purposes, and then the Internet, ultimately became important for business activity.

Almost at the beginning of the ARPANET/Internet, satellite communications became an integral part of the newly established network. However, the range of services offered by the satellite systems and its role in the Internet has changed significantly during the last 30 years. The first use of data transmission over satellite networks enabled only circuit switching, typical for telephony. These systems were able to realize point-to-point connections, established for a long time. Such connections were simply another variant of typical circuit switching connections realized in terrestrial networks (wired or wireless, via radio lines). From the point of view of classical telephone connections, designed for uncompressed voice signal (PCM modulation at the target bit rate of 64 kb/s), these solutions are optimal.

A computer network generates a different type of traffic than a telephonic one. In such a network, we observe instantaneous activity periods (e.g., a client gets data from a server) separated by idle periods (e.g., a server waits for reaction of a client). Also data compression, often used in the Internet technology, has influence on the character of traffic. Compressed data, especially video streams, transmitted in real-time, generate complex traffic (sometimes with bimodal distribution), characterized by large dynamics (Chodorek & Chodorek, 2000; Chodorek & Papir, 2000). As a result, while typical for telephony, permanent channel allocation leads to small bandwidth utilization and, in consequence, to an increase in costs of data transmission. That's why data transmission requires more flexible networks than telephone ones. Especially, the arrhythmic nature of transmitted data should be taken into consideration. One possible solution to this problem is to use packet switching instead of circuit switching. Because the Internet is a packet switching network (or, simply, a packet network), it was necessary to introduce the support of packet switching to satellite communication.

Pioneering work on this field was done by scientists from The University of Hawaii, which in the early 1970s developed the ALOHA (multiaccess, terrestrial radio network), which connected branches of the university, situated at different islands of the archipelago. Successfully implemented, the ALOHA concept was then used to control the connection via a satellite channel linking Hawaii and Nasa-Ames via NASA's ATS-1 satellite. Also in the 1970s, more work was carried out by the Defense Advanced Research Projects Agency (DARPA) with the cooperation of partners from the United States and Europe. This work included methods for sharing a common, broadband satellite channel by traffic originating from several earth stations. As a result, the SATNET satellite network was built. The SATNET connected research centers from the United States and Europe via Intelsat-IV satellite. From 1975-1978, the network was the subject of intensive performance evaluations and tests, and in 1979, it provided services at a fully professional level.

The technological development in the 1980s enabled reception of satellite transmission via small size antennas. Moreover, transmit/receive terminals equipped with small antennas (less than 2.5 m in diameter) also became both technologically realizable and cost-effective. VSAT terminals allow the end-user to transmit various types of information, including data. They were (and still are) commonly used for data transmission (usually for credit card transaction purposes) to and from fuel stations and shops in rural areas. In these areas, the cost of an equivalent terrestrial network would be considerably larger. VSATs often are a basis of rural telephony and the Internet in developing countries with poor terrestrial infrastructure (e.g., in African countries, thousands of VSAT terminals provide communication infrastructure on a large scale (Moroney, 2002)). VSATs are also applied in urban areas, where usage of terrestrial networks could be cheaper. However, VSAT offers uniform service via a dedicated network, characterized by high levels of reliability and availability. Thus, although within the last few years we have observed rapid development of terrestrial wireless networks, VSAT satellite terminals are still attractive to individual users in rural areas, and small and medium enterprises, as well as to big organizations.

Generally, VSAT networks are built in star or mesh topology, although there are networks which can support mesh and star simultaneously.

VSAT networks often have star topology, in which designated earth station, so-called hubs (usually equipped with larger, 6-10 m in diameter, high-gain antenna), manage the network, perform access control, and relay transmissions between other stations. Star topology allows the user to apply VSAT end systems with antennas of around

1 m. If the hub is located at the head office, the vast majority of traffic will be transmitted between the hub and other VSAT earth stations.

Mesh VSAT networks enable individual connections between VSAT terminals, without the hub. Such a topology reduces transmission delay (typically to half a value), which is especially important in the case of real-time transmissions, as for example, Voice over IP (VoIP). Because of the lack of the hub, which acts as a kind of amplifier, all VSAT terminals in mesh networks have to be equipped with larger antennas. Also, one of the mesh network earth stations must control access to the radio channel linking VSAT terminals and the space segment.

In general, VSAT networks both enable bidirectional point-to-point transmission between end-users (One-to-One transmission scheme) and point-to-multipoint data dissemination to many recipients (One-to-Many transmission). Using VSATs allows you to build effective Internet access. Internet via VSAT is used by individual end-users, enterprises, and Internet service providers (ISP).

By merging Internet and VSAT technology, enterprises can build private Intranet networks between branches of a company situated at distant geographical locations. Such a network, independent of the public Internet, assures a high level of security and quality of service.

In the 1990s, a new solution, an alternative to VSAT, has appeared—data networks, which are based on "satellite versions" of DVB and ATSC standards, namely DVB-S, DVB-S2 and ATSC-S. These networks, initially intended for distribution of MPEG-encoded digital radio and television, were later adapted to data transmission, carried out using the IP over MPEG technique.

In the case of DVB-S, DVB-S2, and ATSC-S, space segments have high-power transponders that enable distribution of digital TV signals to end-users equipped with relatively small antenna. Transponders usually are able to transmit two types of downward beams: wide beams, which cover a substantial fraction of a continent, and spot beams, focused on relatively small areas (hundreds of kilometers). DVB-S, DVB-S2, and ATSC-S systems typically offer broadcast services, nonbroadcast services and professional applications services (digital TV contribution, news gathering, TV distribution to terrestrial transmitters, Internet trunking, etc.) (ETSI, 2005a). Broadcast services are intended for transmission of digital content (radio, TV) and associated data (e.g., teletext, EPG, etc.) to a wide group of recipients. Nonbroadcast services and Professional Applications services can be used for data transmission. Nonbroadcast services offer both two types of unidirectional transmission (point-to-point and point-to-multipoint) and bidirectional transmission (carried out by terminals with additional equipment).

Due to large throughputs and the low cost of terminals, DVB-S, DVB-S2, and ATSC-S systems were extended to enable transmission of computer data. As a result, apart from their basic function, the transmission of digital radio and television, these systems can be applied to the Internet access. In these systems, two variants of the Internet access are used: without and with the satellite return channel. The first solution assures only unidirectional satellite communication, and only downlink to the terminal is used. Because Internet access requires, typically, bidirectional communications, return connection is established via terrestrial technology (e.g., via normal dial-up telephone). This solution is very simple and easy to implement, without the necessity of additional infrastructure.

The second solution, Internet access with the satellite return channel, was introduced a few years later than the first one. It is based on digital video broadcast-return channel via satellite (DVB-RCS) ETSI standard, and enables fully bidirectional Internet access via a homogeneous satellite network. DVB-RCS uplink channel has lower throughput than the downlink one, which arose from limited power of transmitters in terminals (which, in turn, was caused by the necessity of building relatively cheap service with low-cost terminals). This asymmetrical link meets the requirement of a typical individual Internet user, who sends less data than they get from the Internet. Usually, Internet access via DVB-S, DVB-S2, or ATSC-S systems is much cheaper than via VSATs. However, Internet access via VSAT assures better quality of service (e.g., higher throughput of the uplink channel and better availability) and better utilization of the satellite channel.

INTERNET TECHNOLOGIES AND APPLICATIONS

In this section, an overview of the Internet technology is presented, from the layered structure, mapped to OSI model, through the Internet Protocol and modern transport protocols, to the Internet applications.

Layered Structure of the Internet

According to the open system interconnections (OSI) reference model, the Internet has a layered structure. The first, physical layer, is an interface between the network node and the physical link (wired or wireless). The second, link layer, is concerned with moving blocks of information, called frames, within the link-local network (e.g., from the earth station to the satellite or from one end of a fibre to the other). The third, network layer, provides transmission of blocks of information, called datagrams, between the source and the destination, via intermediate nodes (routers), which receive frames and resends them toward the receiving node. The fourth, transport layer, of the OSI model creates and maintains separate network connections for each transmission required by applications. In the Internet, transport layer is an interface between the network (seen as a network layer protocol) and an application.

Figure 1. The OSI reference model

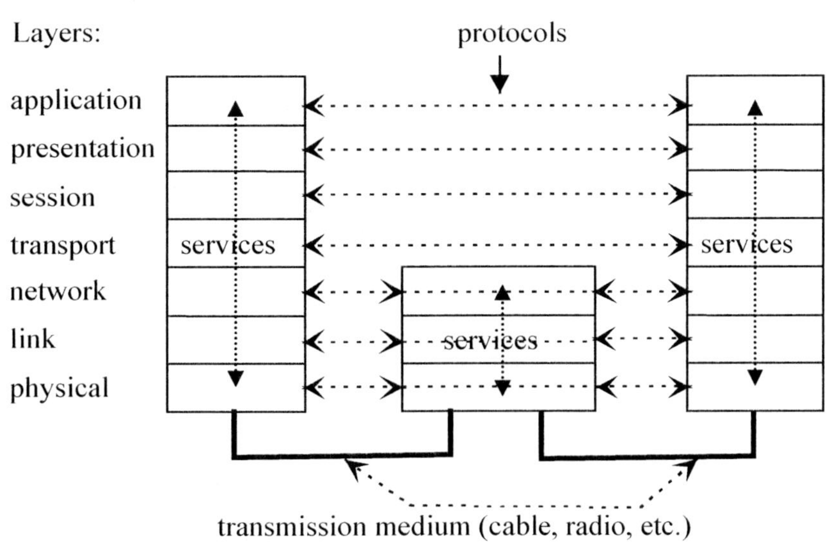

The fifth, session layer, is concerned with session management and control, while the sixth, presentation layer, deals mainly with data compression, encoding, and security. The seventh, application layer, provides application-specific functions, for example, getting objects (text, pictures) for WWW, getting and putting files for file transfer, remote video cassette recorder for video on demand, and so forth.

The OSI model (Figure 1) assumes that horizontal communication (between nodes) is possible only within the layers and vertical communication (between layers) is performed only within a node (both end systems and intermediate). A set of algorithms and mechanisms, which control the horizontal communication together with the format of blocks of information sent within a single layer is called a protocol. The most known Internet protocols are the Internet Protocol (IP) of the network layer, the Transmission Control Protocol (TCP) of the transport layer, Session Initiation Protocol (SIP) of session layer, the Hypertext Transfer Protocol (http), and the File Transfer Protocol (FTP) of the application layer. Internet protocols are standardised by the Internet Engineering Task Force (IETF).

Vertical communication consists in the encapsulation of the upper layer's block of information into the lower layer's data structure. For instance, application layer protocol (e.g., FTP) has to send a file from the source node to the destination node. FTP hands over the file to the TCP transport protocol, which copies successive fragments of a file and encapsulates them into TCP packets (simply by attaching a TCP header to each fragment). The IP protocol encapsulates packet into an IP datagram and the link layer (e.g., the Ethernet) encapsulates the IP datagram into the frame and sends the frame via the physical medium (e.g., twisted-pair cable). In the destination node, the IP datagram is decapsulated from the frame, the TCP's packet is decapsulated from the datagram and, finally, each fragment of a file is decapsulated from the packet. As a result, a copy of the file is transmitted piece by piece via the link and reassembled in the destination node.

The Internet Protocol

The Internet Protocol (IP) is the most known network layer protocol and a common platform, on which the Internet has been built. It provides connectionless, unreliable, best-effort transmission between the sender and one or more receivers via one or more networks. IP protocol is implemented both in end systems (stations or terminals) and in intermediate nodes (routers), located on the network boundary. An IP datagram is transmitted hop-by-hop, from the sender to the receiver, via appropriate routers, which direct the datagram to the proper link, toward the destination.

Nowadays, there are two versions of IP protocol, the older, fourth version (or IPv4), standardized in 1981 (Postel, 1981), and the newer, sixth version (IPv6), standardized in 1995 and changed in 1998 (Deering & Hinden, 1998). These versions differ in header structure (the most known is a difference between size of the address field: 32 bits of IPv4 and 128 bits of IPv6), datagram structure (common and optional parts of an IPv4 header vs. mandatory and optional headers of IPv6) and in functionality. Some functions of IPv4 were omitted (e.g., Record Route), some functions were introduced (e.g., Flow Label), some were changed (e.g., fragmentation and reassembly), and some others stayed unchanged or were changed slightly (as Time to Live of IPv4, mapped to Hop Limit of IPv6). The IP protocol, both IPv4 and IPv6, continuously evolve. As an example, in 1998, security assurance IPSec had appeared, and then it was extended in early 2006.

The IP protocol doesn't assure all functions of the network layer itself. Some functions are provided by auxiliary protocols. These protocols offer signalling and error reporting services, translation of IP addresses to link layer addresses, resource reservation, routing services, and others. The most known auxiliary protocols are, certainly, routing protocols, which find (sub)optimal routers to all possible subnetworks in their working area. Currently, the Internet is divided into large areas, called "autonomous systems." As a result, we have a two-level routing hierarchy:

- Intradomain routing, inside autonomous systems
- Interdomain routing, between autonomous systems

The IP protocol introduces network layer addressing, of global (IPv4, IPv6) or local (IPv6) scope. The IP addressing scheme determines the mode of transmission (point-to-point or point-to-multipoint). There are four types of addressing schemes and corresponding modes of transmission: unicast, anycast, broadcast, and multicast one. The first two addressing schemes describe point-to-point transmission. Unicast addressing, provided both by IPv4 and IPv6, is the most typical addressing scheme and defines individual address, assigned to specific stations. Unicast transmission is carried out between the two stations, both identified by unicast source and destination addresses. Anycast addressing, provided by IPv6 only, is the newest IP addressing scheme and defines a group address, assigned to a specific set of stations. Anycast datagrams are forwarded to one (usually: the nearest) of the stations belonging to the set identified by anycast destination addresses. Anycast addressing scheme is similar to "emergency phone numbers" (112 in Europe and 911 in the United States and Canada).

The two last addressing schemes describe point-to-multipoint transmission. Broadcast addressing, provided by IPv4 only, defines a group address, assigned to every station in a given network. Broadcast datagrams are forwarded from the station identified by unicast source address to all stations in a given network. Multicast addressing, provided both by IPv4 and IPv6, defines a group address, assigned to each station belonging to a given multicast group. Multicast datagrams are delivered from the sender (identified by the unicast source address) to each station identified by a given multicast address. Instead of the broadcast scheme, stations belonging to multicast groups don't have to be located on the same network. A similarity between broadcasting and multicasting caused all IPv4 broadcast transmissions in IPv6 to be replaced by multicasting.

Transport Protocols

Internet users usually associate the Internet, on the one hand, with WWW or e-mail, and on the other with TCP/IP. The last association emphasizes the importance of a transport-layer protocol (or simply transport protocol) to the Internet technology. Applications utilize the functionality of the communication network via the transport service, given by transport protocols. Transport protocols can enhance the functionality of the underlying network (wired or wireless, also satellite), but can also confine it.

Reliable Transport Protocols

A group of protocols, which deal with packet loss or damage, are called "reliable protocols." The primary reliable transport protocol of the current Internet is the Transmission Control Protocol (TCP). TCP is a general-purpose, connection-oriented, unicast transport protocol, designed to provide reliable end-to-end transmission over potentially unreliable networks. It's used by such applications as e-mail, WWW, file transfer (ftp), and so forth. The important feature of TCP is its ability to fairly allocate bandwidth between competing TCP flows, working in the same circumstances.

The three main mechanisms of the TCP protocol are:

- **Flow control:** Prevents the receiver's buffer from overflowing

- **Congestion control:** Deals with congestions, temporary or permanent losses of stability of a network
- **Error control:** Assures reliability of TCP connections

The first two mechanisms slow down the data sending rate both to adjust it to the receiver's ability of reception (flow control) and to adjust it to poor network circumstances (congestion control). The sending rate is limited using two independent windows (a window is an N consecutive data in sequence space that the sender is able to send without prior acknowledgement). The error control consists of the confirmation of proper reception of delivered data (through so-called "positive acknowledgements" or ACKs), error detection (the TCP detect both packet damages and packet losses), and error correction (using selective retransmission, where only undelivered data is retransmitted, or go-back-N retransmission, where all packets in current window are retransmitted).

Although the TCP stays the most popular Internet transport protocol, it is able to carry out only point-to-point transmission. To enable point-to-multipoint data dissemination, several multicast transport protocols have been designed. Nowadays, the three most representative reliable multicast transport protocols are the pragmatic general multicast (PGM), the negative-acknowledgment (NACK)-Oriented Reliable Multicast (NORM) protocol, and the Asynchronous Layered Coding (ALC) protocol.

Real-Time Transport Protocols

Nowadays, there are two most popular transport protocols, able to provide multipoint transmission for applications transmitting real-time data (such as audio or video), over multicast network services. They are the User Datagram Protocol and the Real-time Transport Protocol.

The User Datagram Protocol (UDP) was designed for early IP networks, which were unicast in nature. It was designed as a very simple transport protocol, completing the TCP/IP protocol suite. However, in contrast to TCP, whose mechanisms were optimized from the point of view of unicast transmission, the simplicity of the UDP protocol also allows it to be utilised for multicasting.

Real-time Transport Protocol (RTP) was designed for real-time transmission of multimedia information (Schulzrinne, Casner, Frederick, & Jacobson, 2003). Its practical application covers, among other things, interactive audio/video services and real-time distribution of audio and video data to a large number of recipients. In contrast to UDP, RTP was designed as a multicast protocol, and unicast transmission is treated as a particular case of general multicasting.

Internet Applications

A man in the street usually associates Internet applications with e-mail and WWW. Electronic mail, which in the beginning allowed the user to send simple text messages only, was modified several times during the half of century of the Internet to meet requirements of modern users, both individual and commercial. Nowadays, e-mail enables the exchange of multimedia electronic letters between Internet users and is an important element of business activity for many enterprises. Electronic mail performs the exchange of e-letters using Simple Mail Transfer Protocol (SMTP) as the application layer, which enables both direct message delivery and delivery through intermediate systems. To assure reliable transmission of electronic letters, the TCP protocol is always used in the transport layer.

The second most popular Internet application, and certainly the most frequently used, is the World Wide Web (WWW). The WWW enables easy and intuitive access to multimedia information, stored in the Internet resources. WWW applications use the Hypertext Transfer

Protocol (HTTP) as the application layer, which provides transmission of multimedia data (HTML documents, pictures, files, etc.) between the Web browser and the Web server. The HTTP protocol sends each component of the Web page (e.g., JPEG-encoded image) inside the protocol's messages. The HTTP always cooperate with the TCP protocol of the transport layer.

Historically, the oldest Internet applications, which are still used, are file transfers. File transfers can be implemented either as an element of WWW service, or as a specialized, dedicated file transfer service, ftp. The ftp service uses the File Transfer Protocol (FTP) as the application layer, which was designed to reliably transfer files between two end-systems. As with the SMTP and HTTP, the FTP protocol is built on top of the TCP transport protocol. The ftp applicability usually is not limited to file transfer only and ftp applications also enable making different operations on file systems of connected stations (both, the remote station and the local one). Such operation includes creation of new directories, deleting files, moving files from one location to another, listing content of directory, and so forth.

Although e-mail, ftp, and WWW were the main Internet applications for a number of years, it is possible that in the future they will lose their privileged position in the market place to real-time multimedia services. Real-time multimedia has been talked about for at least 15 years. However, lack of high-speed networks (to send large amount of data in real-time), lack of processing power on home computers (to decode multimedia signals in real-time), and lack of efficient compression standards (to achieve large coefficient of compression without strong degradation of compressed signal) meant that initially they were available only to scientists and a handful of enthusiasts. Nowadays, from a niche market offering, known only to academics and net enthusiasts, modern real-time Internet applications (as Voice over IP—VoIP, teleconferencing, IP television—IPTV) become mature products, objects of interests to businessmen and corporations.

Real-time multimedia services are usually based on International Telecommunications Union (ITU) H.323 standard or IETF standard (also known as SIP-based standard, from the IETF's SIP protocol). Both standards differ in management and control blocks, as well as in the protocol stack for non-real-time applications (like shared whiteboard or slide presentation). IETF's standard recommends the usage of reliable multicast transport protocols (like PGM, NORM, and ALC) for delivery of reliable data, associated with real-time audio/video, while the ITU promotes their own protocols. However, the central idea of a real-time audio/video transmission is the same in the case of both ITU and IETF standards - the IETF's RTP protocol is always applied in the transport layer. It's worth remarking that while the TCP/IP protocol stack is primarily used for classic Internet applications, the RTP/UDP/IP protocol stack is primarily used for real-time Internet.

The third group of Internet applications, partially coinciding with the real-time applications, are multicast applications. Because IP multicast technology allows the sender to easy disseminate information for either a few or a few million Internet users, IP multicasting is an important service for any distribution system. The usage of IP multicasting optimises the utilisation of network resources (available bandwidth), because the same information is forwarded only one time by any link and any node, independent of the number of receivers. The multicast sender sends only one multicast IP datagram, and the network automatically replicates the datagram in the proper nodes, and as a result delivers a copy of the disseminated information to N receivers. In contrast, the classic, unicast sender has to send N copies of disseminated information in separate datagrams. As an example, using a 10 Mbps link during one second, the IP multicast sender is able to send about one thousand different blocks of information, each 1000 bytes in size, to millions of users, while the IP unicast sender is able to send only one block to about one thousand users.

Reliable multicast transmissions, provided they are using reliable multicast transport protocols, allow the user to build effective applications for database replication and update, collaborative tools (used , for example, in telemedicine, tele-working, tele-education), software upgrade services, Web cache updates, and multihome and multisite Web services. Multicast technology is the basis of real-time audio/video applications. It is used to build both typical audio/video group applications, such as multimedia distribution systems (e.g., teleconferencing) and nonmulticast services, such as video on demand or VoIP telephony. In the second case, unicast transmission is treated as a special case of general IP multicasting. The multicast transmission mode is a natural for any type of broadcasting, from IPTV, through online weather map distribution (working in "push" manner), to disaster warning systems. Using multicast transmission in the case of stock quote streaming allows all users to receive information at the same time (within the accuracy of network delays).

CONVERGENCE OF THE INTERNET AND SPACE TECHNOLOGIES

The satellite Internet is a convergence of satellite networks (VSAT, DVB-S, ATSC-S, etc.), belonging to layers 1 and 2 of the OSI reference models, and Internet technologies (layers three to seven). In this section, a convergence of the Internet and space technologies is discussed. This section covers both the problems of Internet access via satellite networks and the challenges and opportunities of the upper (3rd – 7th) layer protocols in heterogeneous terrestrial-satellite networks.

Internet Access via Satellite Networks

Modern satellite distribution systems, such as DVB-S (ETSI, 1997), DVB-S2 (ETSI, 2005a), and ATSC-S (ATSC, 2005), use the MPEG-2 TS (where TS stands for "transport stream") standard of data transmission. Although the word "MPEG" is commonly associated with audio/video compression and encoding, the MPEG-2 standard in fact consists of two issues, transmission systems and compression/encoding algorithms. In the case of satellite data networks, the MPEG-2 TS standard is used for the transmission and multiplexing of many streams, forwarded via a shared delivery channel, while the compression/encoding part of the MPEG standard is not applied.

A single, physical satellite channel is divided among some logical channels using any method of multiplexing (e.g., time division multiplexing (TDM) or frequency division multiplexing (FDM)). Logical channels convey TS Packets, which have fixed lengths of 188 bytes and begin with a 4 byte-long header. The TS Packet header includes the packet identifier (PID), identifying the single stream.

In the case of the satellite Internet, the IP datagrams are encapsulated into TS Packets (usually with prior fragmentation carried out in the network layer). There are two standards of transmission of Internet traffic (and more precisely, IP datagrams) via the MPEG-based broadcast network, both known under the common name of "IP over MPEG." The first standard was specified by both the European Telecommunications Standards Institute (ETSI) (ETSI, 2004) and the advanced Television Systems Committee (ATSC) (ATSC, 2000); the second was specified by the IP over DVB (ipdvb) IETF's Working Group (Fairhurst & Collini-Nocker, 2005; Montpetit, Fairhurst, Clausen, Collini-Nocker, & Linder, 2005). The standards define two alternative methods of transmission of

IP datagrams inside MPEG-2 TS packets, which differ, among other things, in the way they define encapsulation and link-layer addressing.

DVB-S, DVB-S2, and ATSC-S systems provide downlink IP over MPEG transmission. The return connection can be established using any terrestrial technology (e.g., dial-up telephone) or via satellite, through DVB-RCS uplink (using IP over ATM or IP over MPEG technology) (ETSI, 2005b).

Although the IP over MPEG has several advantages (simplicity of solution, homogeneity of transmission inside logical channels, easy coexistence of data and TV transmission, etc.), this solution gives relatively large overheads in the case of bulk data transfer. An alternative solution - the VSAT networks - enables bidirectional connections, optimized from the point of view of data transmission. Thanks to easy encapsulation and variable frame size, VSAT systems are better than DVB-S, DVB-S2, or ATSC-S systems in utilizing satellite links. This is caused by smaller overheads (up to 15-30% smaller) than in the case of IP over MPEG and the possibility of using larger frames (e.g., 1500 B, as in the case of the Ethernet network) than the 188 B of MPEG-2 TS Packets.

Available solutions enable practical online negotiation of parameters for transmission, from the size of the frame conveying the IP datagram to the requested channel capacity. There are solutions that allow the user to configure guaranteed channel capacity with the granularity of 1kbps. It reduces wasted bandwidth (unused fraction of reserved channel capacity) and results in improved efficiency of transmission (e.g., VoIP applications can reserve channel capacity according to the applied method of encoding). It is also possible to configure traffic priorities for given applications. For example, an enterprise might want to restrict the amount of bandwidth available for file transfer in order to guarantee delivery of VoIP through a shared link.

VSAT systems, which apply advanced methods of encoding based on turbo product codes (TPC), are able to achieve a low bit error rate BER of about 10^{-9}, using low powered transmission signals. With additional automatic uplink power control, VSAT systems achieve a relatively high availability of service, independent of weather or atmospheric conditions.

Building Effective Commercial Services Based on Satellite Internet

When building effective (in the sense of optimal utilization of network resources to reduce the costs of the service to maximize profits) commercial services via satellite, there is a need to solve many problems associated with the properties of satellite links. As a result, satellite Internet must face the challenges of:

- Greater latency
- More bandwidth asymmetry
- Higher bit error rates (BER)
- Security issues

The end-to-end latency is a sum of total propagation delay, transmission delay, and queuing delay. In the broadband satellite network, the dominant component is the propagation delay, which ranges from tens to hundreds of milliseconds. For instance, the propagation delay between two stations via satellite, calculated for the LEO, MEO, and GSO altitudes, are of the order of 7, 75, and 260 ms, respectively (Elbert, 2001). The great latency results in low performance of the TCP protocol because of TCP's window-based mechanisms (flow control, congestion control, and error control) that require acknowledgements of correctly received data. In low-latency networks, transmission is carried out continuously, whereas in high-latency environments, transmission typically consists of periods of activity alternate with idle periods. Idle periods are in fact waiting times, where the sender waits for acknowledgements to slide the window. The existence of idle periods is caused by too small rwnd (flow control window) size, which disables TCP to "fill the pipe." As a result, typical rwnd settings theoretically limit TCP throughput to no more than about 2 Mbps (thus, only 2 Mbps channels can be fully utilised) in GSO satellite systems. In practice, achieved throughputs are much lower (analysis carried by the authors shows that settings used by popular operating systems lower TCP's throughput to about 0.5 Mbps and some literature sources report even 0.1 Mbps). To avoid this unintended limitation of achieved throughput, the window scale option introduced by RFC 1323 (Jacobson, Braden, & Borman, 1992) can be used. However, the solution described in RFC 1323 solves the problem only partially, because "satellite" operating systems should be configured differently to the "terrestrial" operating systems, while users in many situations do not know whether the given transmission is routed via satellite or not. Moreover, operating systems are usually configured to one network environment and alternate satellite and terrestrial transmissions require constantly repeated reconfigurations of a system. Thus, the applicability of the window scale option is, in practice, limited only to satellite access networks, while in the case of mixed satellite-terrestrial networks, TCP/IP acceleration (also known as TCP/IP spoofing) is used.

TCP accelerators are intermediate devices (Figure 2) which intercept TCP packets, acknowledge them, and resend them to their destination nodes. As a result, source nodes believe that there are short-distance networks between end-systems. Accelerators buffer all packets unacknowledged by destination nodes. Thus, they are a kind of proxy, able to perform retransmission (both, selective and go-back-N) as well as congestion avoidance and control.

Acceleration can be also applied in other layers than the transport one (Border, Kojo, Griner, Montenegro, & Shelby, 2001). Typically, it is used in the application layer, such as WWW proxies (Web caches, which store local copies of WWW pages originating from different servers). They are used to improve WWW performance in both terrestrial and mixed, terrestrial-satellite environments. Accelerators, located at different layers of wireless (here, satellite) networks are known under the common name performance enhancing proxies (PEPs) (Border et al., 2001).

The second feature of a satellite network is bandwidth asymmetry. Satellite networks are asymmetric in at least two ways:

Figure 2. TCP accelerators

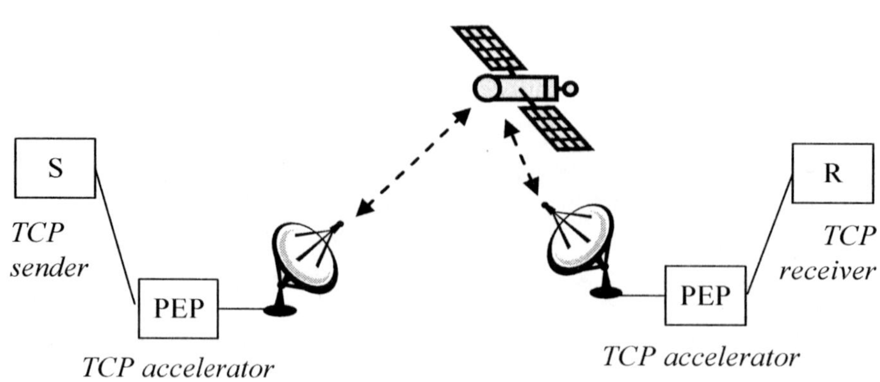

- **Apparent bandwidth asymmetry:** For example, direct broadcast satellite downlink and return via dial-up modem line
- **Unapparent bandwidth asymmetry:** For example, direct broadcast satellite downlink (at Mb/s) and return via slower uplink (at kb/s)

Apparent bandwidth asymmetry has great influence on IP routing, which assumes that links are bidirectional. One possible solution to this problem is a link-layer tunnelling mechanism for unidirectional links (UDL) which emulates a bidirectional link (Duros, Dabbous, Izumiyama, Fujii, & Zhang, 2001). Such emulation is carried out in the second (link) layer, what enables usage not only of routing protocols, but also other auxiliary protocols, as for example, the Address Resolution Protocol (ARP), which converts IP addresses to link layer addresses.

Bandwidth asymmetry is especially important in the case of reliable transport protocols (multicast or unicast). The closed loop formed by such a protocol between the end systems (feedback of error or congestion information through positive or negative acknowledgements) results in the dependence of achieved throughput on network conditions both on the forward and reverse path.

The third, well-known disadvantage of satellite communications is higher bit error rates (BER) than typical cable ones. Typical BER, on the order of 10^{-7} or 10^{-4} in the worst case (Zhang, 2003), is quite enough for analogue voice and video services but unacceptable for data transmission. However, due to new modulation and coding techniques (such as TPC), along with higher powered satellites, normal bit errors are usually much lower and achieve "fibre-like" quality (ITU, 2002). Current satellite systems designed for data transmission generally have BER as low as 10^{-6} or even 10^{-10} (ITU, 2002).

Apart from new modulation and coding techniques, compression-enabled PEP devices deal with higher BERs. They are typically accelerators with the additional function of loseless data compression (Border et al., 2001). Data are compressed by PEP located at the satellite link entry point and decompressed by PEP at the link endpoint. Compression reduces the amount of sent data, which leads to the reduction of packet error rate (PER) when compared with uncompressed data. Typically, PEPs perform TCP and IP header compression (Degermark, Nordgren, & Pink, 1999). During multimedia transmission, RTP, UDP, and IP headers can be carried out (Casner & Jacobson, 1999). IP payload compression can also be used (Shacham, Monsour, Pereira, & Thomas, 1998). Some PEPs enable compression of WWW static and dynamic components (typically pictures) using lossy compression algorithms.

Because the satellite signal is received everywhere on large geographical area, data security is an especially difficult challenge in the satellite Internet. To assure secure transmission, typically two alternative solutions are used: link encryption (e.g., 3DES) and virtual private network (VPN). Link encryption is carried out in the link layer by the satellite network operator and protects only transmission within the satellite link. Virtual private network (VPN), based on IPSec, is performed in the network layer by the user (most often, a corporate user), which create an encrypted tunnel between two security gateways located on the boundary of corporate network. Because security gateways are situated between the end-user and the PEP, TCP acceleration cannot be used (PEP has no access to encrypted TCP headers). Known solutions place lightweight PEPs (software or hardware) before the security gateway.

As described above, there are many challenges in the building of commercial services based on the satellite Internet. However, there are also opportunities which exist in a satellite network such as, and in particular, the natural multicasting.

In the case of the satellite Internet, every receiver located in the area covered by the downward beam always receives each layer-2 frame (which conveys unicast or multicast IP datagrams) sent by a satellite transponder. This frame is then filtered by a network adapter (the so-called "network card") and frames conveying datagrams addressed to another receiver are rejected. Only frames conveying datagrams whose IP address (unicast or multicast) matches the receiver's IP address (unicast) or the address of the multicast group to which the receiver belongs, are accepted for decapsulation.

Multicast technology makes full use of the most important feature of satellite communication, the ability to disseminate information to a large group of users that may span a large geographical area. A combination of IP multicasting and space infrastructure gives the possibility of simultaneous distribution of information to receivers located in different areas and creates opportunities for a new dimension of information technology.

The main advantages of the satellite Internet are:

- Relatively high bandwidth
- High availability—The Internet access is available anywhere in the world, independently of local infrastructure
- Easy and effective IP multicasting

These advantages result in the satellite Internet being especially beneficial for remote area communications (rural communications and services to isolated sites). It allows users in sparsely populated areas to easily access the WWW (probably the most popular Internet service for individuals), electronic mail (e-mail), file transfer (ftp), and so forth. It also enables easy data distribution (e.g., online weather map distribution, stock quote streaming) and access to databases (including e-banking).

The satellite Internet allows inhabitants of isolated communities to improve their standard of living. The broadband Internet available anywhere is especially important for telemedicine and tele-education. Telemedicine applications, intended to support medical, sanitary, and dental practitioners, integrates tele- or audio-conferencing tools with the transmission of specific data (typically high-resolution, uncompressed, or lossless compressed static images or sequence of images). In the case of remote areas, it is difficult (and sometimes impossible) to move patients for diagnosis to large medical centers (which usually offer better diagnostic specialists than local clinics, attended by one doctor or by a nurse). Telemedicine service also helps inhabitants of remote areas get fast medical assistance in case of emergency.

The usage of the satellite Internet for training and distance learning concerns both Internet access, where the Internet is used as an educational tool, and specialized tools for e-learning. These tools are based on videoconferencing applications, which work in a centralized manner (the teacher must have an opportunity to let a student respond and to be able to interrupt that student in return). Videoconferencing is typically associated with data transmission (e.g., drawings—or scanned pictures—during classes in drawing as well as slide presentations and shared whiteboard); however, there is no necessity for the transmission of high-resolution and uncompressed images. While telemedicine conferencing applications often work in the unicast manner, educational tools benefit from the satellite Internet's natural multicasting. Participants, situated in their homes or local educational centers, receive voice and video from the teacher and their colleagues, "sitting" in a virtual classroom. Thanks to IP multicasting, downlink transmission needs only one single connection, while nonmulticast solutions need at least one connection per participant.

FUTURE TRENDS

Future trends of the satellite Internet should be examined on three planes.

The first plane is a network layer. On the one hand, network infrastructure (hardware), functionally located in the network layer, is evolving toward routers onboard satellites (Wood, 2005). On the other, network layer protocols are optimised for satellite communication. Current work in progress includes extensions to the Mobile IPv6 protocol to support mobile receivers moving to access networks with unidirectional satellite links (Menzel, Wagner, & Miloucheva, 2006), as well as translation of IP addresses to link layer addresses in the case of IP over MPEG-2 transmission (Fairhurst & Montpetit, 2006).

The second plane is the transport layer. To improve TCP behavior in high speed (e.g., optical) or long distance (e.g., satellite) networks, limited slow start (Floyd, 2004) and Quick-Start (Floyd, Allman, Jain, & Sarolahti, 2006) mechanisms were introduced. However, the novel trend observed in terrestrial network—reliable multicast transmission—is also reflected in the satellite Internet (Chodorek & Chodorek, 2005).

In the case of the third plane, Internet applications, the technology trend will be the same as in the case of the terrestrial one—migration toward new real-time multimedia services, including IPTV and VoD. IPTV is expected to account for about 10 of the 200 million European households in 2009, that is, more than HDTV (8 million recipients are predicted), while VoD is expected to account for about 22 million households in the European, Middle Eastern, and African region (Reding, 2005). Although these services won't be able to replace "traditional" DVB and ATSC television in the nearest future, they will become a significant part of the audiovisual market.

CONCLUSION

The satellite Internet is a convergence of the satellite network (VSAT, DVB-S, ATSC-S, etc.) and Internet technology. Migration toward Internet technology enables better utilization of satellite communication infrastructure, even in the case of "traditional" satellite services. As an example, VoIP-based telephony enables a larger number of telephone connections in the same link than "traditional" (analogue PSTN or digital ISDN) telephony, which leads to a significant reduction in the cost per telephone call. Moreover, Internet technology makes it possible to utilize the remaining (unused by VoIP) bandwidth for reliable data transmission (e.g., file transfer).

The satellite Internet is a low-cost solution for rural telephony and the Internet in developing countries with poor terrestrial infrastructure. From the other side, in the era of globalization, the satellite Internet gives great opportunity to build global virtual private networks. In contrast to the terrestrial Internet, with decentralized management, satellite Internet assures the same security policy and uniform (and, typically, centralized) management, independent of local determinants (e.g., political factors or disasters). These advantages mean that satellite networks are sometimes used as backup links for terrestrial connections.

REFERENCES

ATSC. (2000). ATSC Data Broadcast Standard (A/90). Washington, DC: Advanced Television Systems Committee.

ATSC. (2005). ATSC Digital Television Standard (A/53). Washington, D.C.: Advanced Television Systems Committee.

Border, J., Kojo, M., Griner, J., Montenegro, G., & Shelby, Z. (2001). Performance Enhancing Proxies Intended to Mitigate Link-Related Degradations. RFC 3135.

Casner, S., & Jacobson, V. (1999). Compressing IP/UDP/RTP Headers for Low-Speed Serial Links. RFC 2508.

Chodorek, A., & Chodorek, R. R. (2000). Characterization of the MPEG-2 video traffic generated by DVD applications. In *Proceedings of the 1st IEEE European Conference on Universal Multiservice Networks* (pp. 62-70). Piscataway, NJ: IEEE Press.

Chodorek, A., & Chodorek, R. R. (2005). PGM-based reliable multicast data dissemination in satellite IP networks. In *Proceedings of the International Workshop on Satellite and Space Communications* (pp. 767-771). New Jersey: IEEE Press.

Chodorek, A., & Papir, Z. (2000). Selfsimilarity of an MPEG-2 video stream generated by DVD applications. In *Protocols for Multimedia Systems (PROMS'2000)* (pp. 389-399). Cracow, Poland: Faculty of Electrical Engineering, Automatics, Computer Science and Electronics Press.

Clarke, A. C. (1945). Extra terrestrial relays. Can rocket stations give world-wide radio coverage? *Wireless World*, 305-308.

Deering, S., & Hinden, R. (1998). Internet Protocol, Version 6 (IPv6) Specification. RFC 2460.

Degermark, M., Nordgren, B., & Pink, S. (1999). IP Header Compression. RFC 2507.

Duros, E., Dabbous, W., Izumiyama, H., Fujii, N., & Zhang, Y. (2001). A Link-Layer Tunneling Mechanism for Unidirectional Links. RFC 3077.

Elbert, B.R. (2001). *The satellite communication ground segment and Earth station handbook*. Boston: Artech House.

ETSI. (1997). *Digital video broadcasting (DVB): Framing structure, channel coding and modulation for 11/12 GHz satellite services* (European Standard Rep. No. EN 300 421 V1.1.2). Sophia Antipolis Cedex, France: European Telecommunications Standards Institute.

ETSI. (2004). *Digital video broadcasting (DVB): DVB specification for data broadcasting* (European Standard Rep. No. EN 301 192 V1.4.1). Sophia Antipolis Cedex, France: European Telecommunications Standards Institute.

ETSI. (2005a). *Digital video broadcasting (DVB): User guidelines for the second generation system for broadcasting, interactive services, news gathering and other broadband satellite applications (DVB-S2)* (Tech. Rep. No. TR 102 376 V1.1.1). Sophia Antipolis Cedex, France: European Telecommunications Standards Institute.

ETSI. (2005b). *Digital video broadcasting (DVB): Interaction channel for satellite distribution systems* (European Standard Rep. No. EN 301 790 V1.4.1). Sophia Antipolis Cedex, France: European Telecommunications Standards Institute.

Fairhurst, G., & Collini-Nocker, B. (2005). Unidirectional Lightweight Encapsulation (ULE) for Transmission of IP Datagrams over an MPEG-2 Transport Stream (TS). RFC 4326.

Fairhurst, G., & Montpetit, M.-J. (2006). Address Resolution for IP Datagrams over MPEG-2 Networks. Internet-Draft, draft-ietf-ipdvb-ar-03.

Floyd, S. (2004). Limited Slow-Start for TCP with Large Congestion Windows. RFC 3742.

ITU. (2002). *Handbook on satellite communications*. New York: Wiley.

Floyd, S., Allman, M., Jain, A., & Sarolahti, P. (2006). Quick-Start for TCP and IP. Internet-Draft, draft-ietf-tsvwg-quickstart-02.

Jacobson, V., Braden, R., & Borman, D. (1992). TCP Extensions for High Performance. RFC 1323.

Menzel, O., Wagner, D., & Miloucheva, I. (2006). Support of mobile receivers with unidirectional links in MIPv6. Internet-Draft, draft-menzel-udlr-mipv6-00.

Montpetit, M.-J., Fairhurst, G., Clausen, H., Collini-Nocker, B., & Linder, H. (2005). A Framework for Transmission of IP Datagrams over MPEG-2 Networks. RFC 4259.

Moroney, S. (2002, April). *Satellite and VSAT: Innovative uses for rural telephony and Internet development.* Paper presented at the Expert Workshop on Improving IP Connectivity in LDCs, Geneva. Retrieved from http://www.itu.intlipdc

Postel, J. (Ed.). (1981). Internet Protocol. RFC 791.

Reding, V. (2005). *Challenges for TV—convergence—IPTV.* Paper presented at the 8th EIM Conference, Vilnius. Retrieved from http://ec.europa.eu/commission_barroso/reoling/olocs/speeches/eim_20051208.pdf

Schulzrinne, H., Casner, S., Frederick, R., & Jacobson, V. (2003). RTP: A Transport Protocol for Real-Time Applications, RFC 3550.

Shacham, A., Monsour, R., Pereira, R., & Thomas, M. (1998). IP Payload Compression Protocol (IPComp). RFC 2393.

Wood, L. (2005). Testing a router onboard a satellite in space. In *Proceedings of the End-to-End Challenges of Broadband via Satellite* (pp. 83-85), IEE Savoy Place, London, England.

Zhang, Y. (Ed.). (2003). *Internetworking and computing over satellite networks.* Boston: Kluwer Academic.

Chapter VII
The Era of Nanosatellites:
Pehuensat Development Status

Juan Jorge Quiroga
Universidad Nacional del Comahue, Argentina

Roberto Fernández
Universidad Nacional del Comahue, Argentina

Jorge Lassig
Universidad Nacional del Comahue, Argentina

ABSTRACT

Nowadays, it is possible to achieve low cost and short production times space missions using satellites with a mass below 10 kg. These small satellites are described as nanosatellites. Current microelectronic technology makes it possible to develop nanosatellites for scientific experiments and relatively complex measurements (as well as for other applications), making it easy for universities and small research groups to have access to space science exploration and to exploit the new economic possibilities that emerge. This chapter describes an experiment developed in Argentina at the Universidad Nacional del Comahue to design and construct a nanosatellite called Pehuensat-1.

INTRODUCTION

Space science is an exciting activity that can greatly impact our society, particularly young students from elementary to university level. A satellite is one of the most complex and fantastic devices an engineer can design, develop, and build because it involves extraordinary challenges.

As a result, this type of project requires many interdisciplinary tasks that quickly generate enthusiasm and commitment for the different phases, which initially comprise of development, design, and construction, and later entail test, launching, operation, and possible recovery of these devices.

In the beginnings of space exploration, most of the missions were small, due mainly to a limited launching capacity, but as the fuels improved and the rockets power increased, more complex projects were possible. Historically, space exploration has tended toward big satellites with complex and expensive missions. The premises at that time seemed to be, the greater, more complex, and expensive, the better.

The methods traditionally used to place satellites, astronauts, and provisions in space are quite complicated as well as expensive and require long term preparation. Although big missions are able to economize through scale discounts, they are also more complex and rigid, due mainly to their long preparation and tuning period. In addition, it also appeared that limitations in the official budgets of the agencies dedicated to this activity caused a change in the space community tendencies; the need to have less expensive and simpler missions. Also, the number of institutions with limited budgets which were devoted to space science grew enormously when it became a curricular activity and the subject of research at many universities. Nowadays, the tendency is toward simple and fast execution space flights. The new paradigm turned slogan is, the faster, smaller, and cheaper, the better. The return to missions with small satellites was also due to advancements in technology. The development of small-scale digital technologies, and the reduction in the dimensions of the mechanic, optic, and electronic components made possible the beginning of the minisatellite age. This new age brings with it new application in fields such as tele-observation not only for monitoring the pursuit of the terrestrial objectives, but also for environmental control (e.g., the premature detection of field and forest fires), and of course in communications.

The great diversity in dimensions and objectives of space projects has led to a classification system and, although there is no unanimity about their intervals, the following criteria is generally accepted (ONU, 1998): Great satellites are those of more than 1000 kg and small satellites are those of mass inferior to that value. Also, the small satellites are divided into smaller categories:

- **Minisatellites:** 100 to 1000 kg
- **Microsatellites:** 10 to 100 kg
- **Nanosatellites**: 1 to 10 kg
- **Picosatellites:** 0.100 to 1 kg are also built
- **Femtosatellites:** Inferior to 100 g is an "open category being explored" category.

Sometimes, the small satellites are designed to operate together in formation (swarm, cluster, or constellation) and they are ideal for interinstitutional joint projects—even involving several countries—with concrete and complementary objectives for each unit.

Vanguard 1, weighing 1,4 kg, was the first nanosatellite, necessarily restricted as already said, by the launcher. Nowadays, this category is mostly limited by the budgets and chosen not only for being—as in our case the only possible option—but also because it offers a particularly tempting field of application for the incipient research groups of the space science.

The Universidad Nacional del Comahue (UNCo), together with other institutions such as the Asociación Argentina de Tecnología Espacial (AATE: Argentine Association for Space Technology), and the radio amateurs satellite corporation (AMSAT) Argentina, is working on space technology with the objective to form human resources in this discipline tending toward a higher curricular level and an improved regional development (De León, 1999; De León & Lassig, 1999; Lassig; Keil, Fernández, & Quiroga, 1999; Quiroga, Fernandez, Keil, Jurasic, Sierra, De Leon, & Alvarez, 2000a).

The tasks are made within the research project framework *Space Applications : Development of outstanding aspects in the electronic technology for space useful loads and microsatellites*. One of the fixed objectives of this initiative is to design, construct, and put into orbit a small satellite, the Pehuensat-1. In the following sections, main characteristics will be discussed, along with the elements to be taken into account in its design and implementation of the project. This chapter will also discuss the usefulness of this type of initiative as a reference for future attempts in nanosatellite launching.

PEHUENSAT-1

This nanosatellite is a 6 kg nanosatellite with a volume of 5 dm3 designed and built at the university (UNCo) for scientific and educational purposes. The purpose of this project is to obtain the know-how to be able to design more complex projects in the near future (Quiroga, Lassig, Keil, Monte, Fernandez, Simone, et al., 2004).

Besides some limits due to the weight and volume of the satellite, there are also other imposed boundary conditions that restrict the characteristics of the project. The main objectives are to gain experience in the collection and transmission of data, evaluate the performance of off-the-shelf components, and succeed in the coordination with the different involved groups.

Students from the different career programs at the Facultad de Ingeniería (Engineering Faculty) have actively participated in the design and construction of the Pehuensat-1, particularly those in electronic engineering who have been part of different research groups consolidating the theoretical knowledge of the career courses from concrete experience. Their work in relation to this project was presented in a variety of congresses and meetings on the subject and their contribution to the material is presented here.

It is foreseen that elementary and high school students of the Comahue region will also participate in this project, receiving and analyzing the satellite signals.

The AATE and AMSAT Argentina participate in the development team and contribute to the development of the technical specifications and the planning of the satellite mission. They collaborate with the Facultad de Ingeniería in the project management and designing. They will jointly operate with the university once the satellite is in orbit. AATE is responsible for the administration, preparation, launching service, and the final integration necessary for the project. On the other hand, AMSAT is responsible for the communications, adding its experience from the first argentine satellite (LU-SAT). It is cooperating to achieve the educational objectives of the project by providing as much communication equipment as possible to the participating schools and allowing the use of the satellite among students and the community of radio hams of Argentina.

In spite of the smaller size and weight, shorter design times, construction, and its consequent minor launching cost, the small satellites are ideal to stimulate and develop the nucleus of specialization in space technology at the national and regional level. In addition, its benefits increase as the electronic components are improved.

Nowadays, there are many universities and institutions that are working in this field that have their own space program and satellites in the nanosat category and smaller. At this point, education is usually the main objective and the satellite a secondary matter. Here are a few examples:

- "Project Starshine," designed by the U.S. Naval Research Laboratory, has brought in volunteers from all over the world, including Argentina, and has been running for a few years. (http://www.azinet.com/starshine)
- In the UTIAS of the University of Toronto, the CanX project has attracted collaborators, and they have already launched picosat CanX-1 and expect to launch this year the nanosat CanX-2. (http://www.utias-sfl.net/nanosatellites/CanXProgram.html)
- In the CubeSat project developed by California Polytechnic State University and Stanford University, more than 60 educative institutions are already participating. Here,

individual groups construct a picosatellite of cubical form with 10 cm per side and a weight of up to 1 kg. (http://littonlab.atl.calpoly.edu)

- The University of Surrey, in the United Kingdom, has his own aerospace company (SSTL) and it developed the first European nanosatellite, the SNAP-1 of an approximate mass of 7 kg. The SNAP structure supports loads up to 4 kg and includes facilities such as propulsion, attitude control, and an onboard computer. (http://www.sstl.co.uk/index.php?loc=47)
- Arianespace announced its intention to send a set of 50 nano/picosatellites of 1 kg in 2007 to space simultaneously to celebrate the 50th anniversary of the first satellite being put into orbit. Each nano/picosat will be dedicated to a concrete scientific experience under the responsibility of research groups of universities or of the worldwide private sector. (http://www.arianespace.com/site/fr/actualite/p04_10_6.htm)

and another widespread student project:

- The student space exploration and technology initiative (SSETI) gathers students of several European countries, and has already put in orbit at the end of 2005 the SSETI Express, a parallelopiped of 0.6 x 0.6 x 0.7 m with a mass of 80 Kg. It has even more ambitious missions planned. (http:// www.sseti.net)

ELECTRONIC DESIGN

The Pehuensat-1 was originally designed in two parts assembled in separated cards, the transmission and operative control system and the energy management system, constituting a platform of tests in which these actions are combined

Figure 1. Onboard computer

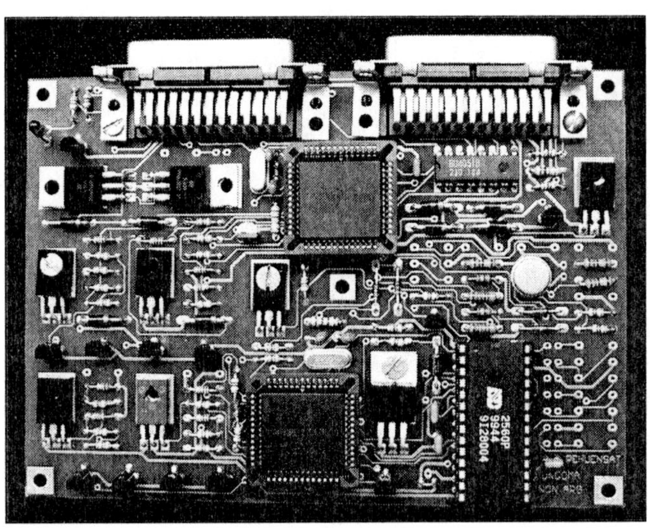

Figure 2. Operative structure of the Pehuensat-1

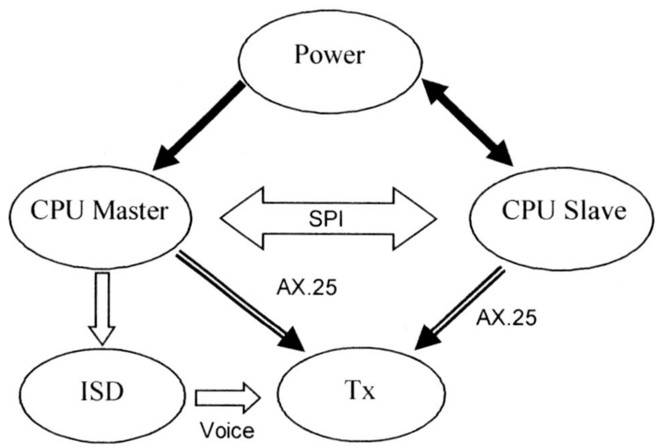

so that the nanosatellite can transmit in packet and voice the state information of the different sensed variables. These variables were chosen with the criterion of being able to have an earthly representative analysis of the operation of the nanosatellite.

These systems were implemented with individual microcontrollers Motorola HC11. The first one, master or main, generates the AX.25 frames for the transmission in packets, conforms the phrases of the voice, and messages and modulates the transmitter. The second one, or slave, makes the sensing of the interesting parameters and administrates the energy of the battery bank.

Both parts communicate to each other through the synchronic serial interface (SPI: serial peripheral interface) of the microcontrollers, and they dialog at regular intervals, exchanging data and operation conditions so that the master can process the telemetry and transmit it to Earth and each microcontroller can verify the operation status of the other and consequently decide.

The reliability of the system is increased by the fact that both microcontrollers have the ability to command the transmitter, because the energy management system also has the routines of the AX.25 transmission and can take total control, although just emitting telemetry in packets and not in voice, in the eventuality that the main system fails.

After some tests considering the intimate relation between both systems with the intention of reducing space, weight, wiring complexity, and probability of faults, it was decided to integrate them in a single plate (Arias, Pelayes, Cajarabilla, Márquez, & Quiroga, 2005).

This main printed circuit card that we denominate on board computer, shown in Figure 1, is the nanosatellite heart. It contains both microcontrollers that not only control the transmission of telemetry and voice, but also runs a system of switches that act on the bank of batteries and multiplexes and adapt the values of the measured parameters.

Operative Structure

The on board computer has basic redundant control (SIHFT: software-implemented hardware fault tolerance) of the keys' parts. This gives the system some tolerance to failures to assure a minimum provision of the actions of the nanosatellite in case of failure of any of the microcontrollers or programs.

a. In case of software failure or hardware failure of the energy management system, the set of switches that act on the rechargeable and dry batteries are closed so that the system has permanent feeding, allowing the limited operation of the activity of the nanosatellite to continue until the batteries have been exhausted.
b. In order to consider the failure of software or hardware of the transmission and operative control system of the satellite, two components have been implemented:

- Software redundancy in the microcontroller energy management system; this redundant program is made up of basic functions of the satellite control, transmitting to Earth, in this case, only packets of telemetry, as was said
- Execution of an additional routine that permanently runs in the microcontroller, supervising the different tasks and acting in case some anomaly in the operation of the master's program occurs.

Besides, a diagnosis procedure was foreseen on Earth in order to be able to finalize details and to activate any component prior to the launching.

After designing the plate, it was assembled, and the tests described later on were conducted.

Diagram in Blocks

Figure 3 shows the on board computer and the other blocks that constitute the electronic system of the Pehuensat-1.

- **Transmission and operative control system of the Satellite (μC Satellite Control):** The satellite control is in charge of running the transmission of the telemetry and controlling the system of voice reproduction by an information storage device (CI ISD 2560), and is the element in which the audio unitary messages (words) have been stored (Alic, Brion, Monte, & Piris Botalla, 2005; Arroyo, Brion, Eggers, Marcellino, Mollo, & Monte, 2003). The telemetry is transmitted in packets to 1200 bauds with radioama-

Figure 3. Diagram in blocks of the Pehuensat-1

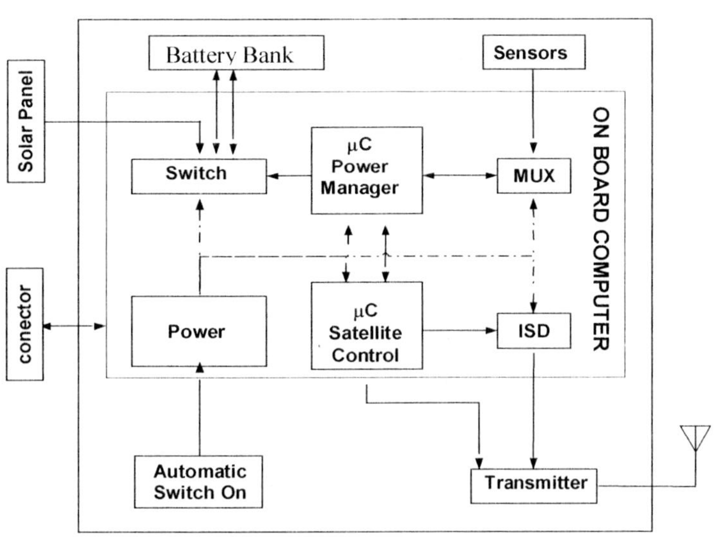

teurs HAM protocol AX.25. The periods of transmission and silence in messages are regulated, depending on the available charge in the batteries.

The educational orientation of direction responsible for the nanosatellite required needed to identify telemetry systems, which could be decoded by simple equipment, for example, there was a requirement to listen to the nanosatellite in a class at an elementary school. That is the reason one voice was one of the forms of telemetry. The other system of telemetry had to allow the data to register directly in the computer in order to quickly evaluate the behavior of the nanosatellite. It was decided, therefore, to transmit information in packet AX.25. This is an adapted protocol of X.25 of the ITU-T (international telecommunication union-telecommunication standardization sector; before CCITT) widely used by the radio hams.

The software is developed in a multitask environment from a basic list which is adapted according to the resources of the microcontrollers. The tasks in which the software and its descriptions have been divided are detailed in Table 1.

With respect to the hardware, the microcontroller works with a clock of 4 MHz, and its resources are fully used because it has a compact design with a low amount of integrated circuits, resulting in a trustworthy system. 2KB of memory EEPROM and 256 bytes of RAM are available. An integrated circuit, MC34064, keeps the CPU in RESET mode during a fixed time in order to allow the stabilization of the power supply and, therefore, avoid possible errors. The entire subsystem is fed with 5 Volts, and normally the microcontroller is in the low consumption mode (WAIT) waiting for interruptions.

The storage of the voice messages in audio format is contained in the integrated circuit ISD 2560. This circuit integrates all the functions of sampling, A/D conversion, and recording of words in cells EEPROM. It also contains all the devices to generate the audio out that once filtered by a low-pass filter (to limit its bandwidth), and a resistive divider (to limit its amplitude) can be injected into the audio input of the transmitter.

Using a connector type DB25, the energy to all the circuit is provided and the CPU is connected to the transmitter. Another DB25 introduces the

Table 1. Master's tasks

Task	Description
Communication μC Energy	Its task is to exchange information with the energy administrator microcontroller regarding the state of the sensors and to notify the operative state of the other CPU.
Telemetry Voice	Responsible for generating the audio messages by reproducing them in the order in which the messages were saved.
Telemetry AX.25	It generates the complete frame AX.25, including the CRC (Cyclic Redundancy Checking), for each packet. It uses to the maximum the resources of hardware available in the microcontroller, generating the signals by means of interruptions of OC in the microcontroller.
Transmission Intervals Analysis	It doses transmission intervals, because the energy is consumed mainly during the transmission. An algorithm evaluates the tension of the rechargeable batteries and the temperatures to determine the schedule for the next transmission.
Transmitter Control	It is responsible for turning the transmitter on sometime prior to the transmission and of controlling the consumption registering the tension fall of the batteries when they turn on.
Diagnosis on Earth	This waits for commands by the port RS232, through which it is possible to permanently save parameters in memory EEPROM, to read the information table about the state of the sensors, and to put in order the execution of tasks. Inclusion of this task makes it easier to diagnose the nanosatellite on Earth.

Figure 4. Cycle of tasks conducted by μC power manager

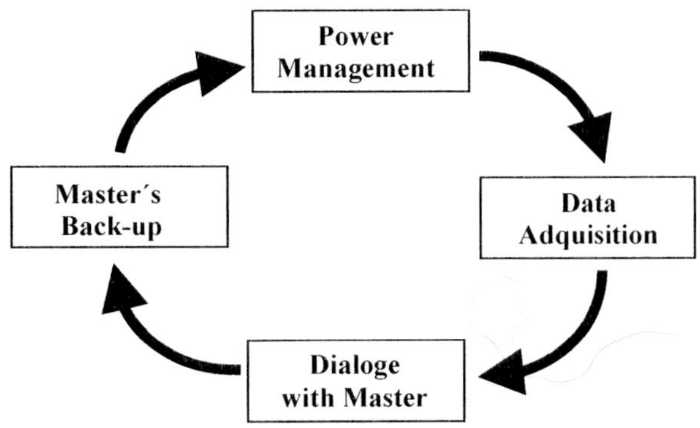

signals of the sensors and allows access of the external connector.

- **Energy management system (μC power manager):** The role of the power manager is the maintenance of the batteries, administering its energy, and selecting the suitable bank not only for its charge by solar panels, but also for its connection to feed the onboard computer and other circuits. It also gathers the data of measurement of the mission basic parameters (temperatures, charge current of the panels, and battery state) that make up the telemetry (Arias, Pelayes, Orejas, Giacomelli, Mare, Odello, & Quiroga, 2003). This information communicates to the main Satellite Control System by means of the serial synchronic interface (SPI), where it is processed for its transmission to Earth.

In addition, this system evaluates the charge state from the different battery banks, and according to the values of measured tension and temperature, it decides which bank is in charge and which one provides the energy to the platform main circuit. In the case of not having enough energy in the rechargeable banks, the software connects to the bank of alkaline batteries.

The energy management system is also prepared, as well, to take control of some of the satellite's basic functions in case of the failure of the Transmission and Operative Control System of the satellite.

The microcontroller software of the energy management system is organized into two basic components: the Main Program and a Periodic Execution Subroutine. The cycle is also based on the multitask programming, as shown in the outlined Figure 4.

- **Main program:** The primary function of the main program is to administer the system's energy based on the analysis of the measurements gathered through the different sensors that make up the measurement system. The flow chart of Figure 5 briefly shows the algorithm of the main program that controls the batteries' charge.

According to the state or charge of each bank of batteries (which is evaluated measuring their tension and temperature) it is put in charge or it is connected, feeding the system. In case of a lack of a bank of NiCd batteries under suitable charge conditions (and only in this case), the software decides to use the alkaline ones.

Figure 5. Batteries charge control subroutine

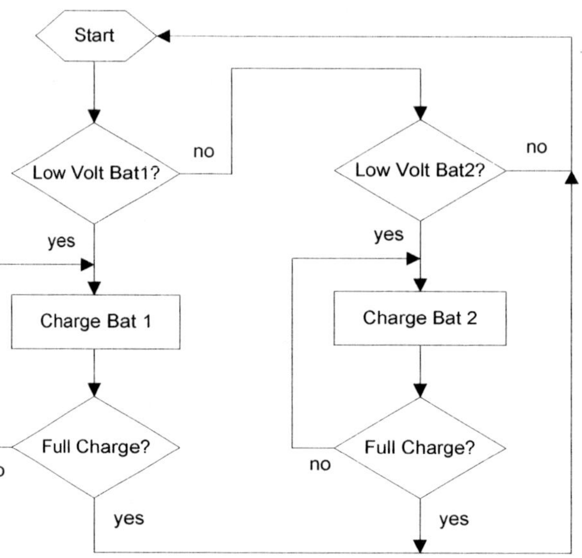

- **Periodic execution subroutine:** This periodic execution subroutine is intermittently executed by means of an interruption in the main program, and at this point it starts up a sequence of subroutines developed to obtain the data that is sent to the control system of the satellite:
  - Sampling of sensors
  - Conversion to binary data
  - Sending of data

The analogical values from the sampling of the measurement sensors are multiplexed, turned to binary in the A/D converter of the microcontroller, and stored in BCD format to be sent later to the control system of the satellite via the synchronic interface (SIP) conforming part of the telemetry package that this last generates.

For a greater efficiency and security (with the expectation of redundancy), the energy control was prepared to take over certain basic functions of the satellite in case of failure in its main computer.

The master failures are detected by watching for the appearance of unusual time gaps of inactivity of the SPI. The secondary microcontroller, thus, takes control and transmits only packets to Earth.

The power source is one of the key systems that defines the expected life of the satellite. Its design was made based on some limitations such as its cost, conditions during its flight, and available material, among others. Nevertheless, the basic objective, the security of operation and the efficiency of the design, were never forgotten.

For this project, we have multiplied the test processes to diminish the probability of material faults, because, due to the low budget, materials of local, standard origin, and low cost were used.

- **Other blocks:** In addition to the activities already described as being made by the microcontrollers, other blocks exist whose tasks are:

a. **System of automatic power-on or starting control:** In this block, the system of power-on of the nanosatellite is present, and because of the electronic detection of acceleration, it will automatically produce the starting of this platform once in orbit (Márquez, Cajarabilla, Quiroga, Arias, & Pelayes, 2005).

   This system has an acceleration sensor of two axes **AXDL250** that activates (wakes up) a timer based on a microcontroller PIC, whose time-out is determined by the launching time and bringing into orbit of the satellite. Some time after detecting the acceleration (launching), the energy is connected to the satellite. The device is duplicated in order to ensure the starting up of the platform.

b. **Sensors:** In order for the energy management system to fulfill its objective, it is of great importance to make a good choice in the parameters that are going to be measured. This way, the selection will optimize the operation and ensure that the success of the mission.

   The voltage of the three battery banks are measured, as well as the charge current from the solar panels and the temperature in six different points. Two of these points correspond to the banks of rechargeable batteries, which, together with their tension, determine when the bank is charged. The other four remaining points of measurement are strategically located in different places in order to balance the interior room temperature of the nanosatellite, with the purpose of being able to value the thermal changes that have occurred on the on board computer plate and the bank of rechargeable batteries.

   A current sensor for the solar panel was placed on the side of the computer, as well as three sensors of the state of charge of the batteries. These measurements are multiplexed because the number of parameters that need to be measured is bigger than the number of analog/digital inputs available from the microcontroller. The temperature sensors activate only when the measurements need to be made. The objective in doing this is to reduce the energy consumption and to avoid heating the sensors, which would introduce errors in measurement.

c. **Solar panel**: To recharge the banks of the batteries of NiCd, a solar panel with 36 cells of 3x5 cm, 400 mA, and 0.5 V, each arranged in series conforming a plate of 25x31 cm are used, located on one of the greater faces of the structure.

   The Pehuensat-1 does not have attitude control and there is no knowledge of the orientation that the solar panel will have in relation to the sun position. This is because the flight, in principle, would be like a parasite of a greater dimension satellite. This flight has not been planned yet, so that is the reason the amount of illumination that the panel will receive is unknown, It is equally probable that it will only be partially permanently illuminated, or even that it remains in the shade during the complete time of flight. In addition, because one of the faces of the structure will be used for its fixation, that leaves only one left for placing the solar cells. All these adverse factors were considered when making the design of the Energy Management System, in order to guarantee a minimum lifetime for this space test platform. To fulfill this objective and provide a source of alternative power, a third bank of nonrechargeable batteries (dry ones for domestic use) serve as back up, and are added to both banks of rechargeable batteries.

d. **Batteries:** The cells used at the banks of rechargeable batteries are made of nickel-cadmium. These cells are characterized by having a hermetic construction, high currents, and long life (in cycles of charge and discharge) (Fernández, Nolly, Sierra, Quiroga, & Monte, 2003; Quiroga, Fernandez, Estevez, Mare, Odello, & Sierra, 2000b).These characteristics make them suitable for space applications. Each bank is constituted by 10 Ni-Cd cells connected in series, having a total of 12 V by bank.

   The cells that are used in the banks of dry batteries are alkaline. Their most outstanding characteristics are a possibility of giving a great density of energy during a brief time, great current capacity, minor self-discharge, and reliability. The bank is constituted by eight alkaline cells connected in a series.

   The choice of the battery types that compose both banks is based on the experience acquired by the group during the previous tests of charge/discharge carried out on batteries of different trade names and different temperatures, for the evaluation of their behavior as well as their feasibility in relation to the mission.

   The mechanical support of the batteries is made of teflon, in order to not only hold them firmly, but also to isolate them thermally to one another.

e. **Switches:** The design criterion used for choosing switches was to achieve the smallest drop of tension with the smallest possible consumption. As a result, a series of tests with different configurations from diverse bipolar transistors and MOSFET's were done. The best result obtained was a decrease of tension of 0.2 V, for a nominal current of 500 mA, which is the maximum estimated consumption of the platform for space testing. The control current of the switch is of 4 mA.

   The switches are commanded by the microcontroller according to the charge state of the batteries and are normal closed, to ensure that the platform be totally fed in case of failure of the energy management microcontroller.

f. **Telemetry: Voice telemetry**—The nanosatellite will emit, as long as there is enough energy available, a message spoken in a radio ham frequency made up of the satellite identification and information of the states of temperature, current, and tension sensors. The message will last between 30 and 35 seconds. It will be emitted in three languages: Spanish, English, and Hindi, in this order. Messages are stored in EEPROM in an integrated circuit ISD (Information Storage Device) of a 60-second capacity. The messages are: identification of the nanosatellite, the digits from 0 to 9, the sign, the decimal point, and the names of the variables.

   With this presaved information, the messages are made up by a subroutine that selects the voice segments. For example, in order to emit a message at a temperature of -12.45 °C temperature, it would be:

TEMPERATURE MINUS ONE TWO POINT FOUR FIVE DEGREES.

The recording of the messages was made using software developed in VisualBasic designed by the work group. The messages are recorded one by one by a presenter in the corresponding language, processed in PC. The interface between the PC and the integrated ISD was designed and constructed with a Motorola HC11E2 microcontroller.

The processing of the messages includes the aggregate of synchronism pulse that is inverted and turned to TTL levels with a NPN transistor and led toward PORTC3 port of the microcontroller. When this pulse is detected, it activates a timer that waits 4 ms and begins the recording of the ISD in the direction received by port RS232 from the PC. The microcontroller also governs the message time, which is of 440 ms for a simple message and 880 ms for double messages. The on/off relation of the transmitter is decided by the microcontroller based on the battery charges 1:3, 1:5, 1:7, and so forth.

Telemetry in AX.25 packets—Adapted X.25 protocol for radioamateurs use. It is transmitted in beacom mode.

License of UNCo LU1YUC CRC-16 CCITT.

ID Destiny: BEACOM

ID Source: LU1YUC

Control: 0x03

PID: 0x04F0

Length of the package: N bytes of information + 17 bytes overhead. All the variables available are transmitted.

Transmission speed: 1200 bauds.

The inclusion of the AX.25 telemetry allows for the registration of the sent variables in order to study the behavior of the solar panels, the batteries, temperatures, and the charge currents of the energy administrator. All the protocol has been completely developed by software in the HC11 microcontroller, and furthermore, the FSK and CRC-16 tones are generated as well. With this design, we managed to reduce the circuit complexity, avoiding the addition of an extra controller to AX.25. This protocol can be decoded by devices currently used like TNC, which are easily connected to a computer.

g. **Transmitter:** The transmitter is a standard equipment of 3W maximum potency, digitally exchangeable to a minimum potency of 250 mW. It has a frequency of 145.825 MHz, and has reconverted to the severe conditions of the space atmosphere, shifting some components (De Zan & Simone, 2005).

The activation of the transmitter is made by a power transistor used as a key. This activation can be done by the

Table 2. Frame of AX.25

Flag	Address	Control	Info	PCS	Flag
7E	12Bytes	8 bits	N°8b	16bits	7E

master or, in case of its failure, the energy management system. To achieve this, diodes have been arranged in order to avoid possible inverse currents or short circuits.

Due to the fact that the CPU has the option of transmitting in formed packet or audio, a capacitive couple has been provided to the design at the audio entrance of the transmitter. Also, by a capacitive couple, the energy management system can make a data transmission in packet format.

h. **Antenna:** Different configurations have been evaluated. The final decision will be made when the transport satellite is defined. (Mingolo, Moya, & Simone, 2005).

i. **External connector:** A circular, female, 19 contact, aluminum connector was placed, in order to have access to the on board computer to perform: terrestrial diagnoses of the computer, programming of each microcontroller, transmission through the operation of the PTT transmitter the audio input and general verification within the parasite charge.

STRUCTURAL DESIGN

The body of the nanosat is constituted by an aluminum T6061 box of aerospace degree of 31x25x7 cm whose interior contains banks of batteries, the onboard computer, transmitter boards, and a scientific experiment to measure the temperature of water freezing in a microgravity atmosphere. In its outer surface it connects to the solar panel and at the opposite face, an aluminum cover screwed in its perimeter. The circular connector and the antenna are located in the laterals.

Figure 6 is a photograph of the nanosatellite during its joint.

Figure 6. Assembly of the Pehuensat-1

TESTS IN SIMULATED SURROUNDING SPACE

One of the most important tasks is to verify the proper operation of the electronic component devices of the nanosat in limited conditions similar to those that it will encounter in the rough outer space (Gilmore, 1994).

The selection of the different tests that help to ensure efficiency of the board computer was an arduous process defined by the final phase of the integration of the nanosatellite. A verification program should be an integral part of the development engineering. The verification program should be useful as the suitable link within the design process and the support logistics and launching, and the main aim should be technical efficiency and low costs.

In this case, the use of low cost components have been incorporated as a condition of the design, which increases the possibilities of failures of the nanosatellite.

The responsibility of the design of the tests has also required a process to lower the cost of each one of the experiences, which allows it to make a proper diagnosis at the level of the required mission exigencies. To achieve this, we resorted to using our own laboratories for making the required tests compatible with the existing infrastructure.

As was previously stated, the use of the devices of commercial category require an exhaustive test program, mainly to assess the satellites' behavior in low and high temperatures within a vacuum to simulate the conditions in which they will work when they are put into orbit.

The experiences selected for such purposes are:

- Vacuum test at high temperature
- Vacuum test at low temperature
- Thermal shock
- Analysis through of X-rays

Next, a brief description of each of them.

Vacuum Test at High Temperature

To develop the test, the microcontroller in charge of the energy management was programmed so that it would easily sense the temperature at two

Figure 7. Vacuum tests camera

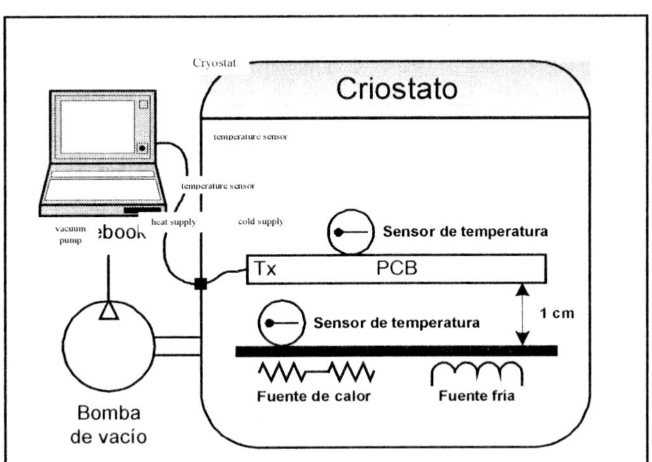

Figura 4. Esquema de ensayo en vacío

different points and transmit the data every second in serial form with protocol RS-232 to a PC. In addition, the state of operation of the system was shown with an indicator LED blinking at the frequency of transmission and acquisition of the data.

The onboard computer was introduced in a cryostat where, after generating emptiness, the temperature was increased by a source of heat in the interior of the camera. One of the temperature sensors was placed in contact with the copper plate of the heat source, and the rest on the PCB to the height of the components, as shown in Figure 7.

The experiment lasted 3 hours 36 minutes and was conducted at an average pressure of $2*10^{-4}$ mbar (getting to minimum of $8*10^{-5}$ mbar, which is the highest temperature that can be attained by the equipment). This pressure would correspond to an approximated height of 110 km. The power provided by the heat source was of 22 W (approximately represents the power to dissipate within the cabinet of the nanosatellite) for more than half of the test; after it diminished to 11 W, where it stayed until the end. The results are presented in Figure 8.

By giving power to the heat source, the temperature of the copper plate began to rise until it arrived at a plateau of 85 °C, and later on it began to drop when the power source was reduced. As the caloric energy presented in the copper plate is transmitted to the area that surrounds it by radiation only, because there is no convection in emptiness, the sensor located on the components presents a curve moved away from the sensor located in contact with the copper plate, reaching a maximum peak of 55 °C.

This test confirmed that the electronics of the onboard computer suitably responds to temperatures of up to 55°C, and that the heating due to batteries, transmitter, structure, and so forth, have less effect on the electronic components in the space atmosphere than in the terrestrial surface.

Figure 8. High temperature vacuum test

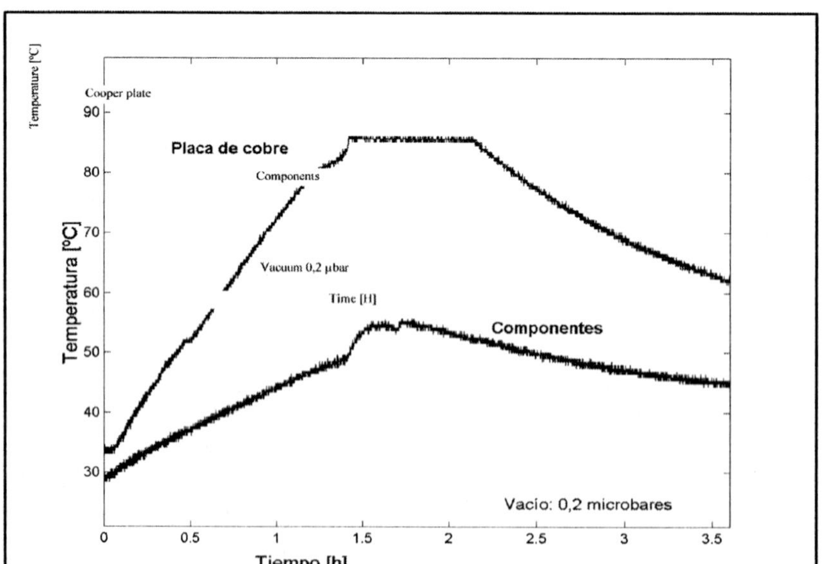

Figura 5. Ensayo en vacío a alta temperatura

Vacuum Test at Low Temperature

This test is performed by replacing the source of heat used in the former with a cold source, again simulating the face opposed to the sun of the nanosatellite structure. The low temperature is obtained by making liquid nitrogen circulate through the serpentine that the cryostat has in contact with the copper plate.

This test lasted 4 ½ hours at a constant pressure of 6 mbar, and the results appear in Figure 9.

The temperature of the copper plate fell linearly to a minimum of -30 ºC and remained there until the nitrogen stopped circulating, at which point it slowly increased. The curve of the temperature of the components presents very smooth variations due to the fact that, in this case, the components are the ones that radiate toward the copper plate (in opposition to the previous case) and have an approximated temperature of only 25ºC. As it is the smaller source of energy, it irradiates less, and that is why the variation of temperature is of only 8º C, whereas in the high temperature test is almost 30 º C in spite of having more emptiness.

Thermal Shock

The test consists of subjecting the onboard computer to abrupt changes of temperature with the objective to study the behavior of the welds and components, that is to say, to detect possible fractures in any one of the elements of the plate. An X-ray of the electronic board was taken prior to performing the test and another one after its finalization by doing a comparative analysis of them with a 3x magnifying glass. The X-rays were taken with X-ray equipment under the following conditions: 90 kV, 6 mA, 1-minute time of exposure, distance between the ionic emitter and film 0.80 m.

The test was done introducing the plate in a furnace at 65ºC, and once it warmed up, the plate was moved to a container cooled with dry ice at a temperature of -40ºC. This heating/cooling cycle was repeated 12 times to cause an important mechanical stress in the welds and components due to the repeated expansion and contraction. The test was carried out with the onboard computer side in operation and was repeated at a greater thermal

Figure 9. Low temperature vacuum test

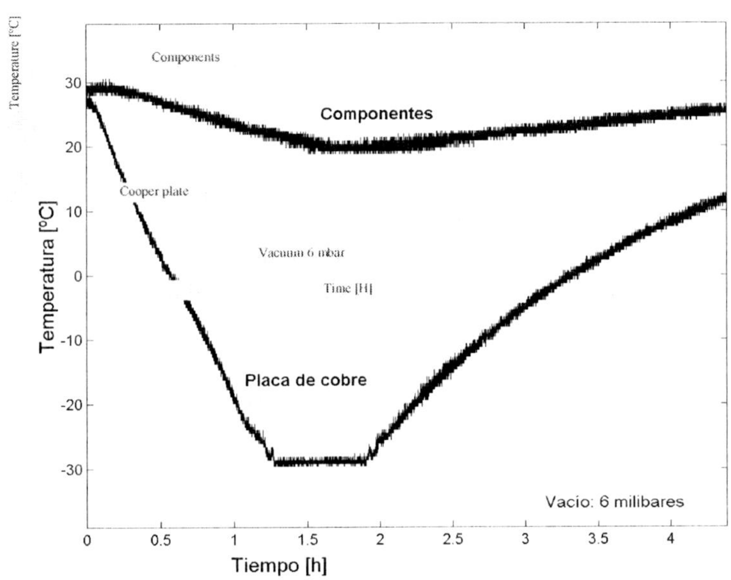

exigency and with the plate disconnected in order to verify only its mechanic answer. For this, the temperature of the furnace was raised up to 100 °C, while reducing the cold source to -60°C, with a methylic alcohol solution with dry ice.

The exposure of the plate to X-rays was made with the microcontroller programmed (EEPROM memory), and then the same went on operating without having any damage or loss of data.

With the developed tests, some of the extreme conditions of operation in which the onboard computer will be put under were emulated within the limits of our reach.

The set of tests that were performed in outer space allowed us to assess the expected behavior of the electronic system, including both hardware and software. Therefore, it is expected that the electronic design will work according to the group's expectations, thus achieving a successful mission.

The results of these tests are still being evaluated, as well as other similar ones made with the plate of the transmitter. New tests are being prepared to diminish the risks of failures in the space, such as subjecting the Pehuensat-1 to electromagnetic vibrations and radiations (EMI).

FINAL RECOMMENDATIONS AND ADVANCE STATE

Space science is as exciting as it is complex. The interdisciplinary characteristic, the atmosphere's aggressiveness, and the null margin for error require attentive and progressive designs in order to ensure final success.

Testing the equipment and components on Earth is a phase of the project that we know we must not scrimp. Test once, test again, and continue to test to be sure that the premature failure (infant mortality) zone is left behind and that the equipment and components are in the deep portion of the bathtub curve (maximum reliability, minimum failure rate) (Fernández, Nolly, Quiroga, Sierra, & Monte, 2005). Extensive testing is worth the trouble, because detection and correction of a fault prior to the launching is surely the difference between success and failure.

To specially verify that the electronic circuits are sufficiently strengthened to resist the space atmosphere, testing should include emptiness, microgravity, extreme thermal variation, radiation resistance, and huge vibrations typical in launching (Lassig, Quiroga, Keil, Fernández, & Jurasics, 2003; Thornton, 1996). In addition, the tests reveal that the economic advantages of the small satellites are remarkable.

The experience reached by the students of the UNCo by the development of the Pehuensat-1 has been very enriching. The Pehuensat-1 nanosatellite has generously fulfilled its educational mission. In spite of the system being simple, it has been designed with more complex future developments in mind. The inclusion of two microcontrollers will allow for the design of the multiprocessing systems with superposition of tasks, where dialog, exchange of information, and resolution of failure conflicts are the framework of design.

With respect to the software, the technique of design through multitask makes it easier for future developments because the "intelligence" of the system is in the communication component negating the existence of a main supervisor program.

The Pehuensat-1 is almost ready for its integration to the transport satellite. It only lacks the definition of the characteristics of the assembly and the transmitting antenna. This is supposed to take place along with the launching, during the second semester of 2006.

FUTURE OF THE SMALL SATELLITES

The low cost involved in the construction and launch of nanosatellites allow developing countries and small research groups to access space missions, gaining very valuable experience, knowledge, and new economic possibilities in this field.

Although small satellites do not represent the solution for all assignments, they offer the possibility to do important scientific experiments and measurements of all types, and they have many other applications, from mere technological demonstrations to remote education and qualification in isolated groups, integrating and complementing missions of greater scale. Even though this is obvious for the industrialized countries where there are already space programs, it is an extremely important alternative for the developing countries and for those with an incipient space technology that can accede to missions and space applications. It is also important for competing with aerospace giants, small companies, universities, and other institutions, for the constructing of small satellites to be used individually in fast and cheap missions or as swarm for more complex missions.

To sum up, in concordance with the diagnosis and recommendations done by the United Nations in the UNISPACE III (ONU, op.cit. 1998), we must try to increase the understanding of the general public regarding the benefits of space technology in order to achieve the economic commitment that encourages sustainable development and strengthens the capacities of countries, especially of developing countries. This will facilitate the use the applications of space exploration for economic and cultural growth, resulting in international cooperation in the improvement and use of space technology and its applications.

We need to motivate and to canalize the interests of young people toward the rewarding activity of space science, in order to produce the human resources necessary for economic development in a world impelled by technology, keeping in mind that small spacecrafts do not equal bad quality technology with short useful life spans. On the contrary, they can mean an advanced technology that offers benefits in relation to the total mass of the spacecraft, that allows it to accomplish valuable missions, for the science along with education and qualification. This is particularly important for developing countries, because it offers them the opportunity to access space missions and generate new business opportunities

REFERENCES

Alic, D., Brion, J. J., Monte, G. E., & Piris Botalla, L. (2005, May). *Nanopehuensat. Telemetría Digital de Datos y Transmisión por Voz.* Paper presented at the Congreso Argentino de Tecnología Espacial, Córdoba, Argentina.

Arias, A., Pelayes, N., Cajarabilla, M., Márquez, F., & Quiroga, J. J. (2005, May). *Nanopehuensat: Integración de la Computadora de Abordo.* Paper presented at the Congreso Argentino de Tecnología Espacial, Córdoba, Argentina.

Arias, A., Pelayes, N., Orejas, M., Giacomelli, M., Mare, J., Odello, L., & Quiroga, J. J. (2003, May). *Administración de Energía y Mediciones de una Plataforma de Ensayo Espacial. Nanopehuensat.* Paper presented at the Congreso Argentino de Tecnología Espacial, Neuquén, Argentina.

Arroyo D., Brion J. J., Eggers, S., Marcellino, A., Mollo, D., & Monte, G. (2003, May). *Nanopehuensat Computadora de Abordo para Transmisión de Datos.* Paper presented at the Congreso Argentino de Tecnología Espacial, Neuquén, Argentina.

De León, P. (1999, July). *Non-governmental and educational space activities in Argentina.* Paper presented at the Unispace III, Space Generation Forum, Vienna, Austria.

De León, P., & Lassig, J. (1999, September). *The impact in the space activities on education and public outreach (The Argentine experience).* Paper presented at the International Simposium on Commercial and Non-governmental Space Activities, Rio De Janeiro, Brazil.

De Zan, H. J., & Simone, D. D. (2005, May). *Estudio y Análisis de las Influencias del Ambiente Espacial en las Comunicaciones Satelitales*. Paper presented at the Congreso Argentino de Tecnología Espacial, Córdoba, Argentina.

Fernández, R., Nolly, O., Sierra, E., Quiroga, J. J., & Monte, G. E. (2003, May). *Fuentes de Energía para Misiones Espaciales*. Paper presented at the Congreso Argentino De Tecnología Espacial, Neuquén, Argentina.

Fernández, R., Nolly, O. A., Quiroga, J. J., Sierra, E. A., & Monte, G. E. (2005, May). *¿Tolerancia O Evitación? ... Esa Es La Cuestión*. Paper presented at the Congreso Argentino de Tecnología Espacial, Córdoba, Argentina.

Gilmore, D. G. (1994). *Satellite thermal control handbook*. Washington, D.C.: The Aerospace Corporation Press.

Lassig, J., Keil, G., Fernández, R., & Quiroga, J. J. (1999, October). *Space education at the National University of Comahue: A means for the integration with multiple purposes*. Paper presented at the 50th International Astronautical Congress, Amsterdam, The Netherlands.

Lassig, J., Quiroga, J. J., Keil, G., Fernández, R., & Jurasics, A. (2003). *Pehuensat. elementos de Diseño de pequeños satélites*. Neuquén, Argentina: Artes Gráficas Limay.

Márquez, F., Cajarabilla, M., Quiroga, J. J., Arias, A., & Pelayes, N. (2005, May). *Nanopehuensat-1: Encendido Automático de Satélite por Detección de Aceleración, con Eliminación de Ruido*. Paper presented at the Congreso Argentino de Tecnología Espacial, Córdoba, Argentina.

Mingolo, F. G., Moya, C. D., & Simone, D. D. (2005, May). *Nanopehuensat-1. Análisis y Diseño de Antenas para Satélites Parásitos*. Paper presented at the Congreso Argentino de Tecnología Espacial, Córdoba, Argentina.

ONU. (1998, May). *Misiones con Pequeños Satélites*. Documento preparado por la Oficina de Asuntos del Espacio Ultraterrestre de la Secretaría, para la Tercera Conferencia de las Naciones Unidas Sobre la Exploración y Utilización del Espacio Ultraterrestre con Fines Pacíficos (UNISPACE III – en junio 99). A/CONF.184/BP/9 del 26 de mayo de 1998.

Quiroga, J. J., Fernández, R., Keil, G., Jurasic, A., Sierra, E. A., De León, P., & Álvarez, A. (2000a, October). *Small satellites Pehuensat Program*. Paper presented at the 51st Internacional Astronautical Congress, Rio De Janeiro, Brazil.

Quiroga, J. J., Fernández R., Estévez S., Mare J., Odello L., & Sierra, E. (2000b, October). *Ensayo de Baterías y Paneles Solares Estándar para Ser Utilizados en Nanosatélites*. Paper presented at the Congreso Argentino De Tecnología Espacial, Buenos Aires, Argentina.

Quiroga, J.J., Lassig J., Keil, G., Monte, G.E., Fernández, R., Simone, D. et al. (2004, October). *Pehuensat-1: Development Status*. Paper presented at the 55th International Astronautical Federation Congress, Vancouver, Canada.

Thornton, E.A. (1996). *Thermal structures for Aerospace applications*. Washington, D.C.: AIAA (American Institute of Aeronautics & Astronautics).

Chapter VIII
Digital Bridges:
Extending ICT to Rural Communities Using Space Technologies

Phillip Olla
Madonna University, Michigan

ABSTRACT

Space technology has advanced rapidly in recent years. Nevertheless, a number of countries still lack the human, technical, and financial resources required to conduct even the most basic space-related activities, such as meteorology, communications natural-resource management, and education. The need to make the benefits of space technology available to all countries has thus grown more urgent with each passing year. This chapter proposes a two-phased approach for using space technology to deliver information communication technologies (ICT) to underserved areas. The first phase involves the definition and implementation of the satellite global infrastructure to provide connectivity to underserved regions. The second phase introduces the concept of a coalition of space Internet providers (COSIP) model. The aim of this model is to encourage the diffusion of space technology delivered by the GBBS infrastructure to the grassroots level. The model defines how Internet capabilities should be introduced to rural underprivileged societies to provide health and educational services in a sustainable manner. This model is a reincarnation of the local information utility (LIU) model that was successfully implemented over a decade ago to aid the diffusion of the Internet to rural American communities. This chapter explains the technology at the foundation of the COSIP model and describes the actors required along with their roles and responsibilities.

INTRODUCTION

Information communication technologies (ICT) are considered the driving force for economic, social, and technical development. In effect, high speed Internet delivers numerous imperative fundamental services such as education, health, telecommuting, electronic commerce, and e-government services, at unparalleled cost and performance conditions (Toumi, 2004).

There has been phenomenal growth in the increase in ICT over the globe; however, the digital divide still exists, posing major challenges to many of the developing countries, which are still grappling with a severe shortage of telephone lines, lack of electricity, and high levels of illiteracy. Although it is important to acknowledge the digital divide, it is more important to focus on the progress that has been made and dispel the myth about the digital divide. The most important document that highlighted the digital divide is the 20-year-old Maitland report. Some of statements that were accurate 20 years ago are now deeply established as global myth. Re-education will be needed to change opinions due to the constant e-mail chains and false reporting from journalists and researchers. An example is "There are more telephones in Tokyo or New York than in the whole of Africa." As of the start of 2004, there were approximately 25 million fixed lines and more than 50 million mobile phones in Africa, which is several times more than the total population of Tokyo and New York. Another urban myth that you may recognize is "half of the world's population have never made a telephone call." Although considerable segments of the world's population do not have access to a telephone, and probably could not afford to make a phone call if a phone was available, the international telecommunication union (ITU) estimates suggest the number is close to one-fifth of the world's population that have no telephone access. Another myth relating to the availability of the Internet is that "there are more Internet users in Iceland than in Africa." This statement originated in the 1999 report *Internet for Development*, and became obsolete in 2004. The Internet has become pervasive in society; however, rural and low-income urban areas are underserved in developing nations, and will probably remain so for the foreseeable future, for economic and structural reasons similar to those that limited expansion of Internet services in rural America over a decade ago (Clement, Holbrook, & Staman, 1996). The important thing to note from these statements is that the rate of adoption of ICT technologies around the globe is accelerating and there is a need for new models, technologies, and networks to introduce new services and applications to those living in underserved areas around the globe.

In recent years the requirements for connectivity and information services has expanded throughout the developing world. However, connectivity is hampered by the time-consuming and costly process of building traditional fixed (i.e., wired) infrastructure. For example, only 3.6 % of the population of Africa has online access (Internet-World-Stats, 2006). The situation in Africa is indicative of the entire developing world: there is a substantial unmet demand for connectivity in developing countries. Further, humanitarian services such as telelearning and telemedicine could effectively and broadly be provided if connectivity costs were lower.

Satellite technology has great potential to reach people living in remote and underdeveloped parts of the globe, and in many instances, it is the only form of technology that can provide connectivity to remote or difficult-to-access regions. Satellite systems play an important role in enhancing the ICT landscape, extending necessary services to the hard-to-reach and bridging the digital divide. It is apparent that satellites, from their vantage points in low, medium, or geosynchronous orbit, dedicate a synoptic view and global coverage for either resources management or for global connectivity. Space technology, through communication and remote sensing satellites, contributes to both the conduit and the contents for the evolving Internet infrastructure. Communication satellites provide necessary interconnection to information sources, thus enabling access to "information services" themselves.

This chapter is structured as follows. The first section provides an overview of the digital divide; the next section discusses the vision of a world without a divide by the creation of a global broadband satellite system (GBSS). The following section explores how a new wireless technology called WiMax can be converged with satellite technology to address the *digital divide*. Prior to the conclusion, this chapter explores the concept of a local partnership approach for digital inclusion called the COSIP model.

MEASURING THE DIGITAL DIVIDE

The use of the term "digital divide" became popular in the 1990s to portray the apparent emerging gap between those who have access to ICT technologies and skills and those who do not. The divide exists due to socio-economic or geographical reasons, and results in limited or no access to services such as the Internet, computers, and communication capabilities. There is a concern that ICT would exacerbate existing inequalities and allow people to be disadvantaged based on their country, geographic location, age, gender, culture, or economic status.

The view that restriction to information access closed doors to economic and social development opportunities is not new. In 1984, the *Missing Link Report* (Kelly, 2005) highlighted that the lack of telecommunication infrastructure in developing countries impeded economic growth. The *Missing Link Report's* scope was limited to access to telephones rather than the current ICT concept. In 1996, the ITU initiated a United Nations project *Right to Communicate*, that was aimed at providing access to basic ICTs for all, with the motivation to reduce information poverty for developing countries. This is a goal of many NGOs, and is also at the heart of plans of the world summit on the information society (WSIS).

Most reports looking at their evolution of ICT over the past decade in developed and developing countries express a view that an overall trend of growing ICT disparities exists between countries. Looking closer at the statistics as published by the different official bodies, such as ITU or the world bank, it is clear that the gap in ICT access between developed and developing countries does exist and is still quite important. It is difficult to comment on the size of the gap, as recent measurement indicators are dependent on specific parameters such as:

- Internet host
- Internet users
- Fixed telephones
- Network speeds
- Costs

Despite these varied parameters that can be considered, one thing that is not in contention is that the availability and the quality of a robust telecommunication infrastructure is the key to a quick and reliable development of ICT in the developing countries.

Although the term digital divide was originally created to differentiate between those who have access to digital information technologies (computers and software) from those that do not, more recently, the meaning has changed to mean access to broadband Internet connections or ICT services. The important term is *connection*. In spite of the rapid evolution of telecommunications, there are still regions of the world that are isolated from major population centres and Internet connections. Over the next decade, the digital divide is expected to grow smaller due to innovative solutions that utilize satellite and wireless technology. To understand the extent of the digital divide, it is useful to identify services and hardware that focus on reducing the digital divide among the four key ICT (Kelly, 2005) domains listed below.

1. **Fixed line telephone networks**: These form the main telecommunication infrastructure. The digital gap fell from 14 times to 5 times greater, in the decade between 1992 and 2003, as economies such as China and Vietnam greatly expanded their fixed-line networks.
2. **Mobile telephones:** The reduction is even more dramatic here. Mobile phones took around 20 years to reach their first billion users; however, it only required 4 years (2002-2005) to surpass the two billion subscriber mark. During the decade, the digital gap was reduced from 30 times to 5 times. Since 2002, mobile phones have outnumbered fixed-lines.
3. **Personal Computers (PC):** Unfortunately, this area is not narrowing as quickly as the communications categories. Although the level of penetration in developing countries has risen from one PC for every 243 inhabitants in 1992 to one for every 29 in 2003, this is still a long way behind the rate of one PC for every 2.2 inhabitants in developed economies. The digital divide is wider in PC ownership than any of the other indicators tracked here. One reason for this is the high cost of acquisition and of ownership (e.g., upgrading memory, software, etc.) of a personal computer. The advent of low-cost computers such as the $100 laptops, together with the widespread adoption of free and open-source software, may help to reduce the digital divide for PCs.
4. **Internet:** The Internet has had the most dramatic reduction in narrowing the digital divide. Between 1992 and 2003, the gap between developed and developing countries narrowed from 41 times more to 9 times more. Although there are fewer estimated Internet users than PCs in developed countries (44.8 and 44.9 per 100 inhabitants, respectively), in developing countries there are more Internet users than PCs (5.1 and 3.4 per 100 inhabitants). This suggests the significance of Internet access from cyber cafes, post-offices, schools, universities, and other public internet access centres (PIACs) in the developing world.

The evidence is clear that for the four ICT domains identified, the digital divide is narrowing as diffusion spreads, and in most cases at an accelerated pace. The concept of a narrowing digital divide is an oxymoron, as the technological changes are also accelerating, and ICT innovations are constantly being invented and deployed. This will lead to the impression that the digital divide is actually expanding faster. As more bandwidth becomes available, the Internet applications to use the bandwidth follow, and is eventually diffused to developing nations, and each succeeding ICT innovation starts the diffusion.

The Importance of Connectivity

The Internet and the World Wide Web symbolize the convergence of media infrastructure and services. Unlike any other medium, it has empowered the average user to become an originator of content and services, in addition to being a consumer. Used primarily in the early days for e-mail, file transfer, and remote login applications, the Internet has graduated to many innovative applications in areas as diverse as distance education, telemedicine, e-commerce, banking, and corporate communication, to name only a few. The browser programs that help the users access the Internet have sophisticated features, such as audio and video streaming and other multimedia applications. At present, the bandwidth constraints are the prime limitation which, once resolved, will allow more innovative, interactive broadband applications (UNESCAP-Report, 2002).

The Internet has had a profound impact on educational pedagogy, and innovative e-learning techniques are being implemented round the globe. However, the digital divide has a profound effect on how these techniques are implemented outside the Western world. Private individuals' homes in developed nations have more bandwidth than most African universities. Further, the type of access American and European households can receive for $100 per month would cost African universities $10,000 for the same time period (Partnership, 2002) African universities are significantly handicapped in their ability to provide modern educational opportunities to their students.

Bandwidth is central to the operation of university life, therefore improving access to a greater number and cheaper bandwidths. Managing it efficiently, and using it appropriately for teaching, learning, and research, has become a critical issue for African universities (Partnership, 2002)

In Africa, most institutions use dial-up, leased lines for connectivity and their average bandwidth is often significantly slower than residential service in the United States. The lines are generally used at 100% capacity, 60% of the time, and the average bandwidth cost is US$5.46/kbps/month (Steiner, Tirivaya, Jensen, & Gakio, 2004). To compound the issues, universities often do not receive the bandwidth promised by their local service provider. Shared lines and lack of monitoring equipment mean African universities pay more for less.

Many end-user service providers already exist in most developing nations, and the major commercial ones, in particular, are expanding in both customer numbers and service offerings. However, rural and low-income markets are underserved and likely to remain so for the foreseeable future. The reasons are similar to those that for a long time limited telephone access to rural and low-income communities in developed nations. Customers are scattered, it is expensive to reach, and often high-priced services are not affordable. Rural, isolated geographical areas offer slim pickings to large commercial service providers.

The challenge is to create models that begin by building on existing local resources, and can be scaled to expand as local markets grow. The community of regional and local service providers needs to come up with imaginative approaches to encourage this expansion. The groups most realistically able to step up to provide these services are the ones with experience in these markets. Local institutions who know their own

Table 1. Bandwidth in African universities (Source: Bandwidth Task Force)

Current bandwidth utilization in selected universities	
Institution	Bandwidth Utilization: Kbps up/Kbps down
University of Dar es Salaam (Tanzania)	256/512
Makerere University (Uganda)	1,280/2,500
Eduardo Mondlane University (Mozambique)	384/1,000
Bayero University (Nigeria)	64/128
Obafemi Awolowo University (Nigeria)	128/256
University of Ibadan (Nigeria)	56/200
University of Jos (Nigeria)	64/128
University of Ghana	512/1,024

communities with backing from local private investments are ideal. This approach is supported by the Organization for Economic Cooperation and Development (OECD). An OECD report concluded official donor assistance has *abandoned support* for communications infrastructure in developing countries (OECD, 2005). Focus has shifted to social investments to reduce poverty. The OECD states that assistance from private sector organizations is required to help cover the shortfall by building, delivering, and operating communications services. The approach discussed in the next section to create a global broadband satellite system is poised to lead the way in this type of initiative by building a communications infrastructure and ensuring a strong partnering between the humanitarian and learning organizations in the local community.

VISION OF A WORLD WITHOUT A DIVIDE: GLOBAL BROADBAND SATELLITE SYSTEM (GBSS)

In comparison to fixed cable solutions, satellite technology provides the advantage of universal coverage, point-to-multipoint transmission capabilities, seamless transmission, independence from terrestrial infrastructure, and rapid deployment. Satellite technology can deliver broadband Internet services to developing countries and to rural and remote areas in developed countries where terrestrial infrastructure is practically nonexistent or its rollout is prohibitive in an affordable and timely basis (ITSO, 2002).

Satellites are considered to be one of the keys to reducing the digital divide, especially in the more isolated areas, in much the same way as they brought telephone services. In fact, as far as Internet access is concerned, there are geographical areas where there is just no alternative to satellites. (EADS-Space, 2005). The ITU has a vision to address the communication infrastructure imbalance affecting underdeveloped countries by implementing a global, connected information and communication society. The initiative would be through an innovative public-private sector partnership that would lead to the establishment of a global broadband satellite system considering the deployment benefits over other communication support infrastructures. Cost permitting, a robust and affordable universal broadband infrastructure could be implemented within a reasonable timeframe. This initiative would lead to the development of a new market for broadband equipment and services through the adoption of a universal technical standard for user terminals, effective access to the geostationary orbital and frequency spectrum resources, and a minimal procompetitive regulatory environment. The idea of GBSS can be traced back to a United Nations Resolution below:

Resolution 1721 (XVI) of the United Nations General Assembly sets forth the principle of the availability of satellite telecommunications to the nations of the world on a global and nondiscriminatory basis.

In the 1960s, while trying to fulfil this resolution, the international community created the International Telecommunications Satellite Organization. The objective of this group was to operate a single global commercial telecommunications satellite system to provide expanded basic telecommunications services to all regions of the world on a nondiscriminatory basis. In theory, their task was to ensure that developed and developing countries would benefit equally from emerging satellite technologies. The same political will was behind the establishment of other government-operated international and regional systems, such as Interspoutnik, Inmarsat, Eutelsat, and Arabsat, subsequently joined by dozens of successful private satellite systems.

When you consider the fact that there are over 200 commercial satellites in geostationary orbit that cover the entire planet, there should not be a digital divide, as there is no region on the planet without reasonable coverage. There are 69 countries, accounting for more than 60% of the world population, currently relying on satellites for their domestic and international telecommunication services. An investigation by the ITU into the reasons behind the mismatch of existing global satellite capacity and the need for communication and Internet services revealed the following (Toumi, 2004).

a. User terminals are expensive and cumbersome. Traditionally based on proprietary standards, satellite systems are not interoperable. This is an obstacle to the economies of scale required for mass production of low-cost equipment.
b. The technical and operational bases guiding allocation of frequency spectrum and geostationary orbital slots for fixed satellite services (FSS) are not optimized for use by inexpensive terminals accessing broadband services. In effect, these allocations were determined based on a small number of sufficiently large terminals (earth stations) coexisting with the terrestrial stations and taking up almost all frequency bands.
c. The *passband* transmission capacity is costly. Satellite operators face tough administrative, technical, and regulatory hurdles to gain access to domestic markets. Restrictions on user terminals, including utilization taxes and fees, complex and costly type approval procedures, reluctance to use the so-called network, and *head-end* or *gateway* stations located outside the national territory, to mention just a few, are governed by agreements satisfactory to governments, operators, equipment manufacturers, and service providers.

Barriers to the Broadband ICT's Access Deployment

Broadband ICT access has been primarily deployed in developed countries in urban areas due to the economics and limitation of existing technologies. In the initial phases of the concept development stage, the service providers and the operators are always inclined to serve the most highly populated regions, which normally translate to rich suburbs and cities, where most of the potential customers reside.

There are two factors that affect the diffusion of ICT. These are called the demand and cost factors. The first aspect driving adoption of ICT in developed countries and urban areas is the hunger for more bandwidth to support multimedia applications. Rural areas in developing countries prefer voice communication technologies and applications with a sluggish progression toward ICTs. Developed countries own telecommunication infrastructure and have the means to pool financial resources to invest and pay for new services, while developing countries lack any form of basic infrastructures such as telecommunication, electricity supply, and roads. The developing nations also experience great difficulty mobilizing the required financial resources. This concept can be described as the *demand factor*.

The next key factor driving ICT adoption is the available technologies. Existing wired or wireless technologies have intrinsic limitations either in performances or in capacity, for example, the 6 km maximum distance from the exchange for ADSL or the line of site (LOS) customer premise equipment (CPE) location from the base station for wireless access. If and when these limitations become eliminated using other complementary backbones or equipment such as WiMax, the additional expense for this new equipment, along with their deployment and operation, directly impacts the business model. This factor is referred to as the *cost factor*. To achieve a cost effective deployment of ICT in developing nations, both the demand and cost factors need to be addressed, as they are major barriers to broadband access. One approach used to address this problem is using an international partnership model.

An International Partnership Approach for Resolving the Global Digital Divide

The ITU recently released a statement that indicated the use of partnership as the key to connecting communities. The initiative is called *connect the world*. It is a global multistakeholder effort established and working within the remit of the WSIS to encourage partnerships to bridge the digital divide. The main objective of this initiative is to provide ICT to people worldwide, of whom making a simple telephone call remains impossible.

Currently, the main driver behind the concept of a GBSS is the WSIS. This initiative definitely has political clout. It was attended by 50 heads of state/government and vice-presidents, 82 ministers, and 26 vice-ministers from 175 countries, as well as high-level representatives from international organizations, the private sector, and civil society who attended the Geneva Phase of WSIS and gave political support to the Geneva Declaration of Principles and Geneva Plan of Action (ITU, 2005) that were adopted on December 12, 2003. More than 11,000 participants from 175 countries attended the summit and related events. These types of initiatives are required because research suggests that over 800,000 villages, or 30% of all villages worldwide, do not have any form of connection. Research also reveals that 942 million people living in the developed nations have five time better access to fixed and mobile services, nine times better access to Internet services, and own 13 times more PCs than 85% of the world's population living in low and lower-middle income countries (Cayla, Cohen, & Guigon, 2005).

The trend for telecommunications' markets has been liberalization and privatization of the telecommunications market to free themselves of their public service obligations. This trend makes it almost impossible to visualize a global broadband service infrastructure owned and financed by the public sector or a solitary operator. Unfortunately, this means that any project aimed at providing universal broadband services must depend on global market forces and the voluntary participation of network and device operators.

Formation of the GBSS Association

The most sensible approach to attain a global broadband service infrastructure is to develop a model similar to the digital mobile system GSM system. There are more than 1.7 billion people who use GSM networks and services. This equates to one quarter of the world's population. GSM is one of the greatest technological success stories of our age, on a par with the Internet. In fact, there are more people with GSM mobile phones than have Internet access. GSM networks are represented by the GSM Association (GSMA). This global trade association represents more than 690 GSM mobile phone operators across 213 territories and countries of the world. In addition, more than 160 manufacturers and suppliers support the association's initiatives as associate members. The primary goals of the GSMA are to ensure that mobile phones and wireless services work globally and are easily accessible, enhancing their value to individual customers and national economies, while creating new business opportunities for operators and their suppliers.

Role of the International Governments

Using the principles of the GSMA as a guiding model, there are fundamental issues that need to be addressed before we can even consider providing satellite high-speed Internet services through individual or community low-cost, small-dish platforms. This is a very elaborate project, and governments and private organizations, along with

Figure 1. Actors for creation of global broadband satellite system

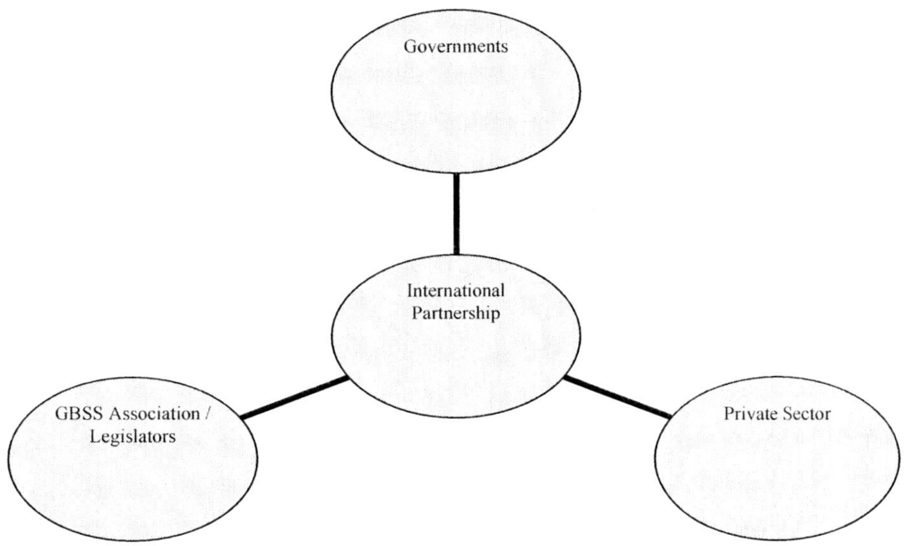

international bodies will have to develop an attractive technical and regulatory framework. Some of the components that fall under the responsibility of the governments would include:

- Identifying internationally harmonized radio-frequency bands and orbital locations that can assure global coverage, suitable for the provision of high-speed Internet services.
- In the likely situation that harmonized frequencies cannot be identified globally in "unplanned" bands, the allotment plans for the direct broadcasting satellite services (AP30 and 30A) and fixed satellite services (AP30B), with some 1500 MHz, will be an appropriate resource for this purpose.
- The GBSS body would need to request that governments for member countries should make available part of their national allocation, without compromising their rights under the Plans.[1]

One of the most important roles governments will also be responsible for is establishing a harmonized and minimal satellite telecommunications regulatory framework that promotes competition and broadband services. The regulatory framework would include the following four responsibilities:

1. Granting *landing rights* to all satellite operators participating in this initiative to provide high-speed Internet services
2. Licensing at least two service providers in each country to provide high-speed Internet services
3. Ensuring the principle of interoperability among all satellite systems providing high-speed services
4. Ensuring competition among operators providing high-speed services
5. Ensuring financial support where market conditions are such as to hinder access to high-speed Internet services

The Role of the Private Sector: Building Infrastructure and Markets

The primary role of the private sector is to develop innovative technical infrastructures and build strong markets for the products and services. The commitment from governments to create a global market for satellite broadband services will introduce new business opportunities for the private sector. It is imperative that the telecommunications industry, especially the satellite operators and device manufactures, should be actively involved in the design and development of a global infrastructure.

To ensure the creation of a harmonized global market, it will be important that GBSS operators have access to prime frequency bands and ideal orbital locations. This will allow them to deliver high-speed services via small-dish and low-cost user terminals. The role that needs to be undertaken by the private sector is described below:

- Harmonize universal technical standards for user terminals to access high-speed Internet service; the aim of this is to promote mass production of simple, low-cost terminals.
- Develop interoperability standards between broadband satellite networks to facilitate easy content exchange; the standards should also take account of ITU Resolution 101 on the development of the Internet protocol (IP) for public telecommunications networks.
- Use the orbital locations and RF spectrum resources identified for the global broadband satellite infrastructure, exclusively to provide broadband services in conformity with the universal technical standard specifications.

This initiative is more than just a system to resolve the digital divide. It also represents opportunities for the expansion of the telecommunications industry, an industry which is constantly investigating new prospects to ensure its survival and progress. Today's satellite telecommunication technologies are capable of providing universal access to high-speed Internet services within a reasonable timeframe at high cost. The GBBS approach will improve the access, provide more innovative features and content and become more affordable. This objective can only be attained by forging a new type of partnership between the public and private sectors. It is important to understand that providing a GBBS is merely providing the backbone network infrastructure. Ground networks will still be required to connect communities. One emerging technology that could be used to distribute the broadband signal over a large area is called world interoperability for microwave access (WiMax).

CONVERGENCE OF SATELLITE-WIMAX TECHNOLOGY FOR ADDRESSING THE DIVIDE

WiMax is a standards-based wireless technology that provides high-throughput broadband connections over long distances. IEEE 802.16 provides Internet connectivity up to 50 km (31 miles) of linear service area range and allows connectivity between users without a direct line of sight. It is also anticipated that WiMax will allow interpenetration of broadband services such as VoIP, video, and Internet access, simultaneously.

The technology industry as a whole is committed to addressing the global digital divide, and there are numerous organizations and initiatives, such as WIMAX Forum, ITU, Intel, Microsoft, and GSMA, that are committed to reducing the technology disparity. The WiMax forum (Cayla et al., 2005) in particular is committed to delivering an innovative wireless technology standard to address the new millennium development goals aiming at a *global partnership for development* (Goal 8), and more specifically through Target 18:

In cooperation with the private sector, make available the benefits of new technologies, especially Information and Communications. And the objective that: everyone can create, access, utilize and share the information and knowledge, enabling individuals, communities and people to achieve their full potential and improve their quality of life in a sustainable manner.

The new WiMax wireless technology will adequately address the following WSIS goals for 2015:

- **Target 1:** To connect villages with ICT and establish Community Access Points, knowing that it is estimated that 1.5 million villages in developing nations remain unconnected to telephone networks
- **Target 10:** To ensure that more than half the world's inhabitants have access to ICT, knowing that the total number of estimated Internet users in 2002 was around 600 million, or just under 10% of the world's population

In areas without preexisting physical cable or telephone networks, WiMax could allow access between anyone within range of each other. Home units the size of a paperback that can provide both phone and network connection points are already available and easy to install. Intel is already working on creating processors that support WiMax. PanAmSat, the satellite communications company recently called satellite-delivered WiMax "the future for handheld devices." At the WiMax Forum Plenary in Vancouver 2006, PanAmSat used WiMax to deliver the first-ever live video sent by satellite to a handheld device. PanAmSat anticipates that mobile phone, PDA, and laptop users will begin to access the Internet over satellite-based WiMax connections, and also foresees the technology as a way to deliver IP-TV throughout.

WiMax is considered to be a cost-effective solution for remote deployment; however, it is definitely not limited to such applications, and may also be an answer to expensive urban deployments. Due to the lack of wired infrastructure in most developing countries, the costs to install a WiMax station in conjunction with an satellite hub will be minuscule in comparison to developing a wired solution. The wide, flat expanses, rugged terrain, and low population density is ideal for WiMax deployment, and its current diametrical range of 30 miles. For countries that have skipped wired infrastructure because of inhibitive costs and unsympathetic geography, WiMax can enhance wireless infrastructure in an inexpensive, decentralized, deployment-friendly, and effective manner (Cayla et al., 2005).

Satellite broadband converged with WiMax technology provides a very strong technological offering, with the potential to cover the globe. Unfortunately, having good technology does not ensure that the people who need it the most can afford the technology or understand the full potential. It is important to have viable and sustainable access models. The model proposed in the next section involves the formation of an alliance called coalition of space internet providers (COSIP) and is based on a proven methodology.

LOCAL PARTNERSHIP APPROACH FOR DIGITAL INCLUSION: OVERVIEW OF COSIP MODEL

This section proposes the concept of using local initiatives to increase demand for new ICT services. The solution proposed involves the formation of an alliance called coalition of space internet providers (COSIP). This model exploits ideas developed over a decade ago to successfully aid the diffusion of Internet connectivity to communities in rural North America. The solution involves identifying a local institution and creating infrastructures that build on existing local resources and expertise. This model can be scaled so as to expand as local markets needs expansion. The COSIP model is based on the Local information utility (LIU) proposed in 1994 (Clement et al., 1996). The idea of a LIU was created by CICNet Rural. The LIU was built around one or more local organizations, such as a school district, a community college, or a public library.

The coalition will provide a means to access health and educational resources to a rural community using one of the described space technology configurations. The technological configuration will be dependant on the technical requirement, size of community, and the purpose of the connection (education, telemedicine, community engagement, etc.). There will also be the opportunity for Internet connectivity to be extended to other groups in the community or to a telecenter which could serve the ICT needs of the local community. The resulting system must provide access to local and community information resources. As experience in localities grows, project stakeholders will need to broaden their information holdings and come to integrate educational, business, cultural, and governmental purposes and information within their structures.

All initiatives similar to COSIP, which aim to implement ICT services to bridge the digital divide, must aim to address the following issues (Gurstein, 2000):

- Provide support for a multiplicity of usage roles involving the creation, dissemination, and retrieval of information
- Address the full range of possible users and the diversity of their life situations
- Recognize the interplay of social and technical dimensions in infrastructure development
- Encompass both conventional and new media

Figure 2. Coalition of Space Information Providers (COSIP) model

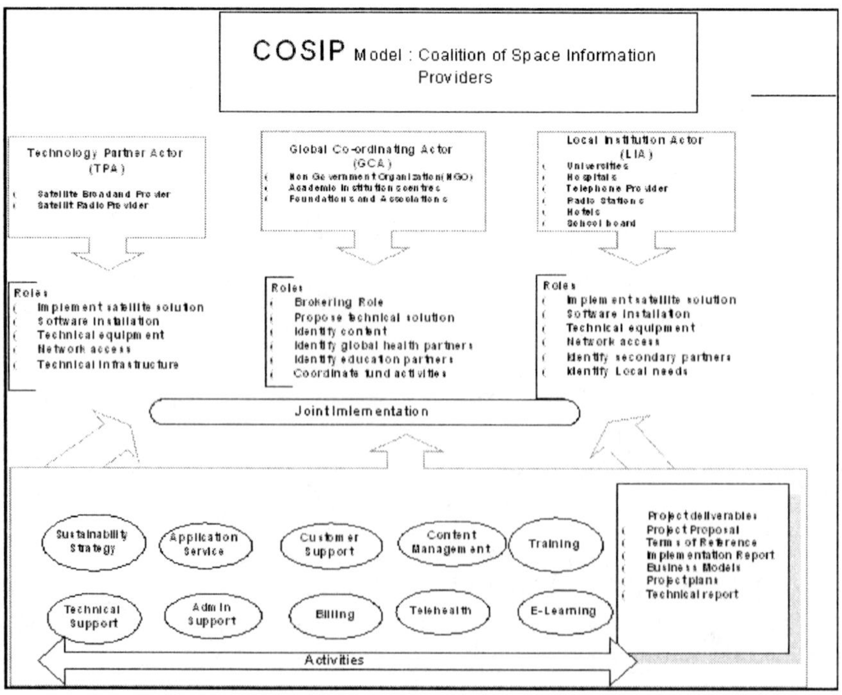

- Highlight *access gaps* and social forces likely to be left out by market forces
- Help to identify essential services

The growth of the Internet as a means of providing communication, education, or health services offers an opportunity for new, agile groups of people to come together and work from widely dispersed geographical locations around the globe. Education is a very important aspect. The perceived potential of computer-based networking to transform primary, secondary, and basic health education will eventually lead to widespread networking for schools. Higher educational institutions are realizing opportunities to extend their campuses by using radio, computer networking, and video technologies to impart educational content, in effect moving the universities closer to the community for the purpose of continuing lifelong education.

Successful implementation of the COSIP Model will certainly contribute to the achievements in the field of ICT, as well as in other priority areas, such as poverty alleviation, managing globalization, and addressing emerging social issues. Using the COSIP model, each interested local community will have to find their own motivations for investing in community networking. The opportunity to improve the economic prospects of many localities, coupled with educational institutions' drive to improve educational services and the ability of local health care providers to form telemedical partnerships with international health care organizations, can serve as overall motivating forces.

This type of initiative will require the creation of a partnership consisting of three focal actors; technology partner actor (TPA), local institutional actor (LIA), and global coordinating actors (GCA). These roles will be explained in more detail in the next section.

COSIP Actors' Roles and Components

To successfully implement the COSIP model, there are three important groups of actors. The local institutional actor, the satellite technology partner, and the gGlobal coordinating actor.

Local Institutional Actor (LIA)

The LIA are normally the primary actor. The primary actor is the focal point of the relationship. The primary actor can be any local institution or group, such as a school district, university, community college, public library, business group, community coalition, or even a local communication service provider such as a telephone, radio station, or television company. The plan should be implemented so as to adapt to local conditions and growth patterns.

Satellite Technology Partner Actor (STP)

The STP will use its considerable technical assets and knowledge to build in the air what does not exist on the ground in developing countries. Satellite and broadband wireless technologies will be blended to provide low cost, secure, and easy to use connectivity to locations not readily accessible by wired, land-based systems.

Traditionally, satellites are used to link a facility in a developed country to a particular location in a developing country. From that location, a wired network is built to provide connectivity to computer users. Often, connectivity is limited to a single location such as a university, hospital, or cyber café. The solution being proposed by the COSIP model builds on this concept to extend the coverage to sites in the vicinity of that community. The coverage can be extended depending on the requirements of the COSIP group members. Satellite and wireless technologies can be combined

Figure 3. Primary location=university secondary location=school (nonprofit making)

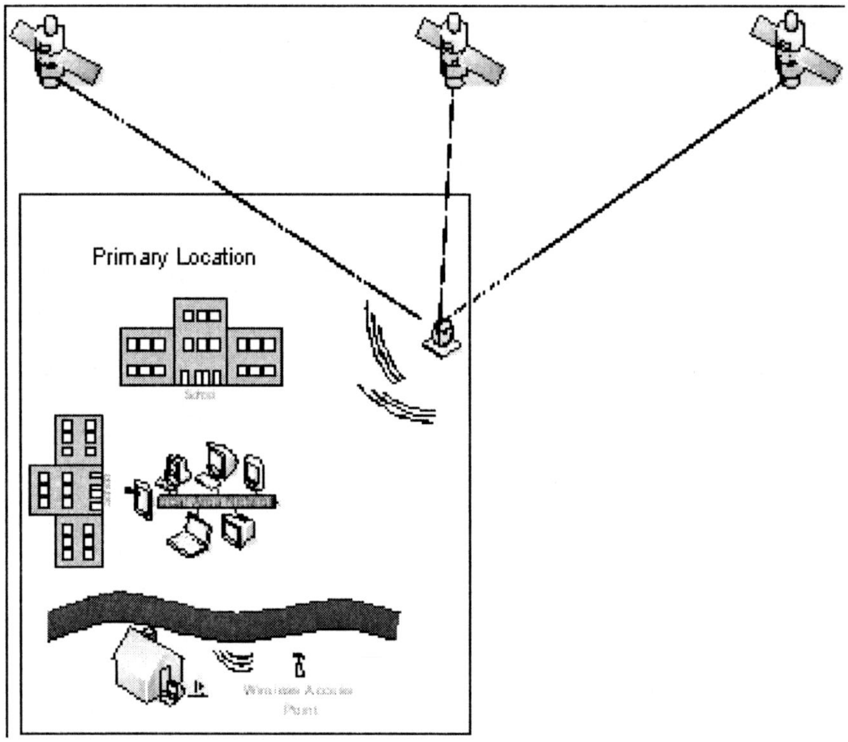

Figure 4. Primary location=university secondary location=school (nonprofit making)

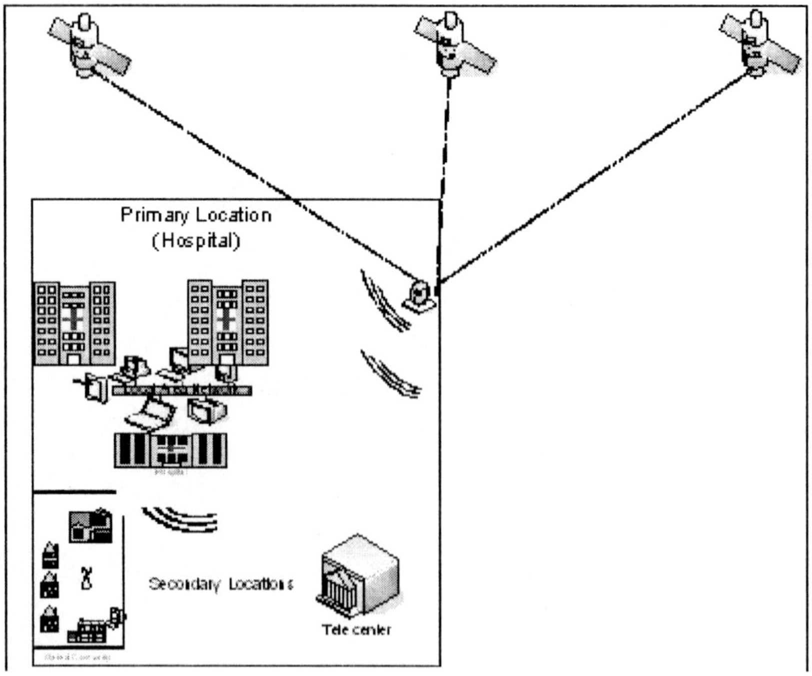

to provide fast, reliable, and secure connectivity by establishing "hotspots" that are independent of location and wired networks. See Figures 3 and 4 for example configurations.

Global Coordinating Actors (GCA)

The GCA takes on the role of a broker, responsible for validating the LIA's claim for resources and matching it with the capabilities of one of the STPs. They will also be responsible for coordinating the local organizations along with regional service providers providing guidance and advice. Any applications for funds will be via this group, and initial payments to the TPA for equipment, training, and support will be handled by the GCA. To achieve the goal and support the alliance objectives, the GCA, in conjunction with service providers, needs to play a brokering role, developing technical, human, and capital resources to help local groups plan and implement realistically.

Components of COSIP

The local institutional actor LIAs may, if they so choose, resell services to other users or user groups in the community. This approach is encouraged to aid sustainability of the project. The LIA also has the potential to create private information exchanges (PIX), purpose-driven connections between local groups to support more efficient means of local resource access. Providing the connectivity is not enough. The core components that must be considered include customer support, user training, the right mix of services, the proper information infrastructure, and content.

Support and Training

Support is critical to the success of the project in the community, and the agreement between regional service providers and local customers should place special emphasis on the customer's likely needs to provide a robust service package. The models implemented must define locally appropriate user training, both entry-level and advanced. Regional service providers should work with the LIA to identify training sources for local trainers or broker referrals to knowledgeable resources, often at postsecondary educational institutions in the region. The LIA should also define a help desk capability to answer service questions from users, and should provide a centralized referral capability for networking consultants, trainers, and other experts.

Application Services

The range of services an LIA will provide can include electronic mail, news, Web and Gopher servers, internet relay chat (IRC), video conferencing, voice over IP (VoIP), mailing list management, and caching applications for Web client and server response speedup. Other services can be provided on the basis of an appropriate bandwidth infrastructure (i.e., higher bandwidth may permit video applications).

Technical Infrastructure

The TPA will need to supply the LIA with a basic package of middleware and services, and should be responsible for their coordination and for conducting ongoing, standards-based research and development. The capabilities should include:

- Authentication services (i.e., verification that a given user is allowed access)
- Authorization services (i.e., verification that a user can invoke a given application)
- Call detail and settlement services (for access from remote LIAs)
- Domain name service
- Accounting and financial support services

If the components above are considered, the chances of success are increased. It is important to consider models such as the COSIP model due to the fact that these initiatives are being implemented in developing nations with limited resources, knowledge, and technical competencies. It is also likely that government and local official are corrupt, which would mandate that the concept of a GCA is utilized to avoid fund mismanagement.

CONCLUSION

The availability of ICT services and applications are considered to be the driving force for economic, social, and technical development. The convenience of a good infrastructure facilitates the delivery of high speed Internet, which transports important services such as education, communication, health, telecommuting, electronic commerce, and e-government services. The development of ICT applications promises a ray of hope to the developing countries; however, the prospect of accelerated growth still seems to be far away. It is imperative to facilitate the emergence of a global broadband satellite system capable of providing high speed Internet access on a global and nondiscriminatory basis, in accordance with the provisions of United Nations General Assembly Resolution 1721 (XVI), as well as the International Telecommunication Union Resolution 64, which mandates nondiscriminatory access to modern telecommunications facilities and services. Once a global system is in place, the next challenge would be to encourage deployment in the local vicinities using models such as the coalition of space Internet providers. This type of approach is necessary to ensure that a sustainable model is implemented at the grass root facilitating the use of appropriate equipment, devices, services, and business models.

REFERENCES

Cayla, G., Cohen, S., & Guigon, D. (2005). *WiMax: An efficient tool to bridge the digital divide.* Retrieved February 22, 2007, from www.wimaxforum.org

Clement, J., Holbrook, P., & Staman, M. (Eds.). (1996, June 25-28). Extending Internet access to rural areas and small communities in the upper Midwestern United States. In *Proceedings of the INET'96 International Conference.* Montreal, Canada, http://www.isoc.org/inet96/proceedings/e5/e5_4.htm

EADS-Space. (2005). *Orbiting Internet satellites to bridge the digital divide.* Retrieved February 22, 2007, from http://www.space.eads.net/press-center/editorial-features/428-en

Gurstein, M. (2000). Community informatics: Enabling community uses of information and communications technology. In M. Gurstein (Ed.), *Community informatics: Enabling community uses of information and communications technology* (pp. 1-30). Hershey, PA: Idea Group Publishing.

Internet Usage Statistics for Africa. (2006). *Africa Internet usage and population stats.* Retrived November 2006, from http://www.internetworldstats.com/stats1.htm

ITSO. (2002). *Global broadband satellite infrastructure initiative.* Retrieved February 22, 2007, from http://www.itso.int/php_docs/BroadbandInitiative-FINAL-english-17Dec02.htm

ITU. (2005). *WSIS outcome document.* Retrieved February 22, 2007, from http://www.itu.int/wsis/outcome/booklet.pdf

Kelly, T. (2005). *Twenty years of measuring the missing link.* International Telecomunication Union. Retrieved February 22, 2007, from http://www.itu.int/osg/spu/sfo/missinglink/index.html

OECD. (2005). *Financing ICTs for development efforts of DAC members*. Organization for Economic Cooperation and Development, Development Assistance Committee. Retrieved February 22, 2007, from www.oecd.org/dataoecd/41/45/34410597.pdf

Partnership. (2002). *Securing the Linchpin ICT for teaching, learning and research*. The Partnership for Higher Education in Africa. Retrieved February 22, 2007, from http://www.foundation-partnership.org/linchpin/index.php

Steiner, R., Tirivaya, N., Jensen, M., & Gakio, K. (2004). *Connectivity in African tertiary institutions*. Retrieved February 22, 2007, from http://www.ifc.org/ifcext/bsn.nsf/AttachmentsByTitle/ATICS_2004_Report/$FILE/ATICS+2004+Report.pdf

Toumi, A. (2004). *A global broadband satellite system to serve development*. International Telecomunication Union. Retrieved February 22, 2007, 2006, from http://www.itu.int/newsarchive/wrc03/documents/broadband.html

UNESCAP—Report. (2002). *Towards a policy framework for integrating space technology applications for sustainable development on the information superhighway*. Retreived November 2006, from http://unescap.org/icstd/SPACE/documents/InfoHighway/4.asp

ENDNOTES

1 UNESCAP—Report. (2002).

Section III
Space Policy and Economics

Chapter IX
Constructing the European Space Policy:
Past, Present, and Future

Lesley Jane Smith
Lueneberg University, Germany

Kay-Uwe Hörl
EADS ASTRIUM, Germany

ABSTRACT

This chapter examines the development and progress of European space policy from its beginnings over a decade ago up to today's perspectives for a European space policy. By outlining the institutional structures and responsibilities between the differing communities of the EU and ESA, it demonstrates the financial parameters behind the European space programmes and highlights accompanying structural difficulties between the institutions. Current European space efforts and the solutions adopted for cooperation are then highlighted within the background of the structures developing in the EU on security issues for Europe. The chapter concludes with a prognosis and summary of Europe's efforts to create a strong European space policy.

INTRODUCTION

This chapter assesses today's European space landscape, with a focused view on the evolution of the European Space Policy, as well as on the ongoing and projected space applications. Alongside this, it sheds some light on the historical parameters that framed Europe's first steps into the space arena, and examines the current European motivations to implement policy and develop space programmes in a particular way. Against this backdrop, the political, institutional, and industrial driving forces moulding the European Space Policy are gauged together with the challenges hindering the establishment of a coherent framework for European space activities.

The early development of space applications required colossal resources which could only be provided by the public sector, which often entertained exclusive ties with the aerospace industry (Pasco, 2003). The declining interest of governments to invest in space applications reveals the waning of the once-valued national security dimension, as well as the lack of profitability within that sector at State-level. As certain space applications increase in use and popularity with industrial and commercial applications, it is "out with the public and in with the private sector."[1] The contribution of the private sector has modified the European space landscape, its expertise helping Europe achieve its goal of independent access to space. Europe has thus succeeded in catering for the majority of space activities with the exception of autonomous human space flight.

Implementation of an effective European Space Policy faces challenges both from within and outside Europe. Curiously, Europe seems to fare better at overcoming challenges posed by other space-faring nations. It enjoys continuous success in both keeping pace and surpassing its *rivals* while promoting cooperation in space activities with countries having varied levels of competence. The difficulties that arise from inside Europe are largely due to the prevailing structures that have framed the European space sector. European space activities have been supported by a multilevel regime of bilateral and multilateral agreements, as well as *ad hoc* European programmes. Although such a regime affords much flexibility, its drawbacks are significant: it lacks the stability, certainty, and the coherence needed for a strong European Space Policy. A coherent European Space Policy requires that Europe realigns its relevant institutions, aiming to centralise its political and technical expertise. While certain European States are members of both the European Space Agency (ESA) and of the European Union (EU),[2] the two organisations are distinct from one another and their institutional structures bear significant differences due to their founding objectives. The political construction of the European Union has followed its own course, with a considerable spectrum of policies, including defence, now coming under areas of shared competence. From an apparently conflicting stance, ESA's constitution provides that the agency should operate for exclusively peaceful purposes.[3]

A number of the challenges threatening the implementation of a sound European space policy have been met, as evidenced by the cooperation between ESA and the EU and the ongoing European programmes, but much remains to be achieved. Outstanding and recurring issues relating to the intricacies of funding, institutional framework, EU enlargement, financial constraints, and European security and defence policy must be addressed.

THE EUROPEAN SPACE SECTOR

The Beginnings

From its early beginnings in space in the 1960s, Europe has come a long way by covering almost all aspects of space activities, and can be proud of its accomplishments so far. At the outset, only a limited number of European States had space capabilities, but their efforts were not comparable to those demonstrated by the United States and the Soviet Union during the Cold War era. The two superpowers were motivated by strong political, ideological, and military interests, and they were prepared to spend the money, intellectual resources, and industrial capability necessary for the pursuit of their goals (Krige & Russo, 2000b). At the time, there were no comparable political motivations in Europe, so that Europe's national space efforts were not designed to be sufficient to compete with the aggressive space programmes that had been developed.

The end of the Cold War has shifted the focus away from military and strategic space operations, such that the substantial investments in military space research and technology have yielded a wide array of systems that space-faring nations have been able to improve upon and reutilise. Europe made use of the marked shift toward increasingly scientific and industrial applications to strengthen its foundations in space activities. Investments have been progressively geared toward those activities that would benefit the public, that is, communications, environmental monitoring and protection, and international security.

International Cooperation in Space Activities

Space activities require significant funding and European public sectors cannot aspire to simultaneously fund national, ESA, and EU programmes. In fact, it is the change in the space landscape in Europe, a period in which companies have been seen to emerge with extended capabilities in system engineering, satellites, launchers, and subsystem manufacturing, which has rendered possible competition with U.S. corporations and increased its prestige as a space power. Dating from the Soviet Union's launch of *Sputnik* in 1957, space activities have conferred international prestige upon successful states. Although the world-scene has much evolved since then, space activities continue to generate similar levels of prestige, spurring countries with space potential to continuously improve their capabilities.

The relatively cordial relations entertained by and between most states foster invitations and offers for international cooperation, while the prestige of contributing to space activities remains a strong motivation factor. The prestige to be acquired is accompanied by the concrete exchange of scientific and technological know-how, results of independent research and, more importantly, the reduction in expenditure. The possibility of using costly space assets for more than one mission to the satisfaction of more than one government alleviates the burden of political justification. Space programmes emerging from Europe and the United States bear numerous similarities in both their structures and their objectives, with cooperation with other countries such as Russia,[4] Canada,[5] China,[6] India,[7] and Japan[8] envisaged or already taking place. The economic, political, and technological benefits of combining their diverse national space exploration initiatives have been recognised by the participating countries.

International Cooperation Strengthening Europe

Space exploration projects require extensive investments and common grounds can potentially be found through consultation, leading to the pooling of international resources upon cooperation. A broadened participation in space programmes can lead to increased burden sharing and makes financial sense. Europe foresees the development of such programmes, offering other states the opportunity to participate in its own space projects often on the basis of reciprocity.

A noteworthy example is that of China's 5% investment[9] in the European Galileo satellite navigation system (Galileo), while agreements regarding participation by India, Israel and Ukraine have been reached and cooperation with further countries is to be considered (European Commission [EC], press release, September 7, 2005). Europe welcomes this investment[10] which contributes to the achievement of its own navigation system, while China gains access to space technology and expertise.[11] Sino-European cooperation in space extends to further fields including telecommunications[12] and Earth observation[13].

Europe is also inclined to consider cooperation in other space projects, with international partners being invited to participate in the global monitoring for environment and security (GMES) initiative. Europe's GMES will be an essential pillar of the coordinated and sustained global

Earth observation system, within the terms of collaboration foreseen at the Earth observation summit.[14]

International cooperation in these programmes reinforces the development of a strong initiative of collaboration in space activities, as evidenced by the joint study on US-European collaboration in space science (European Science Foundation [ESF] & US National Research Council [NRC], 1998). Europe strives to further foster international cooperation not only for the benefits that accrue to all participants, but also in the consolidation of its own space policy. For a number of years, European space has been restrained from projecting its full potential by the multitude of its actors (Dupas, 2001).[15] Besides the European Union (EU) and European Space Agency (ESA), numerous member states have continued with their national programmes through their own national space agencies. Bringing together ESA and the European Commission (EC), the executive arm of the EU, with the establishment of the European Space Strategy (ESS) through the EU Council Resolution (November 16, 2000) therefore marks a tremendous progress toward a sound European space policy framework.

The European space strategy could not address the multitude of issues raised in the definition and implementation of a European space policy. These are addressed progressively by the institutions themselves, with varying degrees of success, while accommodating political and budgetary constraints and security issues as well as private-sector aspirations.

EUROPEAN PROGRAMMES

The European Space Agency itself is responsible for a significant list of achievements within just over a quarter of a century of its existence,[16] and the progress made by the most space-faring European nations during the past decade in defining, planning, and implementing space programmes is considerable. The efforts and successes in all space applications include telecommunications and earth-monitoring satellites, and contributions to manned spaceflight programmes, as well as powerful rocket launches.

The European Union's progressive interest and increasing financial and technological investments in the space industry have paved the way for long-term European space ambitions. The projects to which Europe is allocating its members' contributions have a wide array of objectives and uses, indicating a marked will to strengthen the foundations of its space activities, to enhance scientific knowledge, and to strive toward achieving technology independence. A snapshot of a choice few projects (operational or in the making) provides an insight into the diverse industries and sectors involved.

Galileo Project

The Galileo satellite navigation system will be comprised of a global network of 30 satellites which will provide accurate timing and location information to users on the ground, at sea, and in the air. As well as being globally accessible, the Galileo system is designed to be interoperable with the existing Russian GLONASS and the U.S. GPS systems. A significant and improved deliberate characteristic of the Galileo system is that, unlike the U.S. GPS which is a military-run programme, it will be a civil system. Galileo will be run by a private consortium in a public private partnership (PPP) offering various levels of guaranteed service. Although the impacts of PPPs in the commercial space sector have not yet been fully analysed, they might represent a useful tool to increase private involvement and investment in the development of space systems (Organisation for Economic Co-operation and Development [OECD], 2006).

The Galileo concept bears important political considerations. The strategy for the Galileo satellite navigation system was defined in the early 1990s, and was endorsed a few years later. The initial concerns leading to the development of the Galileo system included possible impairment of Europe's sovereignty and security if its critical navigation systems were not under its own control (EC Communication, February 10, 1999). The concerns contributed to the decision that Galileo would be a civil programme under civil control (EU Council Resolution, April 5, 2001). In parallel, the satellite navigation market has experienced remarkable growth in the past few years, and the forecast for global markets and applications for navigation show that the industry had a global turnover of € 15 billion in 2001, which is predicted to rise to 140 billion euros by 2015 (Galileo Joint Undertaking, http://www.galileoju.com, 2003). Such growth encourages the Galileo initiative to prioritise and pursue applications that will generate funding through users, reaping the benefits for the European market and society.

GMES and Earth Observation Satellites

The global monitoring for environment and security initiative was proposed in 1998 and is a growing environmental research and monitoring effort. Satellites are a unique method for Earth observation and a remarkable source of information for identifying environmental problems at regional and global levels. These characteristics prompt the European Space Agency, the European Union, and their member states, as well as Eumetsat[17], to support and invest in the GMES project. The information obtained can play a central role in determining EU decision-making based on the data gathered for agricultural monitoring, change mapping, water pollution, coastal zone management, and ice monitoring.

In addition to the GMES initiative, there also exists a parallel commitment to begin building replacement satellites for the aging Envisat and ERS-2 radar satellites.[18] Further, Earth observation satellites designed for scientific missions, such as the CryoSat polar-ice-monitoring satellite and the CryoSat-2 (CryoSat recovery mission), are on Europe's space agenda (ESA, News, February 24, 2006).

Launch Vehicle Support Programmes

The development of European launch vehicles started at the beginning of the 1960s, more specifically with the French *Centre National d'Études Spatiales* (CNES). Independent access to space required Europe to have its own launcher as well as launch structures and facilities, and building a European launcher was one of ESA's first objectives. A medium-lift rocket, Ariane, was at the core of endeavours in space transportation for geostationary orbit launches. The Ariane family of rockets has progressively grown, with the Ariane 5 heavy-lift rocket being the latest addition. Furthermore, the Vega small-satellite launcher is Europe's answer to the need to send smaller satellites in low-Earth orbits, mainly for scientific and Earth observation missions. A majority of launches are carried out from Europe's own spaceport (located in French Guyana) which itself benefits from the continuous improvement of its infrastructures. Arianespace is the prime example of international cooperation in this field. Arianespace posted sales of € 1.068 billion in 2005, over 60% higher than the previous year (Arianespace, press release, April 25, 2006). Also because of the joint efforts with Russia, Arianespace not only carried out five Ariane 5 launches from the Guyana Space Centre, but also Starsem performed three Soyuz launches from the Baikonur Cosmodrome, to orbit a total of 11 satellites. Furthering this cooperation, Arianespace will soon launch Soyuz launch vehicles from the Guyana Space Centre.[19]

Space Exploration

Aurora is a European space exploration programme, aiming to formulate and implement a

Table 1. (Source: copyright © by Eurospace 2006 (www.eurospace.org)

European Space Manufacturing Industry - 2005
Geographical Turnover Distribution by Market and by Application (€M)

Country	Distribution by market			Distribution by activity					Total
	Institutional Customers	Commercial Customers	Other/ unidentified	Satellite Applica-tions	Launcher Devt and Produc-tion	Science, μG & Space Infr.	Support & Test activi-ties	Other/ unidenti-fied	
Austria	18,5	5,6	3,9	11,6	4,5	7,5	0,0	4,4	**28,0**
Belgium	51,5	49,8	10,3	38,1	35,0	19,7	10,5	8,3	**111,6**
Denmark	15,7	-	-	1,6	-	4,6	8,7	0,8	**15,7**
Finland	9,4	-	0,1	6,4	-	2,6	0,1	0,4	**9,5**
France	939,7	985,9	11,9	1 252,2	516,8	138,4	19,4	10,7	**1 937,5**
Germany	338,2	247,6	27,8	268,0	166,5	147,5	10,7	20,8	**613,7**
Ireland	2,7	1,8	0,1	0,2	2,1	-	2,2	0,1	**4,6**
Italy	547,8	152,4	32,6	311,3	211,2	165,7	14,0	30,5	**732,8**
Netherlands	56,8	10,2	5,7	24,3	14,6	31,9	0,6	1,4	**72,7**
Norway	17,1	13,7	6,2	17,8	8,1	10,3	-	0,8	**37,0**
Portugal	3,5	-	0,1	2,1	0,0	0,4	1,0	-	**3,6**
Spain	144,2	33,5	2,2	123,6	24,8	27,7	3,0	0,8	**180,0**
Sweden	62,7	26,0	2,2	51,1	29,5	10,2	-	-	**90,8**
Switzerland	41,6	35,7	1,0	11,6	43,9	19,0	0,4	3,3	**78,2**
UK	413,0	81,4	7,2	453,0	4,7	23,2	11,6	8,9	**501,5**
Total	**2 662,5**	**1 643,6**	**111,2**	**2 573,0**	**1 061,7**	**608,9**	**82,3**	**91,5**	**4 417,3**

European long-term plan for the robotic and human exploration of the solar system. By improving on the existing technologies and generating more novel ones, the Aurora programme is also designed to complement the space missions that are planned by ESA as well as those at national levels. This method benefits the European space strategy by working toward the consolidation of a coherent and unified framework for space exploration.

The space missions planned and under development by the European space sector indicate Europe's intentions to become technology-independent while promoting international cooperation. Such intentions are often prompted by strict technology transfer and export control rules imposed by certain states, such as the USA's [U.S. State Department's international traffic in arms regulations (ITAR) and U.S. Commerce Department's export administration regulations (EAR)] and the strong culture of military influence on space activities. European projects are increasingly geared toward technology advancement, knowledge enhancement, and benefiting markets and society. All of this will significantly add to European space assets upon completion. The following overview serves to describe current markets and areas of business.

INSTITUTIONS AND COMPETENCIES

ESA & EU

The European Space Agency is not in all respects equivalent to its probably better known counterpart, the U.S. National Aeronautics and Space

Administration (NASA). While NASA's birth in 1958 was directly related to the pressures of the Cold War[20] with a corresponding agenda, ESA was more the result of a coordinated European space effort from scientists. A main objective in mind prior to the inception of ESA was to create "an International organisation pooling the resources of, say, ten European countries ... to enable the scientists of Europe to make a valuable contribution to the exploration and study of outer space" (Krige & Russo, 2000a). The preamble to the convention which brought ESA officially into existence in 1975 emphasises the desire to strengthen European cooperation, for exclusively peaceful purposes, in space research and technology and their space applications.[21] The European Space Agency is not the agency of the EU, and membership of the former does not equate to membership of the latter.[22]

The European Union has evolved from the European communities and was founded to enhance political, economic, and social cooperation.[23] Member states have set up common institutions to which they delegate some of their sovereignty, so that decisions on specific matters of joint interest can be made at the European level. Europe's initial goals of establishing a common market and approximating economic policies of its member states, increasing stability and raising the standard of living, have been largely achieved with the support of structural policies financed and firmly implemented by the EU itself. The union has seen its power to set European-wide standards and regulations in a growing number of areas across its member states increase with successive policies designed to strengthen economic and social cohesion. As such, the progress and multiplication of space technologies have gradually elevated *space* to a key issue within the *economy and society* policy of the union.

Up until recently, the European Space Agency defined and applied its own policy with regard to scientific missions. However, the EU's growing interest and investment in space technology and applications have brought ESA to reorient its agenda, using its technological expertise to develop the union's space projects. With the EU increasing its number of space programmes mandating the European Space Agency as the technical expert for implementation,[24] the latter has set up the EU and industrial programmes directorate (D/EUI) within its departments. This directorate is exclusively concerned with assignments from the union.

EU and Space: Legal Basis

The EU passes legislation and manages programmes in over 30 areas, ranging from agriculture to external relations to human rights and transport. Space research and applications are relevant to a number of the areas that are determined at the European level, bearing a leverage potential in social, economic, and political projects. Apart from constituting the subject matter for a European policy in its own right, space is also relevant to areas including agriculture, energy, environment, foreign and security policy, information society, research and innovation and transport. The Nice Treaty (2001) provides the legal basis for space or space-based relevant technologies to be used in the implementation of existing EU policies.[25] The treaty establishing a constitution for Europe makes explicit reference to space alongside research and technological development as an area of shared competence,[26] giving the union the authority to implement space programmes, as long as it does not interfere with member states' activities. The constitution document contains further references to space—unprecedented in a European treaty—particularly under section 9.[27]

In EU affairs, space is categorised as part of the research and innovation policy and as such benefits from the recurring framework programmes (FP) for research.[28] The framework programme—a legal and political obligation stemming from the Amsterdam Treaty[29]—is based on the treaty establishing the European union, and serves two main strategic objectives: that of strengthening the scientific and technological bases of industry while encouraging its international competitiveness and that of promoting research activities in support of other EU policies. Europe funds space research through the fFramework programme for research and development and space has acquired a growing importance with the successive FPs.[30] The current sixth framework programme designates aeronautics and space as a thematic priority area to which is allocated € 1182 million (EC, Research, FP6, 2003). Based on the future *European space programme*, the FP7 (2007-2013) should focus on technologies for the exploitation of space in the areas of navigation, Earth observation, and telecommunications, as well as on space transport technology toward the guaranteed independent access to space for Europe, and scientific activities in space (International space station and space exploration).

Europe's space ambitions are too potent to be left solely to structures like the framework programme. During the recent years—and partly resulting from technological progress—consultations, recommendations, and regulations focusing on Europe and Space and aiming to use space applications for the benefit of European citizens have flourished. They address all aspects of the space business (EC COM (2003) 600), European competitiveness (European council meeting, October 16/17, 2003), the 'Digital Divide' (EC COM (2002) 263), radio spectrum policy (EC decision, 676/2002), electronic communications (EU guidelines, SEC (2003) 895), future investment targets in research and development (EC COM (2003) 226)[31], and more significantly, international security, for which the EU presented its ambitions in the European security strategy,[32] which was approved by EU governments.

Common Foreign and Security Policy

The benefits that space can bring to many areas are placed in the limelight, almost (understandably) outshining the advantages of using space assets in support of the EU's common foreign and security policy (CFSP). The European Commission had endorsed the need to consider space as a whole and not just as commercial applications, alongside the urgent requirement to reshape the European union's space policy, taking into account all the strategic interests involved in space activities (EC COM (1998) 34). This was later followed by the European Commission's white paper on the future of Europe in space (2003) which, among other aims, addressed the possibilities of using space technology in the practical implementation of the European security and defence policy (ESDP), a component of the CFSP.[33]

The European Union has important foundations for security and defence, having enhanced its capabilities through institutional/structural reforms. The ESDP was established in order to carry out civil and military crisis management operations (Petersberg Tasks[34]) and to prevent conflicts where the North Atlantic Treaty Organisation is not involved. The Nice Treaty[35] established permanent political and military structures for ESDP, giving a legal basis to the political and security committee (PSC), the European union military committee (EUMC), and the European union military staff (EUMS). Those structures aim to monitor the international situation and exercise political control and strategic direction of crisis management operations under the supervision of the EU council within its agreed competences.[36]

Following the Nice Treaty, the treaty establishing a constitution treaty for Europe explicitly attributes space competencies to the European union.[37]

It simultaneously provides for the establishment of a European Armaments, Research, and Military Capabilities Agency and for the implementation of "any measure needed to strengthen the industrial and technological base of the defence sector, to participate in defining a European capabilities and armaments policy, and to assist the council in evaluating the improvement of military capabilities."[38] The Headline Goal 2010,[39] reflecting the European security strategy, together with the European Defence Agency (EDA)[40], mark the commitment of member states to develop the capabilities to respond with rapid and decisive action by 2010, applying a fully coherent approach to the whole spectrum of crisis management operations covered by the treaty on the European union. The European security research programme (ESRP)[41] initiative should play a major role in the light of the FP7.[42] It will seek to bridge the gap between civil and military research and maximise the benefits of multipurpose aspects of technology. The dual nature of space systems makes their applications relevant to both the civil and defence communities, and the European union intends to avail its space technologies and assets to serve its security and defence policy.

Framework Agreement and Closer Collaboration

The concept and vision for a European space policy were initiated nearly three decades ago[43] in response to the needs of European states engaging in space research, with concrete results being achieved particularly in the past few years. As mentioned above, the multitude of actors involved in the European space sector rendered the process of drafting a coherent framework even more complex. The European Space Agency's members are not all EU member states and vice-versa, a situation that has also resulted from the union's enlargement. ESA members pursued their own individual national space objectives while in parallel participating in the agency's programmes in parallel; balancing the pursuit of national space objectives with that of joint European programmes and objectives. This method led to a remarkable series of successful space projects framed by a multilevel regime of bilateral and multilateral agreements. However, the enlarging European union with a broadening spectrum of policies already including space[44] have pushed this regime of *ad hoc* planning and management of programmes to its limits. The European Commission expressed the view that the political framework of the European union is the only adequate one to provide the appropriate conditions to reap the benefits of an extended European space policy that should also benefit the union's new member states (EC, White Paper, 2003). Altogether, various documents, propositions and reports[45] have led toward a coming together of ESA and the European community, for the achievement of a truly European space policy (Smith & Hörl, 2004). The EU and ESA have adopted the path of rapprochement in order to combine their political, social, and technological expertise and centralise available resources and funding, essential for the establishment and development of European space activities.

The lengthy process of bringing the European Space Agency and the European union to work together "officially" started with the Green and White Papers on Space in 2001, progressing with the ratification in 2004 of the ESA-EU Framework Agreement[46] (FA), which provides the legal basis for cooperation. The Framework Agreement concluded a long period of negotiation between the two organisations and represents a milestone in Europe's progress toward becoming a world leader in knowledge-based research.[47] The Framework Agreement has created a system of cooperation over a period of 4 years, allowing for a strengthening of relations between the two parties while investing in concrete projects. The main purpose of the agreement is to ensure that European space activities benefit European cohesion and economic growth, leading to a wider political, economic, scientific, environmental, and social framework more directly at the service of European citizens.

The Framework Agreement attempts to deal with *all* aspects relevant to cooperation between ESA and the EU. Critics have put forward that the relatively short document consisting of a preamble and 13 articles is not adequate to govern the relationship between two institutions having conflicting constitutional provisions and principles.[48] However, the FA cannot and does not purport to bear a solution to the institutional and operational divergences that divide the two organisations. Its provisions address the areas and overall objectives of cooperation, the rules governing the implementation of joint programmes, and the establishment of a *Space Council*, as well as the exchange of personnel, public relations, dispute settlement, and final dispositions. The cooperation agreement aims to ensure that European space activities benefit European cohesion and are carried out for the service of European citizens.

The agreement seals a commitment for the two parties to work together for the implementation of space projects that are beneficial to both, based on their specific complementary expertise. Cooperation should avoid unnecessary duplication and waste of efforts, optimising available resources. The relative lack of success experienced by *ad hoc* structures[49] had subsequently resulted in the setting-up of joint undertakings and special-purpose institutions designed to support cooperation: the benefits to be accrued from a defined collaboration between ESA and the EU outweigh the disadvantages, if any, which could arise. There is no shortage of suggestions for potential models that could be adopted for a more integrated collaboration between the two organisations. However, none of them appear to be readily acceptable to both.

BOUNDARY CONDITIONS

The crafting and implementation of Europe's space policy are influenced by and subject to conditions which themselves flow from member states. Just as Europe benefits from the early development of space activities achieved across its founding states, the prevailing political will of the various governments of the day has a potential impact on policies, including space projects. Establishing coherent policies across intergovernmental and national agencies can be a lengthy process. From as early as 1966, it had been generally agreed that Europe suffered from having too many space organisations, and that some attempt should be made to coordinate the work of European Space Research Organisation (ESRO), European Launcher Development Organisation (ELDO), and *Conférence Européenne des Télécommunications par Satellites* (CETS), as well as to make a more cost-effective use of national facilities and programmes (Krige & Russo, 2000b). For several years now Europe's major space agencies—CNES, *Deutsche Zentrum für Luft-und Raumfahrt* (DLR), *Agenzia Spaziale Italiana* (ASI), as well as ESA, have found themselves in a critical phase of adjustment and reorientation. There may be relational issues, but the troubles stem from a shift in demand and expectation (Von Kries, 2003)—directly influencing the allocation of resources)—and from an expanding European union.

Financing Space Activities

Space programmes conducted by the various national agencies in Europe[50] are subjected to the available funding and priorities set by the governing political administrations. The EU and ESA are also affected by similar financial constraints because their budgets consist of their members' contributions. For instance, after deliberations,

ESA member states and Canada agreed to provide 95% of the requested budget for 2006-2010,[51] with this result being considered as more than satisfactory and hailed as a "great success."

Space programmes run by ESA are divided into two categories; mandatory and optional. Mandatory activities consist of space science programmes and are funded by all the member states based on their respective gross national product. In addition, Member States are free to decide on the optional programmes they wish to participate in and the extent of their contribution. The European Space Agency's *modus operandi* is based on geographical return, or the *juste retour* principle: the Convention of one of ESA's predecessor, ESRO, awarded contracts competitively to the successful bidder (Krige & Russo, 2000a). Following the request that member states be guaranteed a return for their European space effort, the principle of just return was agreed upon and implemented in 1962, and subsequently to be included in the Convention of the European Space Agency. ESA now awards contributing member states with industrial contracts equivalent to their financial participation.

The EU budget is financed by the union's own resources (paid by member states) and other revenue, with space being catered for under the framework programme for research and development. The EU annual budget is decided by the Parliament and Council on the basis of proposals by the commission as to how revenues and expenditures are to be distributed among various areas of activity in the EU. The parliament president's signature finalises the approval of the agreed budget. Once the budget has been adjusted and the European Parliament has given its approval, the commission may request payment of contributions from the member states.

Financial resources deemed necessary for the fulfilment of space projects but which are not always made available hinder the application of the European space policy. Financial constraints, coupled with concerns over space budget duplication, motivate calls for increased coordination and suggestions for an ever closer working relationship (more integrated than the Framework Agreement) between the European union and the European Space Agency (Gaubert, 2006).

Closer Collaboration

The Framework Agreement remains a facilitating instrument in the EU-ESA relationship, and provides for models of cooperation[52] which the institutions may adopt in the implementation of joint projects. As a necessary condition for an integrated space policy, closer cooperation is desirable; not the least because the current setup lacks dependable decision-making or dispute resolution procedures. Although short of being the ideal solution, the four main models proposed by the Framework Agreement and described as follows do overcome institutional divergences to some extent.

i. **The management by ESA of EC space-related activities:** ESA can act as the implementing agency of the EC, upon request of the latter to undertake a given space-related operation as per its policies. ESA will have no decision-making powers, following the procedural guidelines from the EC and applying its own technical and scientific expertise. Because ESA will be acting for the EC, it will not apply the geographical return principle but is bound to apply the EC's procurement rules.
ii. **The participation by the EC in an optional programme of ESA:** In accordance with Article V.I.b of the ESA Convention, international organisations may participate in optional programmes of the European Space Agency. The European Commission could participate in the space programmes it supports, akin to other ESA Member States. Although the agreement does not detail the arrangements that are to frame such coopera-

tions, it does leave the decision to be made by a two-third majority of the participating states. The designated projects would be cofinanced by the EC and the participating ESA Member States. Whereas the ESA Convention includes methods to calculate the contribution to be brought by participating states, a different scale may have to be agreed upon between the EC and ESA.

Decision-making for the optional programmes is carried out by all the participating Member States through their representatives on a board, each carrying one vote. Participation of the EC in ESA's optional programmes will require prior arrangement of its voting rights.

iii. **The creation of bodies for the implementation of specific programmes:** The EC and ESA can jointly set up special purpose committees, whose task will be to supervise designated projects until completion. A concrete example is that of the Galileo Joint Undertaking (GJU): the GJU was founded by the EC and ESA to supervise the Galileo project for an initial term of 4 years "to ensure the unity of the administration and the financial control of the project" for the research, development, and demonstration phase of the Galileo programme, and to mobilise funds assigned to that programme. Its role is to oversee the development of the space and ground segments of the project, foster the development of applications and services, and prepare markets for Galileo services. It is also to prepare the selection of a commercial operator for the follow-up phases of the programme.

This model is an original solution and points at such bodies for pursuing initiatives complementary to research and development activities, such as the provision of services and the management of infrastructures. Once the purpose is defined, specific arrangements come into play to determine the institutional and organisational structure of the body to be created.

iv. **Activities which are coordinated, implemented, and funded by both parties:** Of course, the EC and ESA could opt for each of them to retain maximum flexibility in determining the specific arrangements for common projects. However, this will require specific requirements to be made regarding every single aspect of the joint activities, from the respective role of each party to financing the project. This model of cooperation fails to provide guiding principles aimed at facilitating negotiation.

Each of the above models raises important constraints for one or the other organisation's mode of operating, and on occasion necessitating constitutional amendments. As far as the known projects are concerned,[53] ESA is the implementing agency of the EU. Agreement remains to be reached on the path of institutional realignment of the EU and ESA. If the agency is to become the union's space agency, ESA's founding convention must be amended; removing "exclusively peaceful purposes"[54] references and provisions for *juste retour.*[55]

As seen above, the procurement of contracts according to the principle of *juste retour* provides a major incentive for member states to participate in optional programmes. However, such a practice is unknown to EU law and is incompatible with EU competition law in particular and with the World Trade Organisation (WTO) rules as implemented in the Government Procurement Agreement (EU, 1994). Article 87 of the treaty establishing the European Community Treaty (EC Treaty, 2002) limits state aids, whereas ESA member states have the assurance that their financial contributions to the optional programmes will return to their national industries as business contracts. It is unlikely that ESA's industrial policy rules would systematically benefit from the exceptions from the general prohibition of state aids contained in the EC Treaty.[56] In parallel, space industry is an important employer for highly qualified staff and it is understandable that European policy strives to support this sector. The success of this model will thus also depend on the specific arrangements that can be made to determine and apply different procurement rules to the EC contributions and to that of ESA member states. In any event, the situation of legal uncertainty must be supplanted by more consistent procedures in the long run if ESA is to progress as the technical arm of the EU.

The models of cooperation suggested from various sources[57] and embodied in the Framework Agreement point toward ESA becoming the implementing agency of the EU or the EU becoming a member of ESA, indicating that a coherent European space policy must be implemented under the aegis of one organisation.

Security and Defence

Europe recognises space as a market for new technologies, but appears hesitant regarding the priority that it should be given as a potential security and defence issue. Involving defence decision-makers in space at the European level is not an easy task and the majority of such programmes remains funded and used at the national level. But leaving European military space systems to national agencies and agreements outside the ESA framework would contradict the policy of consolidation of the European space effort (Creola, 2001). While the common space strategy addresses the role of space systems in the common European security and defence policy, ESA's *exclusively peaceful purposes* clause is one of the reasons preventing ESA members from supporting specifically military space technology as part of its activities, leaving them to instead fund defence projects nationally. But ESA already contributes indirectly (or directly) to ESDP/CFSP by developing space technology for the EU: dual-use nature of space activities means that deciding between civilian and military use is merely a political and not a technological decision.

In the European space context, *security* is generally taken in its wider sense,[58] but GMES already covers the security and protection aspects related to environmental threats.[59] Improving European military capabilities is vital for an effective common ESDP and space assets constitute an essential segment of the military means needed. Space-based capabilities are reliable and integral to any nation's national security operational doctrines and at the very least, make ground segments more effective (Kolovos, 2002). Much of the required technology, including optical reconnaissance and radar observation satellites, is in existence and operation within European states,[60] which may eventually agree to make them available to the EU.[61] In the meantime however, defence remains intergovernmental and cannot be handed over to the commission as such.

Member states' reluctance to allow security and defence issues that may affect them be subjected to laws other than national ones, is clear from their response to EU proposals concerning defence procurement. The European Commission has published a green paper on defence procurement (2004) in which it puts forward its proposals for developing a transparent and competitive European defence equipment market (EDEM). The green paper was anticipated in the commission communication "towards a European union defence equipment policy" (2003) and its aim is to improve cross-border competition in certain areas of defence procurement. Article 296 of the treaty establishing the European community allows member states to derogate from the rules of the internal market for the procurement of "arms, munitions and war material" if their "essential security interests" are at stake. Most governments have regarded the derogation clause as an automatism, passing almost all defence contracts on the basis of national procurement laws which differ considerably from country to country.

Such tendencies have subsisted in Europe for a long time, but the establishment of the European Defence Agency in 2004 illustrates member states' determination to combine the development of military capabilities with new approaches in the fields of defence research, armaments cooperation, and defence markets (EUISS, 2005). Member states had tasked the EDA with exploring the possibilities of drawing up an intergovernmental code of conduct to foster intraEuropean competition within the scope of Art. 296 EC Treaty. The resulting Code of conduct on defence procurement[62] will be the basis of the new European defence equipment market which will be launched on July 2006.[63] The regime was approved by defence ministers to cover defence equipment purchases which are exempt from the normal cross-border competition rules of the EU single market.

The shifting perceptions about defence equipment in general result from Europe's desire to develop capabilities for strategic decision-making and action, independent of other states. A growing number of European countries agree with the need to pursue military-related space programmes and recognise that while defence procurement is a sensitive area, it nonetheless remains subject to business realities and market forces, such as competitiveness and profitability. Without constituting the end goal in itself, competition in the European defence equipment market is desirable, at least to avoid unnecessary duplication of costly resources. However, the success and benefits of the new European defence equipment market cannot be assessed in the near future and the willingness of European states to include military space technologies in the EDEM remains to be seen.

EU Enlargement

In the course of its enlargement, the EU views space technologies as indispensable for the implementation of numerous policies across Europe. In many respects, space technologies are unique in their capabilities to collect and distribute information at any time and any place for every citizen (Verheugen, 2005). Ten new member states have joined the European union since May 2004 and further enlargement is on the agenda.[64] It is likely that, following in the footsteps of new member states and even of candidate countries,[65] more countries will join ESA in years to come.

Nevertheless, the issue of new entrants in the EU impacts the implementation of the European space policy at various levels, particularly at the political and financial levels. The difference in ESA and EU membership and the marked disparity in space infrastructures between the "old" and the "new" states hinder the even implementation of a space policy. Most of the new entrants have little or no space infrastructures or technological expertise. In the wake of enlargement, reducing disparities means also that the European union must act as a catalyst to raise the standards, and evaluate the best practices (EC Communication, Financial Perspectives, 2004). Within new states, it would be possible and even desirable to create a demand for space-related user applications, as well as industries for space-related components, which would generate social and economic benefits.

OUTLOOK AND CONCLUSION

In spite of a fragmented approach, the following characteristics—independent European access to space, Europe's leading position in Earth observation, navigation, science, contributions to manned space flight such as the ISS, and international cooperation and European science programmes—vouch for the important challenges that have already been overcome as part of the European Space Strategy. But the debate remains rife with scenarios of the future relationship between ESA and the EU. There is indeed justification for further rapprochement between the two institutions: operational uncertainties delay the harmonised implementation of a truly European space policy. ESA's many years of experience and independence from the EU make it the best entity to implement European space policy and the common space strategy crafted by both institutions marks their commitment. At the highest political level, the strategy recognises the importance of space for the welfare of EU member states. It also establishes ESA as *the* space agency of the enlarged Europe. Unfortunately, this effort of cooperation is not accompanied by the necessary long-term financial commitment from the EU as a whole, in contrast to individual, member states.

To successfully implement a unique space policy across its member states, the EU should confront the challenges arising from within its own borders that pertain to finance, institutional framework, security and defence commitments, and enlargement. Many EU members contribute to the ESA budget while maintaining their own national space agencies: a coherent space policy should seek to channel available funds to avoid duplication of efforts at many levels. The institutional framework in place, for instance, is not conducive to an optimal centralisation of financial, technological, and political capabilities involved. A decision defining the roles of the European Space Agency and the EC in the implementation of the space policy must be made. Traditionally, governments had interpreted their obligations to mean that ESA could not run programmes with a military content, but even this definition has been revised. It is agreed that ESA can develop space systems involving monitoring and surveillance satellites, which may be used in nonaggressive military activities.

Europe has evolving security interests and commitments - partly due to its enlargement - and should define a more appropriate security doctrine with a new focus on space and security, including the use of its space assets. The EU must address the issues of how to involve the new member states in space activities and especially how to harness the variety of technological developments, and economic and political capabilities. Ultimately, it is the responsibility of the European political decision-makers to create an environment conducive to full use of space as an indispensable tool (Plattard, 2005) in achieving the objectives already formulated at the highest European political level.

NOTES

The content of this paper reflects the views of the authors and not necessarily those of their affiliated institutions. The authors wish to thank Ms. Kamlesh Gungaphul-Brocard D.E.S.S. in preparing this paper. E-mail address: kamlesh_gungaphul@yahoo.co.uk

REFERENCES

Arianespace. (2006, April 25). *Arianespace reports 2005 results: A very successful year* (Press release). Retrieved March 2, 2007, from http://www.arianespace.com/site/news/releases/06_04_25_release_index.html

Bildt, J., Peyrelevade, J., & Späth, L. (2000). *Towards a space agency for the European Union. Report for the Director-General of the European Space Agency.* Retrieved March 2, 2007, from http://esamultimedia.esa.int/docs/annex2_wisemen.pdf

Creola, P. (2001). Some comments on the ESA/EU space strategy. *Space Policy, 17,* 87-90.

Dupas, A., Janichewski, S., Kries, W.v., & Schrogl, K.-U. (2001). A Franco-German view of Europe's ambition in space for the 21st century. *Space Policy, 17,* 103-110.

European Commission. *Financial programming and budget: Financing the budget.* Retrieved

March 2, 2007 from http://ec.europa.eu/budget/budget_detail/financing_en.htm

European Commission. *Sixth Framework Programme: 2002-2006*. Retrieved March 2, 2007, from http://europa.eu.int/comm/research/fp6/index_en.cfm?p=0_instruments

European Commission. (1999, February 10). *Communication on Galileo*. COM (1999) 54 final. Not published in the Official Journal.

European Commission. (2002). *eEurope 2005 Action Plan: An Information Society for all (Digital Divide)*. Communication from the Commission to the Council, the European Parliament, the European Economic and Social Committee and the Committee of the Regions. COM (2002) 263 final.

European Commission. (2002). *Strategic Aerospace Review for the 21st century (STAR 21)—Creating a coherent market and policy framework for a vital European industry*. Report by the European advisory group on aerospace.

European Commission. (2003). *A coherent framework for aerospace—A response to the STAR 21 report*. Communication from the Commission to the Council, the European Parliament, the European Economic and Social Committee and the Committee of the Regions.COM (2003) 600 final.

European Commission. (2003). *European Defence—Industrial and Market Issues—Towards an EU defence Equipment Policy*. Communication from the Commission to the Council, the European Parliament, the European Economic and Social Committee and the Committee of the Regions. COM (2003) 113, final.

European Commission. (2003). *Guidelines in criteria and modalities of implementation of Structural Funds in support of electronic communication (broadband access)*. Commission Staff working paper. SEC (2003) 895 EU Com July 2003.

European Commission. (2003). *Investing in research: An action plan for Europe*. Communication from the Commission to the Council, the European Parliament, the European Economic and Social Committee and the Committee of the Regions. COM (2003) 226 final.

European Commission. (2003). *White Paper on Space: A new European frontier for an expanding Union—An action plan for implementing the European space policy*. COM (2003) 673, released 11 November 2003.

European Commission. (2004). Commission decision on the implementation of the preparatory action on the enhancement of the European industrial potential in the field of security research 2004-2006. C (2004) 249 final.

European Commission. (2004). *Financial Perspectives 2007-2013*. Communication from the Commission to the Council and the European Parliament. COM (2004) 487 final.

European Commission. (2004). Green Paper on Defence Procurement. COM (2004) 608, final.

European Commission. (2005). *Report of the Panel of Experts on Space and Security*. Retrieved March 2, 2007, from: http://ec.europa.eu/comm/space/news/article_2262.pdf

European Space Agency. Convention for the Establishment of a European Space Agency, Paris, done 30 May 1975, entered into force 30 October 1980; 14 ILM 864 (1975).

European Space Agency. (2006, February 24). *ESA confirms CryoSat recovery mission*. Retrieved March 2, 2007, from http://www.esa.int/SPECIALS/Cryosat/SEM7WFMVGJE_0.html

European Space Agency. *The European Space Exploration Programme Aurora*. Retrieved March 2, 2007, from http://www.esa.int/SPECIALS/Aurora/ESA9LZPV16D_0.html

European Union. (1992, July 29). Treaty on European Union. *Official Journal of the European Communities, C 191.*

European Union. (1997, November 10). Treaty of Amsterdam amending the Treaty on European Union: The treaties establishing the European communities and certain related acts. *Official Journal of the European Communities, C 340.*

European Union. (2000, November 16). Council resolution on a European space strategy. *Official Journal of the European Communities, C 37, 23/12/2000.*

European Union. (2001, March 10). Treaty of Nice amending the Treaty on European Union, the treaties establishing the european communities and certain related acts. *Official Journal of the European Communities, C 80.*

European Union (2001, January 22). Council decision setting up the political and security committee 2001/78/CFSP. *Official Journal of the European Communities L 27/1,* 30 January 2001.

European Union. (January 22, 2001). Council decision setting up the military committee of the European Union 2001/79/CFSP. *Official Journal of the European Communities L 27/4,* 30 January 2001.

European Union. (January 22, 2001). Council decision on the establishment of the military staff of the European Union, 2001/80/CFSP. *Official Journal of the European Communities L 27/7,* 30 January 2001.

European Union. (2002). A regulatory framework for radio spectrum policy in the European Community (towards a harmonised market for space services in the EU. Decision of the European Parliament and of the Council, 676/2002/EC. *Official Journal of the European Communities L 108, 24/04/2002.*

European Union. (2002). The treaty establishing the European community. *Official Journal of the European Communities C 325,* 24 December 2002.

European Union. (2003). Council decision on the signing of the Framework Agreement between the European community and the European Space Agency. 12858/03 RECH 152 OC 589.

European Union. (2004). *Headline Goal 2010.* Retrieved March 2, 2007, from http://ue.eu.int/uedocs/cmsUpload/2010%20Headline%20Goal.pdf

European Union. (2004). Treaty establishing a constitution for Europe. *Official Journal of the European Communities C 310,* 16 December 2004.

European Union. (September 24, 2004). Competitiveness (Internal market, industry and research). Press release of the 2605th Council meeting. Brussels. Press 269. 12487/04.

European Union. (2005, September 7). *The Galileo family is further expanding: EU and India seal their agreement.* Press release IP/05/1105.

European Union Institute for Security Studies. (May 2005). *Defence Procurement in the European Union—The current debate* (Report of an EUISS Task Force. ISBN No. 92-9198-073-0).

Galileo Joint Undertaking. (2003). Business in satellite navigation: An overview of market developments and emerging applications. Retrieved March 2, 2007, from http://www.sbf.admin.ch/htm/services/publikationen/international/raumfahrt/business_in_satnav_en_march2003.pdf

Gaubert, A. (2006). Is there really any duplication in Europe's space activities? *Space Policy, 22,* 1-2.

Kolovos, A. (2002). Why Europe needs space as part of its security and defence policy. *Space Policy, 18,* 257-261.

Krige, J., & Russo, A. (2000a). The story of ESRO and ELDO 1958-1973. *A history of the*

European Space Agency 1958-1987 (vol. I). Retrieved March 2, 2007, from http://www.esa.int/esapub/sp/sp1235/sp1235v1web.pdf

Krige, J., & Russo, A. (2000b).The story of ESA 1973-1987. *A history of the European Space Agency 1958-1987* (vol. II). Retrieved March 2, 2007, from http://www.esa.int/esapub/sp/sp1235/sp1235v1web.pdf

Lindstrom, G., & Gasparini, G. (2003). *The Galileo satellite navigation system and its security implications* (Occasional Paper No. 44). Paris: European Union Institute for Security Studies.

Murray, A. (2004, March 8). *The Lisbon Scorecard IV: The status of economic reform in the enlarging EU* (Press release). Centre for European Reform (CER).

National Aeronautics and Space Administration. *A brief history of NASA: Launching NASA*. Retrieved March 2, 2007, from http://history.nasa.gov/factsheet.htm

Organisation for Economic Co-operation and Development. (2006). *Global forum on space economics: Descriptive overview.* Retrieved March 2, 2007, from http://www.oecd.org/dataoecd/4/17/36468669.pdf

Pasco, X. (2003). A new role for a new millennium? The changing nature of space activities. *Space Policy, 19,* 15-22.

Plattard, S. (2005). Going further: Propositions to build up a European Space Policy. *Eurofuture, European Space Policy Institute,* 16-19.

Smith, L. J., & Hörl, K.-U. (2004). Institutional challenges for space activities in Europe. In *Proceedings of the 47th International Institute of Space Law Colloquium*, Vancouver. IAC-04-IAA.5.13.1.04.

Verheugen, G. (2005, February 17). Speech on the importance of space. In *Proceedings of the International Space Conference*, Brussels. Retrieved March 2, 2007, from http://europa.eu.int/rapid/pressReleasesAction.do?reference=SPEECH/05/99&format=HTML

Von der Dunk, F. (2003). Towards one captain on the European spaceship: Why the EU should join ESA. *Space Policy, 19,* 83-86.

Von Kries, W. (2003). *Which future for Europe's Space Policy, 19*, 157-161.

ENDNOTES

1 This creates a number of constraints to private actors: whereas the public sector could assume all risks associated to the often dangerous space activities with a high potential for liability, private companies cannot accept such exposure. Institutional customers should therefore, when placing contracts to industry, recognise and respect this boundary condition.

2 Austria, Belgium, Denmark, Finland, France, Germany, Greece, Ireland, Italy, Luxemburg, the Netherlands, Norway, Portugal, Spain, Sweden, Switzerland, and the United Kingdom are ESA members, while Canada (since 1979), Hungary, the Czech Republic (since 2003) and Romania (since February 2006) can participate in certain projects under cooperation agreements.

3 Convention for the Establishment of a European Space Agency (hereafter ESA Convention). The ESA Convention refers to the organisation's promotion of space for "exclusively peaceful purposes" in the Preamble as well as at Art. II. (ESA, 1975).

4 Signature of a cooperation agreement on March 10, 2006, between the Federal Space Agency of the Russian Federation (Roscosmos) and ESA to enhance bilateral cooperation in space activities encompassing space applications, access to space, space exploration, and the use of the International

Space Station and space technologies development.

5 A number of cooperation agreements governing space activities have been concluded between Canada and the EU, including the declaration signed in 2002 by the Heads of State at the Canada-EU Summit in Toledo, reaffirming the shared intention of the two parties to advance cooperation in the field of Science and Technology.

6 See examples below, notes 9, 12, 13.

7 The Chandrayaan-1 Indian lunar space probe to be launched in 2007 will carry ESA-made instruments. See the Web site of the Indian Space Research Organisation (*www.isro.org*) for more details.

8 ESA is participating in the AKARI mission (Japanese infrared sky surveyor), which was launched on February 21, 2006, from the Uchinoura space centre. AKARI is the result of an international effort, developed by the Japan Aerospace Exploration Agency (JAXA) and the participation of Asian and European universities and Research Councils.

9 China has agreed to invest a total of € 200 million in the global consortium.

10 EU countries have constraints imposed by the Stability and Growth Pact (SGP) regarding public deficits (which are to be kept within the reference value of 3% of Gross Domestic Product) and as such are not able to drastically increase their public space budgets. The SGP was agreed on July 7, 1997, and consists of a Resolution and two Council Regulations. See Resolution of the European Council on the Stability and Growth Pact (Amsterdam, June 17, 1997) OJ C 236 of 02.08.1997, Council Regulation on the implementation of the excessive deficit procedure (EC) No 1467/97 OJ L 209 of 02.08.1997, and Council Regulation on strengthening of the surveillance of budgetary positions and the surveillance and coordination of economic policies (EC) No 1466/97 OJ L 209 of 02.08.1997. While there is agreement about the political benefits of international cooperation, worries subsist within the industry with regard to the uncontrolled transfer of intellectual property. See Lindstrom and Gasparini (2003), for further discussion.

11 China is unable to gain such access to U.S. space expertise because of the hefty export restrictions, like U.S. State Department's International Traffic in Arms Regulations (ITAR) and U.S. Commerce Department's Export Administration Regulations (EAR).

12 China's Sinosat 2 communications satellite was manufactured in Europe.

13 The Dragon Programme is a joint undertaking between ESA, the Ministry of Science and Technology of China, and the National Remote Sensing Centre of China. Its purpose is to stimulate increased scientific exploitation of ESA space resources within China, as well as to stimulate increased scientific cooperation in the field of Earth Observation science and technology between China and Europe.

14 At the Earth Observation Summit of July 31, 2003, in Washington, DC, more than 30 nations, international organisations, and institutions agreed to work together to forge a global, coordinated, and sustained system of Earth observations (www.earthobsesrvationsummit.gov).

15 At the international level, space actors include EUMETSAT (The European Organisation for the Exploitation of Meteorological Satellites) and EUTELSAT (The European Telecommunications Satellite Organisation), an intergovernmental organisation until 2001, when it was incorporated as a private company. These two specialised user organisations prompt different mem-

berships, strategies, and decision-making mechanisms.

16 ESA was established in 1975 with 11 founding members.

17 See above, note 15.

18 Europe's environment satellites. Envisat was launched in March 2002 with an expected operational cycle of 5 years.

19 See above, note 4.

20 After World War II, the United States and the Soviet Union were engaged in the Cold War, a broad contest over the ideologies and allegiances of the nonaligned nations. During this period, space exploration emerged as a major area of contest and became known as the space race.

21 See above, note 3.

22 EU Member States which are not also ESA members are: Cyprus, Czech Republic, Estonia, Hungary, Latvia, Lithuania, Malta, Poland, Slovakia, and Slovenia. The Czech Republic and Hungary are ESA European cooperating states by virtue of cooperation agreements.

23 Formally created in 1992 with the Treaty on European Union (Maastricht Treaty)

24 Space-related projects benefit from increasing investment; from € 200 million under the 3rd Framework Programme (1991-1994), to € 1 182 million under the current FP6 for Aeronautics and Space.

25 Treaty amending the Treaty on European Union, the Treaties Establishing the European Communities and Certain Related Acts (2001). See in particular Art. 70, 154, 157, 163-173.

26 See Art I-14 of the Treaty establishing a Constitution for Europe (2004).

27 See above, note 26 Art III-121 and III-254.

28 Since 1984, Research and Innovation activities of the EU are grouped in one Framework Programme.

29 Art 166(1) of the Treaty of Amsterdam (1997)

30 FP3 (1991-1994), FP4 (1995-1998), and FP5 (1998-2002) included satellite communication and Earth Observation. Since FP5, € 350 million has been allocated to space and satellite navigation.

31 3 % GDP in R&D by 2010 – Lisbon Goal. For a review of the Lisbon process, see Murray (2004).

32 EU Security Strategy endorsed by the EU Council, Dec. 2003. Various programmes in this field are underway; see for example, the ASTRO+ project. It is part of the European Commission's Preparatory Action for Security Research (PASR) initiative and is a first step to conclusively demonstrate the crucial contribution of the space segment in the field of security. It should pave the way toward a further research and technology programme, aiming to establish synergies with other Research & Technology, commercial and defence applications, and to create networking between space industry and research, to consolidate multidisciplinary applications.

33 Actions in the framework of the ESDP, as well as in the framework of the CFSP, are based on intergovernmental cooperation due to their sensitive character, and decisions in the European Union Council are taken unanimously.

34 The Petersberg Tasks include humanitarian, rescue, peacekeeping, peacemaking, and international crisis management roles developed to meet the new global problems that have arisen since the end of the Cold War. They are implemented by the Western European Union (WEU) at the express request of the EU.

35 See above, note 25.

36 See European Commission Council Decisions 2001/78/CFSP and 2001/79/CFSP. The

Western European Union has been replaced by the EU itself as an operational force in the defence field.

37 See above, note 26, Art. I-14(3) in the areas of research and technology, and Art. III-254 for the implementation of the European Space Policy.

38 See above, note 26, Art I-41(3).

39 Headline Goal 2010 approved by the General Affairs and External Relations Council on May 17, 2004, endorsed by the European Council of June 17-18, 2004. Retrieved on April 24, 2006, from: http://ue.eu.int/uedocs/cmsUpload/2010%20Headline%20Goal.pdf

40 The EDA was created on July 12, 2004. (www.eda.europa.eu).

41 A preparatory action on the "Enhancement of the European industrial potential in the field of Security research 2004-2006"

42 Budget allocation of € 65 million. Retrieved on March 20, 2006, from http://fp6uk.ost.gov.uk/fp7/page.aspx?sp=2572

43 Community participation in space research OJ (C 127), 21.5.1979, European space policy OJ (C 260), 12.10.1981, European space policy OJ (C 190), 20.7.1987, European space policy OJ (C 305), 25.11.1991, Community and space OJ (C 205), 25.7.1994.

44 See above, note 26.

45 These include the report on "Strategic Aerospace Review for the 21st century" (STAR, July 21, 2002), and the Report for the Director-General of the European Space Agency (Bildt, Peyrelevade, & Späth, 2000).

46 Council decision 12858/03 RECH 152 OC 589, on the signing of the Framework Agreement between the EC and ESA.

47 A goal enunciated in the Commission's White Paper on Space (2003).

48 See below for an analysis of the EU's competition laws vs. ESA's geographical return principles.

49 The Galileo Programme was agreed upon initially on an *ad hoc* basis, which rapidly caused complications and resulted in a 2-year delay due to the absence of a common decision-making structure and disagreements regarding the final contributions of ESA Member States.

50 ESA/EU members tally 11 national space agencies: Austrian Aeronautics and Space Agency (ALR), Belgian Federal Science Policy, Danish National Space Centre, Finnish National Technology Agency (Tekes), French *Centre National d'Études Spatiales* (CNES), German Aerospace Centre (DLR), Italian Space Agency (ASI), Portuguese *Gabinete de Relações Internacionais da Ciência e do Ensino Superior* (GRICES), Spanish *Centro para el Desarrollo Tecnológico Industrial* (CDTI), Swedish National Space Board (SNSB), and the British National Space Centre (BNSC).

51 ESA Council Meeting in Berlin, December 5-6, 2005

52 See above, note 47, Art. 4-7 Framework Agreement.

53 Galileo and GMES, among others

54 See above, note 3.

55 See above, note 3, Art. IV.

56 Art. 87(3) EC Treaty. For further discussion, see Smith & Hörl (2004).

57 For earlier discussions on this subject, see among others, Von der Dunk (2003).

58 "Security refers to combating all threats that might affect our population, our institutions, our environment, our economic infrastructure and socio-economic interests. Many threats are man-made … but the impact of major natural catastrophes [indicate that] the early developments of early warning and crisis management tools for natural disasters are also crucial"(Report of the Panel of Experts on Space and Security, 2005).

59 See above, note 14.

[60] For instance, France's Helios-2 satellites, Germany's SAR-Lupe radar observation satellite, and Spain's dedicated military satellite, Spainsat.

[61] Any assets and facilities made available will certainly be subject to satisfactory compensation.

[62] The Code of Conduct on Defence Procurement of the EU Member States participating in the European Defence Agency. The Code, approved on November 21, 2005, is a voluntary intergovernmental regime.

[63] With the participation of a majority of member states, excluding Denmark, Hungary, and Spain.

[64] Bulgaria, Croatia, the Former Yugoslav Republic of Macedonia, Romania, and Turkey are candidate countries.

[65] See above, note 2.

Chapter X
Application of Satellite Earth Observation for Improving the Implementation of Multilateral Environmental Agreements

Ikuko Kuriyama
Japan Aerospace Exploration Agency, Japan

ABSTRACT

This chapter introduces the potential of satellite Earth observation (EO) as a tool for improving the implementation of multilateral environmental agreements (MEAs). It provides the technical and legal characteristics of EO and discusses the unique advantages of EO in collecting the environmental information which is a key for effective implementation of MEAs. It also studies the challenges and future steps for application of EO into the MEAs process. Emerging trends and recent initiatives are introduced for reader's future consideration. By showing the issues surrounding such an application, the author hopes to contribute to the further promotion of EO application to the effective implementation of MEAs.

INTRODUCTION

Multilateral environmental agreements (MEAs) have been under development for the last three decades. Now, the key challenges we face are to improve their effectiveness by enhancing implementation and compliance with their obligations. Because there seems to be a trend that recent MEAs require the best and most comprehensive scientific information regarding Earth's environment and the impact of human activities on that environment, the demand for environmental information to implement MEA obligations is continuously increasing.

Current advancement of space-based Earth observation (EO), together with information and modeling technologies, are likely to meet these requirements. Provision of more precise and comprehensive environmental data from EO satellites is expected within the present and subsequent decades. Interest in applying EO to implement MEAs is growing accordingly, and several related initiatives are in progress. However, challenges still remain in the practical application. The issues to be resolved are both technical and institutional.

This chapter provides a broad overview of the issues related to the application of EO for effective implementation of MEAs. The focus is primarily on institutional aspects; a detailed discussion of technical specification is beyond the scope of this chapter. The chapter will (1) generally review the characteristics of EO technology, and examine its potential in support of effective implementation of and compliance with MEAs, (2) propose possible solutions and future steps to facilitate actual application of EO in the MEA process, and (3) provide an insight on this issue for readers in the IT community.

MULTILATERAL ENVIRONMENTAL AGREEMENTS AND ENVIRONMENTAL INFORMATION

Multilateral environmental agreements (MEAs) are international agreements which provide the framework of international cooperation among the members to protect the environment (UNEP-DEPI, 2002). MEAs have rapidly developed within three decades in number and content. There are over 200 existing MEAs ranging from climate change, combating desertification, and protection of biodiversity.

Environmental information is the essential feature of MEAs. Sands (1995) observed that improving the availability of environmental information is one of their well-established objectives. He also noted that this information is recognized as a prerequisite for attaining the main goals of MEAs, which are effective national and international environmental management, protection, and cooperation (p. 627).

Indeed, MEAs entail various obligations that call for the collection, exchange, and analysis of environmental data or information. Such obligations include collection, exchange, and dissemination of data and information, notification, consultation, monitoring, reporting, verification, assessment, fact-finding, inspection, research, and public access. Examples of the obligations in major MEA from 1971 to 1995 are found in the study of Lee A. Kimball (1996). As Kimball observed, the scope and volume of information gathered pursuant to the obligations of MEAs continue to grow, and the format and guidelines become more elaborate as obligations develop (p. 854).

A good example is the United Nations Framework Convention on Climate Change (UNFCCC) and its Kyoto protocol. All Parties of the UNFCCC are required to develop and submit annual national inventories of anthropogenic emissions by sources, and removal by sinks (any process, activity, or mechanism which removes a greenhouse gas, an aerosol or a precursor of a greenhouse gas from the atmosphere) of all six greenhouse gases (CO_2, CH_4, N_2O, HFCs, PFCs, SF_6) (UNFCCC, Article 1.8, Article 4.1, Article 12.1.). Article 3 of the Kyoto Protocol sets the specific reduction target for Annex I Parties (developed countries) on overall emission of such gases, and each Party shall quantify direct human-induced land-use changes and forestry activities including "afforestation, reforestation and deforestation since 1990 (ARD)," and measure the net changes in greenhouse gas emissions and removals as verifiable changes in carbon stocks in each commitment period. Furthermore, each Annex I Party shall provide data to establish its level of carbon stocks in 1990 to enable an estimate of its changes in carbon stocks in subsequent years (Kyoto Protocol, Article 3.4.). The Parties shall promote and cooperate in systematic observation and development of data archives related to climate systems and in full, open, and prompt exchanges of relevant scientific data (UNFCCC, Article 4.1 (g), (h), Article 5; Kyoto protocol, Article 10(d).).

As information and data on environment and human activities are increasingly required, lack of appropriate information or lack of a capacity to collect information becomes critical causes of noncompliance of MEA's. Failure to report has a particularly impact on effective implementation of MEAs. It is because compliance-securing mechanism and institutional reviews under MEA's are often based upon national reporting. Thus, a noncompliance response requires improvement of or assistance in collecting and reporting data or transferring information. Improving the availability of information will help states' compliance with procedural or institutional obligations, and thus lead to states' compliance with the more substantive and detailed obligations in MEAs.

The overall objective must be to increase the amount of information available, improve its quality, allow it to be broadly disseminated among all relevant members of the international community, and ensure that it is used to guide decision-making at all national and international levels (Sands, 1995, p. 627). If EO can provide an appropriate method of collecting data or be a source of information required by MEAs, it may be identified as an innovative method for improving the implementation and compliance of MEAs. The relevance of EO should be confirmed by not only technical, but also institutional point of views. One basic criteria for EO to be used in the implementation and compliance of MEAs:

- Political and technical reliability (noninfringement of sovereignty issue, allowing scientific uncertainty but based on best scientific knowledge available, adequacy in accuracy, resolution, global coverage, frequency, timeliness, and environmental parameters)
- Equality and fairness in accessibility, availability, and content (balancing and taking into account the needs of all countries, cooperation and assistance to fill the gaps, objectiveness)
- Easiness (ease to manage, access, and use)
- Cost-effectiveness and efficiency
- Comparability (homogeneous, provide common reference, same and harmonized standard/conditions for all, and long term archive)
- Environmental decision-making relevance (Kuriyama, 2002)

POTENTIAL AND RELEVANCE OF EARTH OBSERVATION FOR THE IMPLEMENTATION OF MULTILATERAL ENVIRONMENTAL AGREEMENTS

Advantages of Earth Observation's Technical Characteristics

Space-based Earth observation (EO) (i.e., remote sensing by satellite) is the technology for observing the Earth's surface by using space systems. Currently, there are more than 60 EO satellites collecting data which allow us to obtain information about the Earth and human activities (UNEP-GRID, n.d.). Being sensed from space, the data from EO satellites inherently possess characteristics relevant to supporting MEA implementation. The primary EO technical features that provide advantages include:

- Higher speed and relative ease in acquiring data
- Increased frequency of data collection by use of a single instrument

- Collecting multiple kinds of data at the same time over large areas
- Spatial continuity (see UNOOSA, 1998, p. 10)

These technical capabilities provide EO data with objectiveness, global and repetitive coverage (capability of temporal analysis), and comparability (providing common reference, homogeneous) without physically infringing on national borders. Many of the data issues (e.g., lack of data, difference in method and format) found in implementation obligations in MEAs, such as reporting, assessment, and monitoring could be resolved by using EO data. Moreover, EO offers particularly excellent support in:

- Remote and difficult-to-access areas
- Areas undergoing rapid change
- Countries with poor infrastructure and limited transportation
- Areas with natural hazards and political sensitivities
- Constructing a broad overview or a detailed map of a large area (see UNEP-GRID)

The case of Mesopotamian marshland illustrates the above advantages. The desiccation of the vast marshlands located in southern Iraq and southwestern Iran was revealed in March 2000 by an assessment study of the United Nations Environment Programme (UNEP) based on analysis of EO data (UNEP, 2001b). Comparative analysis of Landsat (U.S. land remote sensing system) imagery from 1973-76 and 2000 enable a quantified assessment of the changes in marshland habitat extent (pp. 30-32). The study found that in total at least 7,600 km^2 of primary wetlands (excluding the seasonal and temporary flooded areas) disappeared between 1973 and 2000. Of the original 3,121 km^2 domain of the central marshes in 1973, only 98 km^2 or 3% remain in 2000. The study concluded that the disappearance of marshland was mainly driven by massive drainage works undertaken in the wake of civil unrest following the Gulf War in 1991. In addition, a small remaining area is also under a high risk of disappearing due to upstream activities in Iran, Kuwait, and Turkey (p. 36).

In principle, it could be argued that satellite data may provide the only practical source of information to survey large areas covered with marsh, particularly those located in developing countries and also in a region with difficult political realities. The assessment study (UNEP, 2001b) demonstrated that the availability of satellite data back in the 1970s compensated for insufficient information and enabled us to detect changes in the marshland. Changes in the marshland occurred particularly rapidly between 1991 and 1995. Using EO increased the speed of multitemporal analyses, and therefore helped in making quick and focused decisions. Moreover, EO data provided scientific facts for the negotiation of international agreements after the findings.

In addition to its technical capabilities mentioned above, recent movements and developments may provide a thrust for future application of EO to MEAs. As the UN report (United Nations, 1999) mentioned, "An extremely valuable tool is already available and will be greatly improved over the next decade" (p. 23, para 30). Capability of EO satellites has dramatically improved recently, and availability and accessibility of these data are increasing. For example, high-resolution commercial satellites, such as IKONOS, can distinguish one-meter objects. The hyperspectral capability, which is under development, can even identify hundreds of different parameters of the Earth. More and more developing countries are actively involved in EO activities. Several developing countries such as India, Brazil, and Argentina are using their own satellites. Regional organization such as the Economic and Social Council of Asia and the Pacific (ESCAP) is also actively working in promotion of EO in the region (Davis, Hoban, & Penhoet, 1999, p. 1109). The contribution from the recent development in information technology, including Internet and geographic information systems (GIS), can enable the EO application more effectively. Technical characteristics of EO, such as digital data, along with coherent and objective quality, make successful integration with information technology and increase our ability to process data. In the Mesopotamian marshland assessment (UNEP, 2001b), GIS was also employed with EO data to integrate and visualize the results using maps. In particular, the proliferation of the Internet dramatically increased our accessibility of EO data. The examples of these positive trends can be summarized:

- Development of new sensors with superior capability, such as higher resolution and multispectral sensors, and advances in information analysis and processing
- Development and operation of higher numbers of EO satellites with various sensors, including by private sector companies and developing countries
- A drop in the price of data, thanks to market competition and technology development
- Advances in new technology, particularly information technology such as the Internet and GIS, and effective integration of these technologies with EO technology
- Availability of the public funded data on the Web, sometimes in near real-time (Kuriyama, 2005)

It had been noted that EO has the following disadvantages. It requires in-situ or grand truth data for validation. EO is very costly. It doesn't eliminate the obligations such as reporting. However, above positive trends result in:

- Easier management and use of data
- Increased data accessibility and availability
- Increased accuracy and timeliness of data
- Lower cost of data and products
- A decrease in the capacity gap (Kuriyama, 2005)

Thus, the technical capability and characteristics of EO may enable:

- States and relevant organizations to reduce the burden of reporting, notification, assessment, monitoring, and other data management tasks and their cost
- States, relevant organizations, and the general public to increase their access to and analysis of the data and monitor compliance from the regional to the global scale
- States and relevant organizations to cooperate in optimal and equitable solutions and decision making on environmental issues, risks, and conflicts arising during the implementation of MEAs (Kuriyama, 2005)

The above indicates that EO can facilitate the implementation of obligations in MEAs.

Implications of the Legal Framework of Earth Observation

A legal framework applicable to EO activities is provided by the laws and regulations regarding space activities and remote sensing. The United Nations Principles Relating to Remote Sensing of Earth from Outer Space (Remote Sensing [RS] Principles) is the primary international legal document directly related to EO activities from space, including handling of collected data (Principle I (e)). The remote sensing principles are in a resolution adopted by the UN General Assembly and are initially nonbinding; however, several scholars have already recognized their customary status (See generally, Gabrynowicz, 1999). Another legal source of law for EO is the Treaty on Principles Governing the Activities of States in the Exploration and Use of Outer Space, including the Moon and Other Celestial Bodies (Outer Space Treaty). The treaty addresses the general rules for exploration and use of outer space and provides a legal basis for EO from space systems, which is one of the activities included in "use of outer space." In addition to international legal instruments, national regulations, multilateral and bilateral interagency agreements and arrangements, and various relevant practices form a part of the implementation framework of EO activities. However, the legal characteristics of EO activities presented in this subsection are based on an analysis of the international documents introduced above.

It is found that the following characteristics are contained in the legal framework of EO:

- The objective of EO is to protect Earth's environment (e.g., RS Principles, Principle I, X).
- Free EO activities are permitted as a legitimate measure for acquiring data on Earth's environment (RS Principles, Principles IV; Outer Space Treaty, Art. I)
- Balancing States' interests is carefully maintained to avoid the sovereignty issue (RS Principles, Principles IV; Outer Space Treaty, Art. IX).
- Promotion of international cooperation is a basis of EO (RS Principles, Principle V, VIII, XIII; Outer Space Treaty, Art. I, III, X, XI).
- EO is for the benefit and in the interest of all countries (RS Principles, Principle II, IV; Outer Space Treaty, Art. I).
- The needs of developing countries are considered (RS Principles, Principle II, XII, XIII).
- Data exchange and technology transfer are encouraged (e.g., RS Principles, Principle VI. VII, VIII, XII).
- Nondiscriminatory and reasonable cost access to data is required (RS Principle, Principle XII) (Kuriyama, 2005).

It could be argued that the above characteristics, as a whole, implies EO has legal characteristic relevant and complementary to MEA application to the extent that a state follows and complies with the remote sensing principles in implementing EO activities, due to the following reasons.

First, the legality of EO activities without prior consent of the sensed States has special implications for the application of EO data to MEAs. EO can avoid the impact of sovereignty or political sensitivities, at least at the stage of data collection and dissemination, by substituting for conventional measures such as inspections, which are subject to a State's approval. The increase of availability of the information offered by EO could make noncompliance transparent. This would pressure the relevant States to confront enforcing facts and comply with their obligations, such as prior notification or consultation in MEAs, which can be hindered by a State's unwillingness to comply. The fact that EO is a legitimate method for collecting and distributing data also has an advantage in instilling public confidence in the use of the data, particularly with the compliance mechanism of MEAs. It is because the legitimate procedure of gathering data, as well as the accuracy and reliability of the data, is required as basic information for enforcing States to comply with their obligations.

Second, MEAs and EO frameworks have common fundamental objectives and governing norms. The remote sensing principles contain many norms for the public good (Gabrynowicz, 1999, p. 100): international cooperation, equity, equality, capacity building, and being "for the benefit and in the interests of all countries." Equivalent principles can be recognized in basic international environmental law instruments, such as the Rio Declaration on Environment and Development (Rio Declaration, e.g., Preamble, Principle 3, 4, 6, 7, 9, 10). The nonterritorial and public good characteristics of outer space could facilitate cooperation "in the spirit of global partnership" (Rio Declaration, Principle 7) and could provide a suitable and objective basis for MEAs to address global environmental issues.

Third, it is notable that a basic data distribution policy favorable to the application of MEAs is embodied in the EO's legal framework. The idea of equal and reasonable cost for data access is provided in the text of international legal documents. Assured accessibility of information has a positive implication for implementation and compliance of MEA's, which are dependent on availability of information.

THE CHALLENGES OF EARTH OBSERVATION FOR THE IMPLEMENTATION OF MULTILATERAL ENVIRONMENTAL AGREEMENTS

Remaining Challenge in Applying Earth Observation to Multilateral Environmental Agreements

Against the background of a positive trend in EO activities, the interest in using EO technology for implementing and compliance with MEAs has increased recently. There are already actual instances such as the Vienna Convention for the Protection of Ozone Layer (Vienna Convention) and the Montreal Protocol on Substances that Deplete the Ozone Layer (Montreal Protocol) that apply EO technology to monitoring obligations under MEAs (e.g., Vienna Convention, art. 2.2.(a), 3, "systematic observation"). Moreover, in the response to the Kyoto Protocol, the use of EO technology for measuring environmental parameters is attracting attention, which is likely to spur this trend. Recent workshops (e.g., American Institute for Astronautics and Aeronautics (AIAA), 2000; Center for International Earth Science Information Network (CIESIN), 2000; Kalpakis, Markowitz, & Jamar, 2002; Penders & Ronald, 2002; Rosenqvist, Imhoff, Milne, & Dobson, 1999;) have studied the potential contribution of EO for development and implementation of MEAs, and have started to provide the potential forum for the linkage of EO and MEAs. Findings of the workshops (CIESIN, 2000; AIAA, 2000) show that EO may contribute to every process of MEAs: from prenegotiation to compliance enforcement. Accordingly, the importance of EO data in the context of MEAs is also gradually gaining the recognition among environmental policy makers and regulators (e.g., CIESIN, 2000; United Nations University, 1999; (Remarks by David B. Sandalow, Assistant Secretary of State).

However, the general application of EO to the MEA process is still challenging, regardless of whether EO and EO data have great potential or will only occasionally satisfy the demand for information necessary for effective the implementation of MEAs. There are still several issues to be resolved before practical application is realized at this stage.

First, EO is still not readily usable, robust, and accepted technology for use in MEA. While RS Principles provide the general framework, each EO data provider has its own data policy relating to data use. Increase in protection of intellectual property rights has a negative impact on free and wide utilization and distribution of data. There is also a communication gap between data providers, that is, the EO community (observing agencies or organizations) and users, that is, the MEA community (environmental regulators and government). Analysis and interpretation of EO data are required before obtaining the necessary information or products for reporting obligations. While the technologies are highly developed, it is still needed for trained experts to use EO data effectively. Research into the proper methodology for analyzing data is also required. The new flood of data from EO does not match collection of in-situ corroborating data and assessment capability. Gaps exist between the developed countries and developing countries regarding accessibility and availability of information, technologies, knowledge, and expertise. Furthermore, EO satellites are frequently experimental. Difficulty in finding data with appropriate specifications (e.g., resolution, acquired time, coverage) remains. EO programs are also often poorly funded. Long-term operations and data archiving, which are important for assessment obligations, are not guaranteed.

Second, institutional linkages are still missing. It can be argued that EO itself is a legitimate activity and consistent with international law. However, in order for EO to be introduced as a measure for collecting data or as a reliable data source, and in particular to be part of the compliance mechanism, it is necessary for EO to be authorized in MEAs, by subsequent amendment, or through the conference of the Parties (COP) decision, as appropriate and consistent with applicable international law (UNEP, 2001a, Annex II). The COP has the function of monitoring, updating, revising, and enforcing MEAs as the primary policy-making organization in most MEAs. For example, COP possesses a broad legislative authority under the UNFCCC to adopt and amend protocols and to make "the decisions necessary to promote effective implementation of the convention" (UNFCCC, art. 7.2.). In addition, the subsidiary body for scientific and technological advice (SBSTA) functions to "identify innovative, efficient, state-of-the-art technologies and know-how and advise on the ways and means of promoting development and/or transferring technologies" (UNFCCC, art. 9.2.(c)). Thus, use of EO should be recognized and accepted by these institutional organizations, along with the proper development of compliance-securing strategies based on scientific data.

Next Steps for Achieving Practical Application

The above findings suggest that establishment of both reliable systems and an institutional framework is necessary for EO to fully utilize the benefits and support implementation of and compliance with MEA obligations. Actions that facilitate establishing the operational system and authorizing the use of EO in the MEA process are highly desirable. The long-term objectives that will enable the successful application of EO to MEA implementation can be summarized as follows:

- Establishment of operational global observation systems (including EO and in situ observations) integrated with assessment and research programs that meet the information needs of MEAs
- Establishment of an institutional data center (including a clearinghouse mechanism) and network systems that enable all Parties to exchange, share, and archive reliable data and information (including EO data, in situ data, and relevant legal information) necessary for implementing obligations under MEAs (e.g., reporting, assessment, notification, monitoring)
- Establishment of institutional linkages between the above systems and the MEA implementation process (Kuriyama, 2002)

The necessary actions for the above purposes are as follows:

- Promote and develop EO programs and data utilization and archiving plans to better respond to the requirements of MEAs and environmental decision making
- Coordinate data policies properly to promote broad data sharing and exchange, taking into account the business opportunities of the private sector
- Study the technical and economic feasibility of EO technology for the implementation of MEA obligations, and demonstrate its usability through pilot projects
- Promote further coordination and integration of observation, assessment, and research programs as a global system through existing and new strategies and initiatives

- Conduct capacity-building and technology-transfer programs for both legal and EO technology for developing countries
- Develop proper mechanisms in MEAs, using a scientific basis for monitoring compliance, that enables them to make use of EO data
- Establish a framework to provide formal regular dialog among MEA communities (e.g., COP, SBSTA) and EO communities (EO system and data suppliers) (Kuriyama, 2005)

PROJECT AND INITIATIVES FACILITATING EARTH OBSERVATION APPLICATION

Pilot Project and Services for Multilateral Environmental Agreements

A number of projects are presently underway to verify EO data use in MEAs, based on the increasing interest in this application. Recently, space agencies, such as European Space Agency (ESA), Japan Aerospace Exploration Agency (JAXA), and National Aeronautics and Space Administration (NASA) are particularly active in promoting use of EO for supporting MEAs. These projects are an attempt to fill current gaps and are being prepared for practical application.

Activities of ESA seem to present the comprehensive and organizational approach to the issue. ESA launched with the "Treaty Enforcement Services using Earth Observation" (TESEO) initiative (ESA, n.d.; ESA, 2005b) back in March 2001 to study the application of EO for MEA support. The main objective of TESEO is to explore the capabilities of EO technology to support different national and international bodies involved in the implementation of key MEAs such as UNFCCC/Kyoto Protocol, the United Nations Convention to Combat Desertification (UNCCD), and Ramser Convention on Wetlands (Ramser Convention). The study results of TESEO showed some of the benefits that EO can provide in support of the implementation of these MEAs (ESA, 2005b). Based on the interest in the user community stimulated by the TESEO studies, ESA initiated a series of service development and demonstration projects in order to consolidate key information services for worldwide users (ESA, 2005b).

Regarding UNFCCC/Kyoto Protocol, for example, ESA started with the TESEO carbon project in which the information needs of main bodies involved in implementation of the convention were exhaustively collected, and a set of novel EO-based services were developed as prototype (ESA, 2005b). The Parties to the Kyoto Protocol are required to report the land use, land-use change, and forestry (LULUCF) for consideration of the estimation of greenhouse gas emission (see the second section of this chapter). Following TESEO project, ESA has been demonstrating how EO can provide support in filling the gap in LULUCF in a cost-effective manner under the two dedicated EO-based projects: The Kyoto Inventory and Forest Monitoring (ESA, 2004). The Kyoto Inventory project started in 2002 under the agency's data user programme. In the project, satellite data including ERS, Landsat, Proba, and SPOT were utilized to produce forest maps, land use, and land use change maps covering 1990, 1997, and 2002 across more than 20,000 square kilometers of European territory (ESA, 2005a). The availability of the adequate satellite data, such as Landsat and SPOT, for more than two decades allows analysis of land use and forest status in 1990, which is the reference year for reducing greenhouse gases (ESA, 2004). The Forest Mapping project, having begun in 2003 with a consolidation phase, is now providing a fully operational forest and land use monitoring system offering standardized-information products mainly based on EO (ESA, 2005a). The LULUCF reporting service under the forest mapping is pan-European coverage and has been already delivered to several European countries such as

Germany, Greece, France, Poland, and Sweden. The service is also working with some Asian and African countries in the evaluation of clean development mechanism (CDM) afforestation project in developing countries (ESA, 2005a).

Such pilot EO projects like ESA's, which are specially tailored to the needs of MEAs, can provide the following positive effects:

- Demonstrate the capability of the EO satellite to the MEA community
- Create further demand for applications of data
- Define the needs of MEAs to be met by EO
- Increase satellite capability and enhance knowledge regarding EO methodology (Kuriyama, 2002)

It is important to include EO data users, such as environmental regulators, as partners in the project in order to reflect their needs in planning new systems and to attain full application of EO to MEA processes. In this regard, it is noteworthy that national end-users responsible reporting under Kyoto Protocol were recruited for definition of product and services in the Kyoto Inventory project (ESA, 2005a). It is also necessary to demonstrate workable end-to-end solutions and the technical feasibility of implementation mechanisms, including selecting the appropriate media for data exchange and reporting. The involvement of developing countries and wide publication of the results should also be considered.

Group on Earth Observation and Global Earth Observation System of Systems

Because no single country can cover and respond to all MEA needs, the improvement of data availability naturally requires international cooperation in data collecting and sharing. Binding countries through a coordinated and integrated approach can create an institutional base for information exchange, coordination, decisions, and cooperation on relevant issues, including data policies and pilot projects. This approach may also allot authority to the EO community, which is important for increasing the reliability and credibility of EO in the MEA communities, such as COP and SBSTA. In this regard, the group on Earth observation (GEO) can be observed as a promising initiative for achieving the successful application of EO to MEAs.

The GEO is an intergovernmental initiative which leads a worldwide effort to build and institute the global Earth observation system of systems (GEOSS) (GEO, 2006a). It comprises more than 60 member countries, European commission, and more than 40 participating international organizations such as the World Meteorological Organization (WMO) and UNEP. This globally coordinated and integrated approach has been implemented through the activities of the Committee on Earth Observation Satellites (CEOS) and integrated global observing strategies (IGOS) (CEOS, 2002; IGOS, 2002). The difference between the GEO initiative and its precedents is the involvement and commitment of high-level representatives. GEO was established by a series of three ministerial-level Earth observation summit to further pursue the action plan of the G-8 Evian Summit calling for strengthening international cooperation in global observations of the environment (GEO, 2005a). The first Earth observation summit met in Washington, DC, in July 2003, and adopted a declaration establishing the *ad hoc* group on Earth observation to draft the 10-Year Implementation Plan (Ad hoc GEO, 2003). The second Earth Observation summit held in Tokyo, Japan, in April 2004, adopted a framework document which defines the scope and intent of a global Earth observation system of systems (GEOSS) (GEO, 2004). At the third Earth observation sSummit held in Brussels in February 2005, the GEOSS 10-Year implementation plan was endorsed, and GEO was formally established to carry out the plan (GEO, 2005b).

The 10-Year implementation plan provides the initial steps taken by nations, intergovernmental, international and regional organizations to realize GEOSS (GEO, 2005b, p. 1). According to the plan, the purpose of GEOSS is to achieve comprehensive, coordinated and sustained observations of the Earth system, and GEOSS will be "a system of systems" consisting of existing and future Earth observations covering not only satellite EO but also in-situ and air-bone observations (GEO, 2005b, p. 1; 3.2). In the plan, it is expected that the GEOSS will provide the institutional mechanism to secure the coordination and strengthening the existing Earth observation systems (3.2). It will extend across the processing cycle, from primary observation to information production which meets the information needs of sound decision making and delivers the benefits to society initially in the nine areas such as disaster, climate, and ecosystems (1; 3.2).

GEOSS, collectively, has several functional components, including addressing common user requirements, acquiring observation data, processing data into useful products, exchanging, disseminating and archiving data, meta data and products, and monitor performance (GEO, 2005b). Regarding the data management (5.2), within 2 years, the target of GEOSS is set to facilitate the development and availability of shared data, meta data, and products commonly required across the societal areas. Within 6 years, the implementation of GEOSS is anticipated to establish international information sharing and dissemination drawing on existing capabilities through appropriate technologies, such as Internet-based services, and by supporting common standards. The GEOSS data sharing principle is defined as full and open exchange of data and products within GEOSS, recognizing relevant international instruments and national policies and legislation (5.4). GEOSS will also focus on the capacity building of developing countries and develop infrastructure resources to meet research and operational requirements (5.6).

The objectives and directions of the GEO initiative, particularly the establishment of GEOSS, as envisaged or intended by the 10-year implementation plan, would appear to be compatible with the necessary conditions and systems for EO application to the MEA process examined in the previous section of this chapter. Notably, the bases for most of the necessary actions (see the fourth section of this chapter) are likely to be available in this existing initiative. In addition, the implementation plan originally expects GEOSS to serve for MEAs. It provides that GEOSS will further the implementation of international environmental treaty obligations (3.1). Besides, those assisting with the implementation of MEAs are defined as the user of GEOSS (4.2). Therefore, Molly Macauley's view (2004) regarding "the possibility of GEOSS serving as an international system for compliance monitoring of MEAs" (p. 6) could be realized in the future, while the various factors in realizing GEOSS—including governance, architecture of the systems, operational procedures, the role of the private sector and a proper data policy—are still under the evolving process.

EMERGING TREND AND ITS PROSPECTS

Satellite Mission Dedicated for Multilateral Environmental Agreements Support

With the increasing interest in EO application to MEAs, the satellite mission directly targeted to support the information needs of MEAs is emerging. The Japanese mission, the greenhouse gases observing satellite, GOSAT, is one such example. GOSAT is a satellite to be used to monitor the carbon dioxide (CO_2) and methane (CH_4) globally from orbit, aiming to support the international efforts to prevent global warming, such as the Kyoto protocol (JAXA, 2005). Although the number of ground-based CO_2 monitoring stations is increasing, there are not enough at this time. Furthermore, the distribution of the stations is not sufficiently uniform to estimate CO_2 sinks and sources around the world. Therefore, it is important and valuable to acquire homogeneous satellite data from frequent and global observations.

It is a joint project of the space agency, Japan Aerospace Exploration Agency (JAXA), and the user agencies, the ministry of environment (MOE) and National Institute for Environmental Studies (NIES). JAXA is responsible for satellite development, launch, and satellite operation. JAXA and MOE are in charge of the sensor development, and MOE and NIES are responsible for satellite data utilization. It is scheduled to be launched in August 2008, with a view of the first commitment period (2008 to 2012) of the Kyoto protocol.

The criteria and target of the MEAs support mission are conditioned based on the best compromise between the information needs of MEAs and challenge in technology development. The objectives of the GOSAT mission are defined as contributing to environmental administration by estimating the greenhouse gases (GHGs) source and sink in subcontinental scale and verifying the reduction of GHG's emission, which is required by the Kyoto Protocol, along with the advancement of EO technologies for future missions (JAXA, 2005). The targets of the mission are observation of CO_2 density in 3-month average with 1% (4ppmv) relative accuracy in subcontinental spatial resolution during the first commitment period. Other targets involve reducing errors by half by identifying the GHGs source and sink in subcontinental scale with the data obtained by GOSAT in conjunction with the data gathered by the ground instruments (JAXA, 2005). A more challenging successor with an accuracy of 1 ppmv on a national scale is under consideration, which is expected to support the national reporting under Kyoto Protocol. It is expected that the satellite mission dedicated to support to MEAs such as GOSAT would be increased in the future.

Linkage between Group on Earth Observation and Multilateral Environmental Agreements

Establishing GEOSS as envisioned in the implementation plan is not an easy target, but rather it would be a great experimental endeavor for the global community. Because GEO is established on a voluntary and legally nonbinding basis, the progress of the establishment of GEOSS is based on the voluntary contributions from member countries and participating organizations, which is not necessarily steady, as planned in the implementation plan. However, we could see an excellent opportunity to design GEOSS to support effective implementation of and compliance with MEAs.

In fact, the first clue for GEOSS to move forward in this direction can be observed in the latest development of the MEAs and GEO process. The work under the GEO and MEAs has begun to be implemented in a coordinated manner. Recently, GEO released the GEO 2007-09 work plan (GEO, 2007c) comprising 97 tasks and organizational activities, which provides the detail of planned activities to be performed during 2007-09 to achieve the initial targets of GEOSS defined in the 10-year implementation plan and its reference document (GEO, 2005c). In relation to the societal benefit area of climate, the Work Plan (pp. 10-11) incorporates the key action items defined in the "Implementation Plan for the Global Observing System for the Climate in Support of the UNFCCC" (GCOS, 2004) developed under the framework of the convention (UNFCCC, 2003). The GCOS implementation plan represents a commonly agreed basis for GEO actions in the climate area. On the other hand, the recent meetings of COP and SBSTA of UNFCCC welcomed the endorsement of the 10-year implementation plan at the third Earth observation summit, and expressed the support to promote the GEOSS (e.g., UNFCCC, 2005a). It also requested the continued close coordination in the implementation of these two plans (UNFCCC, 2005b, para 10). Further involvement of the MEA community in GEO initiatives should be proposed for reflecting the requirements of the MEA side in every dimension of the GEOSS design, including observation parameter setting and data utilization planning.

In addition, it appears that GEOSS could contribute to the synergic implementation of the several types of MEAs. In 4.1 of the 10-year implementation plan (GEO, 2005b), GEOSS is expected to facilitate the development and provision of common products for information needs common to many societal benefit areas such as land cover and land use, and a geodetic reference frame for Earth observation. By providing the common data-sets which meets the information needs common to various MEAs, GEOSS truly could be an institutional base for application of EO into MEAs in the future. For instance, GEOSS could act as a forum for practical coordination regarding the common data format for reporting obligations upon the request of and in cooperation with the MEAs bodies, such as SBSTA.

In order to attain the above objective, establishing a regular and closer linkage between GEO and the MEA regime, such as COP and SBSTA, should be considered. Currently, MEAs secretariats, such as UNFCCC and the convention on biodiversity (UNCBD), have become the participating organizations of GEO, which should be extended to other MEAs. In addition, proper development of MEAs, such as introduction of the compliance mechanism and enhancement of related capacity building, would also be required, along with GEOSS development. Specific authorization of the use of EO and a definition of the eligibility of EO data should also be fulfilled in the MEA regime. For example, the further development of the IPCC good practice guideline is expected to clarify the use of EO data in the UNFCCC regime (VTT, GAMMA Remote Sensing and Consulting AG, European Forest Institute, & Stra Enso Forest Consulting, 2003). The GEOSS could gain excellent potential if the efforts on the MEA side match the GEO initiative.

Involvement of the Private Sector

Because GEO is established as an intergovernmental initiative, involvement of the business sector to the activities of GEO seems not fully exploited yet in its initial phase. However, as the activities of GEO move to the actual implementation of GEOSS 10-year plan from the development of its organizational and governance structures, the opportunities for the business sector to participate in the GEO process are increasing.

The committee activities offer such an opportunity. GEO now has subsidiary structures including four committees and a working group to address the specific issues of GEOSS implementation (GEO, 2006b). The experts affiliated to private companies are also participating to the committee activities as the representatives of the member countries or participating organizations, such as the Institute of Electrical and Electronics Engineers, Inc. (IEEE) and the Open Geospatial Consortium, Inc. (OGC). The issues relating to system architecture, including standard setting and interoperability arrangement, and data management, are discussed under the architecture and data committee. These areas are where the private sectors can become main players and contribute to the GEOSS establishment because these issues may have direct impact on private companies doing business in such areas as software development, satellite manufacturing, and sensor development. In every dimension of GEOSS implementation, they are also supporting the implementation of the work plan through contracts with governmental agencies.

Operational services dedicated for the application of MEAs could also provide the various opportunities for private sector. For example, business opportunities, especially for IT business and EO data provider, such as data analysis, development of higher level processed data-sets, and operational system planning, can be foreseen. Because operational services require continued provision of data, there emerges a need to fill the shortages of data from public satellites with the data from private sectors. In order to transform EO data to useful information for users, EO data should be integrated with various data, including socioeconomic data, and analyzed by the specialists. Knowledge and expertise are required for large volume data management, computing, and software. Development of user-friendly data systems need the know-how from the private sectors. IT technology can help to fill the gap between user needs (accessibility, usability) and the services provided by EO sectors. Indeed, private companies start working together with space agencies to develop their services. For example, an international consortium led by an Italian company has been implementing the ESA's Kyoto Inventory project (see the fifth section of this chapter; ESA, 2005a). It is recommended that IT sectors to analyze the data needs of MEAs and develop the business plan for operational services to provide the solution for such needs.

CONCLUSION

The use of EO in MEAs has great potential for contributing to the effective implementation of and compliance with MEAs. The technical capability and legal norms of EO can meet many basic requirements for implementing the obligations of MEAs. EO's technical characteristics, such as objectivity, homogeneity and repetitive global coverage, are unique advantages that conventional systems cannot offer. Recent technological and market developments in EO are likely to eliminate EO's limitations in cost-effectiveness and usability. It will ease States' burden in implementing obligations under MEAs, and provide a basis for optimal and equitable decision making. The legal framework provides EO with legitimacy for collecting and disseminating environmental data globally. EO data can make the whole MEA process transparent and objective, which would facilitate States' compliance. Thus, it could be argued that EO can complement the legal and physical constraints of MEAs. EO is not a panacea and cannot solve all the issues related to data gaps or compliance with MEAs. However, obligations in MEAs are becoming more substantial and demanding of information. EO appears to be a unique and highly relevant system that can contribute to the implementation of and compliance with MEAs.

However, the general application of EO to the MEA process has yet to be achieved. Before such applications are realized, it will be necessary to establish robust and well-accepted global EO systems with an appropriate data distribution method that is systematically integrated with the assessment and research systems and institutionally integrated with MEA implementation processes.

Initiatives that could facilitate the use of EO in MEA implementation are already underway. In particular, the establishment of GEOSS envisioned in the 10-year implementation plan shows promising potential for achieving practical application. It is expected that the proposed GEOSS will contribute to the establishment of the forum or institutional mechanism for coordinating and integrating the existing initiatives and promoting a formal dialogue between the EO and MEA communities. In implementing the GEOSS, there is also a possibility of involvement of business sectors. Definition of architecture of GEOSS, including interoperability and data management issues, along with planning and providing operational EO services for MEAs information needs, would be the areas where private sectors can play a valuable role. Application of EO for MEAs provides the opportunities for creating innovative initiatives and new services not only for government, but also for private sector, including IT business.

NOTE

The views expressed herein are those of the author and may not reflect the views of the agency to which she belongs. They are mainly based on the Individual Project Report Submitted to the International Space University in partial fulfillment of the requirements of the Degree of Master of Space Studies, May 2002, incorporating recent progress in this issue.

REFERENCES

Ad hoc Group on Earth Observation (Ad hoc GEO). (2003). *The Ad hoc Group on Earth observation (GEO): Terms of reference.* Retrieved March 5, 2007, from http://earthobservations.org/

American Institute for Astronautics and Aeronautics (AIAA). (2001, March). The working group on the contribution of space systems to the development and implementation of multilateral environmental agreement (Working Group Rep.). In *Proceedings of the 6th International Space Cooperation Workshop*, Seville, Spain.

Center for International Earth Science Information Network (CIESIN), Columbia University. (2000, December). *Remote sensing and environmental treaties: Building more effective linkages* (Report of a Workshop). Washington, DC: Author.

Committee on Earth Observation Satellite (CEOS). (2002). *CEOS: Overview.* Retrieved March 5, 2007, from http://www.ceos.org/

Convention for the Protection of the Ozone Layer (Vienna Convention), March 22, 1985, 1513 U.N.T.S 293

Convention on Biological Diversity. (UNCBD). June 5, 1992, 1760 U.N.T.S. 79

Convention on Wetlands of International Importance Especially as Waterfowl Habit (Ramser Convention), February 2, 1971, 11.I.L.M.963

Davis, C., Hoban, S., & Penhoet, B. (1999). Moving pictures: How satellite, the Internet, and international environmental law can help promote sustainable development. *Statson Law Review, 28*, 1153.

European Space Agency (ESA). (n.d.). *Treaty enforcement services using Earth observation (TESEO).* Retrieved March 5, 2007, from http://styx.esrin.esa.it:5000/teseo/about.html

European Space Agency (ESA). (2004, December 20). *ESA presents space solutions at post-Kyoto climate summit.* Retrieved March 5, 2007, from http://www.esa.int/esaEO/

European Space Agency (ESA). (2005a, November 11). *Satellite support Kyoto Protocol through forest mapping service.* Retrieved March 2007, from http://www.esa.int/esaEO/

European Space Agency (ESA). (2005b). *How can environmental conventions benefit from space?* [CD-ROM].

Gabrynowicz, J.I. (1999). Expanding global remote sensing services: Three fundamental considerations. In *Proceedings of the Workshop on Space Law in the Twenty-first Century, UNISPACE III* (pp. 100–104). Vienna, Austria: United Nations.

Global Climate Observing System (GCOS). (2003, October). *The Implementation plan for the global observing system for the climate in support of the UNFCCC.* Retrieved March 5, 2007, from http://www.wmo.int/web/gcos/gcoshome.html

Group on Earth Observation (GEO). (2004, April). *From observation to action-achieving comprehensive, coordinated, and sustained Earth observations for the benefit of humankind, framework for a 10-year implementation plan (Framework Document),* as adopted by Earth Observation Summit, Tokyo, Japan.

Group on Earth Observation (GEO). (2005a). *Earth Observation Summit: Science and technology for sustainable development G8 action plan.* Retrieved March 5, 2007, from http://earthobservations.org/summit/g8_action.html

Group on Earth Observation (GEO). (2005b, February). *The global Earth observation system of systems (GEOSS) 10-year implementation plan (10-year implementation plan),* as endorsed by Earth Observation Summit III, Brussels, Belgium.

Group on Earth Observation (GEO). (2005c, February). *GEOSS 10-year implementation plan reference document* (Reference Document). Retrieved March 5, 2007, from http://www.earthobservations.org/doc/10-Year Plan Reference Document (GEO R 1000) .html

Group on Earth Observation (GEO). (2006a). *About GEO.* Retrieved March 5, 2007, from http://www.earthobservations.org/about/about_GEO.html

Group on Earth Observation (GEO). (2006b). *About GEO: Committees & working groups.* Retrieved March 5, 2007, from http://www.earthobservations.org/about/entities/about_cmtes_wgs.html

Group on Earth Observation (GEO). (2007c). *GEO 2007-2009 work plan.* Retrieved May 20, 2007, from http://www.earthobservations.org/docs/wp0709.v4.pdf

Integrated Global Observing Strategy (IGOS). (2002). *IGOS brochure.* Retrieved March 5, 2007, from http://ioc.unesco.org/

Japan Aerospace Exploration Agency (JAXA). (2005, November). Greenhouse gases monitoring from the GOSAT satellite. In *Proceedings of the Microwave Workshops and Exhibition,* Yokohama, Japan.

Kalpakis, K., Markowitz, K., & Jamar, S. (n.d.). *Environmental legal information systems (ERIS): A view from space: Digital Earth applications in environmental law and resource management.* Retrieved March 5, 2007, from http://earthpace.com/conference/confsumindex.html

Kimball, L.A. (1996). Treaty implementation: Scientific and technical advice enters a new stage. *Studies in transnational legal policy, 28.*

Kuriyama, I. (2002). *The contribution of Earth observation in the implementation and compliance of multilateral environmental agreements.* Unpublished master's thesis, International Space University, Strasbourg, France.

Kuriyama, I. (2005). Supporting multilateral environmental agreement with satellite Earth observation. *Space Policy, 21*(2), 151-160.

Kyoto Protocol to the United Nations Framework Convention on Climate Change (Kyoto Protocol), FCCC/CP/1997/L.7/ Add.1, December 11, 1997.

Macauley, K. M. (2004, June). *Is the vision of the Earth observation summit realizable?* A case study prepared for the American Meteorological Society 2004 summer policy colloquium.

Montreal Protocol on Substances that Deplete the Ozone Layer (Montreal Protocol), September 16, 1987, 1522 U.N.T.S. 3

Penders, M., & Ronald, D. (n.d.). *Environmental Protection International: The hanging chads of international environmental law: Towards the better use of technology and information management to detect violations of international agreements, combat transnational crime, and reform environmental law and trade, Publication #1.* Retrieved March 6, 2007, from http://www.epinetwork.com/publication_1.html

Principles Relating to Remote Sensing of the Earth from Outer Space (Remote Sensing [RS] Principles), G.A. Res. 41/65, 42 UN GAOR Annex (95th plenary meeting) at 2 UN Doc. A/RES/41/65 (1987).

Rio Declaration on Environment and Development (Rio Declaration).(1992, June 13). U.N. Doc/A/CONF.151/26.

Rosenqvist, A. (Ed.). (2003, July). Support to multi-national environmental conventions and terrestrial carbon cycle science by ALOS and ADEOS-II—the Kyoto & Carbon Initiative. In *Proceedings of the International Geoscience and Remote Sensing Symposium (IGARSS'03)*, Toulouse, France.

Rosenqvist, A., Imhoff, M., Milne, A., & Dobson, C. (Ed.). (1999, October). *Remote sensing and Kyoto Protocol : A review of available and future technology for monitoring treaty compliance* (Workshop report). Michigan.

Sands, P. (1995). *Principles of international environmental law I: Frameworks, standards and implementation*. Manchester, United Kingdom: Manchester University Press.

Treaty on Principles Governing the Activities of States in the Exploration and Use of Outer Space, including the Moon and Other Celestial Bodies (Outer Space Treaty), January, 27, 1967, 610 U.N.T.S. 205

United Nations. (1999). *Report of the Third United Nations Conference on the Exploration and Peaceful Use of Outer Space (UNISPACE III)* (A/CONF.184/6), Vienna, Austria: Author

United Nations Convention to Combat Desertification in Countries Experiencing Serious Drought and /or Desertification, Particularly in Africa (UNCCD), October 14, 1994, 1954 U.N.T.S.3

United Nations Environmental Programme (UNEP). (2001a). *The Guidelines on Compliance with and Enforcement of Multilateral Environmental Agreements*, adopted by the 7th special sessions of the Governing Council of United Nations Environmental Programme/Global Ministerial environment Forum in its decision SS. VII/4 of February 15, 2002 in UNEP(DEPI)/MEAs/WG.1/3, Annex II.

United Nations Environmental Programme (UNEP). (2001b). Partow, H. *The Mesopotamian Marshlands: Demise of an Ecosystem*. Nairobi, Kenya: Division of Early Warning and Assessment, United Nations Environment Programme.

United Nations Environment Programme, Division of Environmental Policy Implementation (UNEP-DEPI). (2002). *Compliance and enforcement*. Retrieved March 6, 2007, from http://www.unep.org/DEPI/Compliance-and-Enforcement/

United Nations Environment Programme, Division of Early Warning and Assessment, Global Resource Information Database (UNEP-GRID)—North America. (n.d.) *Satellite systems for land cover monitoring: Navigating through the maze* [Brochure].

United Nations Framework Convention on Climate Change (UNFCCC), May 9, 1992, 1771 U.N.T.S. 107

United Nations Framework Convention on Climate Change (UNFCCC), the Conference of the Parties (COP). (2003). *Global observing systems for climate*. Decision 11/CP.9 in FCCC/CP/2003/6/Add.1

United Nations Framework Convention on Climate Change (UNFCCC), the Subsidiary Bodies for Scientific and Technological Advice (SBSTA). (2005a). *Research needs relating to the convention*. UNFCCC/SBSTA/2005/4. P.14, para 71-83.

United Nations Framework Convention on Climate Change (UNFCCC), the Subsidiary Bodies for Scientific and Technological Advice (SBSTA). (2005b). *Research and systematic observation*. Draft Conclusions proposed by the Chair. UNFCCC/SBSTA/2005/L.17

United Nations Office of Outer Space Affairs (UNOOSA). (1998, May). *Management of Earth resources* (Background paper 3, A/CONF.184/BP/3).

United Nations University. (1999, July). *Inter-linkages: Synergies and coordination between multilateral environmental agreements* [Report of a conference]. Tokyo, Japan: Author

VTT, GAMMA Remote Sensing Research and Consulting AG, European Forest Institute, & Stra Enso Forest Consulting. (2003, May). *Treaty enforcement services using Earth observation (TESEO) theme carbon—KYOTO Protocol to the UNFCCC* (Final Report, ESA/ESRIN Contract No. 15557/01/I-LG). Espoo, Finland.

Chapter XI
Extraterrestrial Space Regimes and Macroprojects:
A Review of Socioeconomic and Political Issues

Dimitris J. Kraniou
Point Park University, USA

ABSTRACT

This chapter examines macroprojects to be deployed in outer space. A feasibility study is used to analyze the deployment of such projects in extraterrestrial realms. Moreover, the author argues that these projects will have substantial socioeconomic and political impacts on the international community of nations. Deploying permanent human facilities in space, mining planetary surfaces, asteroids, and a host of other activities will require the use of macroprojects. These macroprojects will be complex by nature. They will require the use of human and technical networks for their completion. All that can be done, and it can be accomplished by using the skills and talents of people coming from a variety of ethnic, racial, and cultural backgrounds.

INTRODUCTION

There are untold riches in outer space (Lainas, 2005; Lewis, 1997; Nelson, 2001). Everything from satellite orbits and microgravity fields, to the mineral resources of planetary and asteroid surfaces, can be harnessed for economic purposes. The road to outer space is the new frontier in the path of advancing human welfare. It presents an outlet from the closed system defined by our planetary environment replete with its finite resources.

Though the possibilities are tremendous, problems are quite apparent. As the developments witnessed during the formulation of international regimes governing the Law of the Seas, the Antarctic Treaty, the Outer Space Treaty, and so forth, are wont to remind us, global commons raise difficult, and possibly insurmountable, obstacles.

One can argue that this routine posturing during negotiations for treaty formalization is attributable to a positioning process in international relations. This is because negotiations at that level are geared toward enabling individual nations to maximize their future pay-offs from the panoply of activities unfolding in space. Outer space law would thus play a very crucial role in delineating parameters on issues and concerns that pertain to the allocation of such resources.

Outer space law and modern economic principles can enable us to successfully overcome the dichotomy provided by the *res communis* over the *res nullius* arguments voiced over the years. Under ancient Roman law of property, items were subdivided into two categories; the *res nullius* and the *res communis*.

R*es nullius* basically means things or items (property) that do not belong to anyone. This implies that property or items that fit the scope of this definition cannot be appropriated by anyone. The exact opposite rationale is derived by the concept of *res communis.* These are basically things or items that belong to the whole community. Thus, unlike *res nullius* that cannot be owned, the state, the group, or the global community can own *res communis.*

The *res nullius* and the *res communis* dichotomy present a problematic approach when dealing with outer space resources. The technologically advanced countries employ the res nullius argument. They favor this approach because it gives them the flexibility to appropriate resources by using superior economic and technological means. On the other hand, the developing and underdeveloped community of nations favors the *res communis* argument. These countries opt for a communal approach, and solutions to resource problems that can be handled by transnational institutions and organizations.

The aforementioned issues have been debated at international forums, and there appears to be no likely end in sight to this debate. On the contrary, it can be expected that future similar debates will increase as outer space becomes a more intensified arena for commercial and other activities.

THE CONCEPT OF MACROPROJECTS

It will not be long before we deploy colossal industrial infrastructures in extraterrestrial realms to appropriate resources. Planetary and asteroidal resources will be targeted. A lot of these will be classified as *Macroprojects* (Hori, 1990; Horwitch, 1990; Sykes, 1990; Weiss, 1988). I will attempt to provide a working definition of this term. *Macroprojects* have a substantial temporal component. In terms of years of completing their corresponding deployment, their time frame may fluctuate from 5 to 10 years or more. Such projects are collaborative in nature. This is the case because of their complexity and diversity. They involve specialized personnel with expertise in all functions of management, in finance, human resources, technologies, and engineering. They ideally employ astute individuals with integrative abilities. Similarly, such projects have sizable societal and cultural impacts.

Representative examples of terrestrial macroprojects are the pyramids, the Panama Canal, the Manhattan program, the Aliesca pipeline, the Apollo program, and so forth.

The financial resources required for the completion of such projects are quite substantial. Billions of dollars are used for such projects. Groups of corporations participate in such projects so as to diversify their strategic priorities and minimize their risk exposure. It is generally expected that members of the multinational corporate community also participate in these corporate groups. Further reduction of the risk element in undertaking such macroprojects is accomplished by the involvement of governmental or quasi-governmental organizations in these corporate groups. The historical analogies to that may resemble the construction of the railroads, and that of NASA and the space program. Moreover, it is expected that international organizations and transnational institutions will play a role in extraterrestrial macroprojects. Please see the analysis of the following section.

THE ROLE OF INTERNATIONAL REGIMES

It will be instructive to provide a working definition of the term *regime*. I will draw from the work of academics representing a variety of disciplines. The analysis below is not meant to be exhaustive. It will provide us with a frame of reference to be used so as to make further connections with the Outer Space regime.

Krasner interprets *regimes* as a "set of implicit or explicit principles, norm, rules, and decision-making around which actors expectations converge in a given area of international relations" (Young, 1986, p.105). In response to ambiguity associated with this definition, Krasner made an effort to define the parameters of a regime, by providing the following (Young, p. 106).

Principles are beliefs of fact, causation, and rectitude. *Norms* are standards of behavior defined in terms of rights and obligations.

Rules are specific prescriptions or proscriptions for action. Decision-making procedures are prevailing practices for making and implementing collective choice. These modifications, along with the original definition, were able to provide some insight to regime activities.

Keohane and others add another dimension to the regime analysis by introducing aspects of *cooperation and power* (i.e., political power, and hegemonic cycle in relations to stability cycles), while focusing on regime formation. Stein and Keohane imply that regime formation is a process of bargaining or negotiation in which members of a group perceive a group agreement, or a *contract zone*, and gradually come to terms with a set of mutually satisfactory institutional arrangements (Young, p.110). In their view, the cooperation process of reaching an agreement is a perquisite for regime formation.

Actors may very well have different reasons for seeking regime formation. Keohane suggests that inspiration may come from the response of a political market failure. Actors may be primarily preoccupied with their self-interest. They are *rational egoists*, and therefore hoping to maximize their own long-term gains. Goals directed toward the *common good* may be a distant second. In either case, there is ground for cooperation.

Kindleberger looked extensively at the *hegemonic cycle* (the rising and declining of a hegemony) and its impacts on the global economic system. One of the outcomes of Kindleberger's research was the factor of *stability* attributed to the hegemon. Moreover, Kindleberger suggests that a *regime* would carry out this stabilizer role in lieu of the degenerating hegemon.

Keohane, drawing on Kindleberger's earlier work, suggests that the presence of a dominant great power is of great importance, perhaps even necessary, not only to the initial formation of regimes, but also to the maintenance of institutional arrangements over time (Young, p. 112). Keohane suggested that *regimes* and *hegemonic power* could coexist and in fact contribute to *cooperation* (Young, p. 113).

Young's definition of *regimes* is perhaps the most comprehensive and decisive of all. Young (1986) states that regimes are *social institutions*. "Social Institutions are recognized practices consisting of easily identifiable roles, coupled with collections of rules or conventions governing relations among the occupants of these roles" (Young, p. 107).

He proceeds to define *institutions* as practices composed of recognized roles coupled with sets of rules or conventions governing relations among the occupants of these role, and *organizations* as physical entities possessing offices, personnel, equipment, budgets, and so forth (Young, p. 108).

Regimes help to infuse order in the global system. The regime's *jurisdictional* role is above that of an organization. There are a large number of interaction/combinations of the above, (regime only, regime and institutions, regime and organizations, and regime, institutions, and organizations) involving many different areas, and concerning many issues. That is illustrated by *Young's 2x2 matrix*.

Thus, arrangements can be complex. This becomes more evident by just looking at the wide spectrum of *functional scope, geographic domain,* and *membership* which regimes exhibit. These include monetary issues, international arrangements for whaling, polar bear agreements, Antarctica, Outer Space, and so forth.

In addition, even though there is a degree of flexibility associated with regimes, we should not forget that regimes are susceptible to changing environments. Young points out that there are all kinds of conditions and factors that may contribute to that change. Regimes may change due to political, economic, technological, social cultural changes, and even moral developments (Young, 1989). In the extraterrestrial arena, the instruments of *civil society*, that is, organizations and institutions, will play an important role.

POLICY ISSUES ON SPACE RESOURCES

Given the previously defined parameters, one can construct an analytical framework to guide policy decisions of the U.S. private and public sectors toward extraterrestrial space. The rationale is that extraterrestrial space is vital to our national interest. It combines tremendous potential and resources and its facets touch all aspects of our socioeconomic, political, and security concerns.

Some additional definitions are in order here. In this manuscript, extraterrestrial space means the space whose lower layer is at a height of approximately 100 miles from the surface of the Earth. At this height, artificial satellites will continue orbiting the terrestrial globe without the danger of being burned up from atmospheric friction.

The upper layer of this space is located at a height where a geosynchronous or geostationary orbit[1] is achieved, or at approximately 22,500 miles. I have coined the term *First Space Regime*[2] (or *Spare I*) for the aforementioned space. Hence, the lower layer of this regime will be referred to as geosynchronous Earth orbit (GEO) or (GSO).

The extraterrestrial Spare I contains the following resources:

a. **The vacuum:** This is a transversable medium of immense dimensions which is being employed by us for placing there technological devices of civilian (including commercial manufacturing and industrial activities) and military use.
b. **Gravitational forces:** These forces range from those having high intensity (i.e., those around planetary objects), to infinitesimally small or almost nonexistent ones (i.e., the fields of empty space) that will be the focus of our concern.
c. **Energy:** Here, I will emphasize exclusively the solar energy that might be utilized by satellite systems for terrestrial use.
d. **The GSO (as defined previously):** This orbital slot also can be used for a multitude of activities.

The above resources have the potential for a multiplicity of applications ranging from purely security to economic matters. For example, advanced

Table 1.

		Organizations: *Yes*	Organizations: *No*
Regimes	*Yes*	Civil Society	International Anarchy
Regimes	*No*	Freestanding Organization	International State of Nature

satellite systems with security applications can be deployed in space. Their efficiency and maneuverability will increase there due to the vacuum and infinitesimal gravitational forces. Scenarios involve the deployment of space lasers, particle beam weaponry, or other improved versions of presently deployed conventional systems.

Similarly, the vacuum and infinitesimally small gravitational fields of space can be used to improve the manufacturing efficiency of chemical substances, semiconductors, and electronic components, among others.

Moreover, we can utilize the GSO for the deployment of solar power satellites that will capture solar energy and beam it to Earth in the form ofmicrowaves. There, it will be transformed to electricity and used for our energy needs.

All the above undertakings will have a significant input on the international arena. It is expected that these will affect economic, security, and political issues that relate tothe East-West rivalry, and to the North-South welfare concerns.

I advocate that extraterrestrial space should be utilized because of its resources and their significance. Moreover, I think thattheneedtoenhancethe security, resource (energy), economic, technological, and political bases of the United States will shape our attitudestowardspaceutilization. Here is another way to visualize this point.

Let us assume that vector Y represents extraterrestrial space utilization. Then:

Y = extraterrestrial space utilization.

Let us also assume that vector X represents the United States' need to promote space utilization projects. Then:

X = U.S. need to promote space utilization projects.

Consequently, we will have: Y = [F] X, meaning that the utilization of extraterrestrial space will depend on the U.S. need to promote space utilization projects.

Vector X, signifying the U.S. need to promote extraterrestrial space utilization projects, will be shaped by the U.S. need to promote her security, resource-energy, economic, technological, and political (domestic and international) concerns.

Thus vector X can be decomposed to:

x_1 = Security concerns
x_2 = Resource-energy concerns
x_3 = Economic concerns
x_4 = Technological concerns
x_5 = Political (domestic and international) concerns

Therefore, the need to promote extraterrestrial space utilization will depend on the importance and interaction of the security, resource-energy-economic, technological, and political variables among themselves, or:

$Y = [F].\ (x_1, x_2, x_{3)}\ x_4, x_5)$.

Extraterrestrial space (vector Y) will be utilized because of its resources. Therefore, vector Y can be decomposed to its elements:

y_1 = Vacuum of space
y_2 = Space gravitational forces
y_3 = Solar energy
y_4 = GSO

Thus, we have the following matrix that shows how the components of vector Y will interact with vector X.

The Components of Vectors Y, X

It should be emphasized that the activities indicated in the cells of the previous matrix will occur at different orbital slots and altitudes.

In terms of altitude from the surface of the Earth, the following can be said. Space com-

Table 2.

	X_1 security	X_2 Resources, Energy	X_3 Economic	X_4 Technological	X_5 Political	
					Domestic	International
* y_1=Vacuum of Space	Deployment of advanced satellite systems (laser, particle beam weapons)/enhanced maneuverability/ "High Ground" advantage	Vacuum of space will facilitate the utilization of extraterrestrial solar energy "manouvcr:,Lilita/"Nich	Application of commercial manufacturing and industrial activities; improve economic bases of US; Increase economic power of U.S.	Increase national power; advance technological and scientific position; increase industrial productivity; reap economic payoff		Enhance power & prestige
y_2=Gravitational forces	As above	Small gravitational forces of space will facilitate the utilization of extraterrestrial solar energy.	As above	As above		As above
y_3=Solar energy	Solar energy utilization will enhance our security by lessening our dependence on foreign energy suppliers	Solar energy utilization will diversify and improve our resource bases	Solar energy utilization will halt the outflow of dollars to foreign suppliers; It will improve our balance of payments	Improve Solar Technology; make extraterrestrial solar competitive with other energy sources	To promote national well being; to uplift national spirit; to mitigate domestic political pressure among interest groups, environmentalists, anti-nuclear activists, and so forth	As above
y_4= GSO	The GSO can be used for security applications (i.e., SPS deployed there can function as security means/ "High Ground" concept)	SPSs deployed at GSO will harness solar energy	Economic flows will result to the users of the GSO(present cases)	Space(solar) technology will enable us to improve our technological abilities to operate & maintain SPS and other macrostructures in SPARE I		As above

mercial manufacturing and industrial activities will occur at low Earth orbit (LEO).

Such activities will include the manufacture of advanced semiconductor items, other electronic components, and biochemical and pharmaceutical substances. These will be undertaken in space because the absence of gravity and the space vacuum will increase the purity of materials or chemical substances. For example, shuttle experiments carried in space by McDonnell Douglas/Johnson & Johnson have demonstrated the ability to purify materials 700 times more efficiently than on Earth (Waldrop, 1983).

The security satellites, depending on their functions, will be placed at different altitudes. For example, surveillance satellites will be placed at low circular orbits. At that altitude, they will be vulnerable to attack. Nevertheless, to minimize the vulnerability of the surveillance satellites, one can utilize *low elliptical orbits*. Moreover, security satellites placed at *geosynchronous orbits* are presently relatively safe from attack. Finally, the deployment of solar power satellites will occur at geosynchronous orbits. Thus, these will represent the most distant activities to undertake within SPARE I.

Another point is worth stating. The activities mentioned above need not occur simultaneously. One might represent the outgrowth, or spin off, of another activity. For example, the implementation of security-oriented space projects may represent the springboard for the other space activities envisioned for SPARE I. The rationale is that the United States will emphasize security-oriented space projects in order to improve her security and power bases, or because *presently* the undertaking of macroprojects in space, such as solar powered satellites (SPSs), for purely economic reasons might be perceived as a risky proposition.

Here, I do not examine the whole range of space resources. For example, under space energy resources, one could have listed electromagnetic radiation or cosmic rays from deep space, for which we have exhibited only scientific experimental curiosity. Their study might provide important clues for future utilization, but present understanding of them precludes that.

Moreover, another future objective of our space exodus will involve massive utilization of extraterrestrial material resources, for activities conducted in space, or for transfer to Earth when their value to mass ratio is high (as it might happen with cobalt, etc.). Nevertheless, I have chosen not to examine in depth this variable here, though some works (Johnson & Hobrow, 1977; O'Neill et. al., 1979) have considered their important aspects.

My rationale for the exclusion of the extraterrestrial material resources is explained by the following. It is believed that their extensive handling will take place well after the utilization of the vacuum of space, solar energy, and gravitational forces will have occurred. But it might be possible to witness smaller scale material utilization from the lunar surface—such as the usage of lunar silica for solar cell construction or lunar material in general, for security considerations—occurring simultaneously with the establishment of solar power satellites (SPS). Generally, though, it is believed that extensive plans for extraterrestrial material resource utilization will temporally noncoincidental with the resources I have chosen to examine here

Similarly, another space resource I will not examine is the electromagnetic spectrum. This has been treated extensively in a number of occasions (Coase, 1959; Levin, 1971; Meckling, 1968). Today, major discussions concern its allocation to states of unequal economic and technological status, and the demands are fashioned after those involving the Law of the Sea Negotiations.

The reasons that made me follow the above distinction of space resources include the following:

a. **Realities of the *state of the arts***: Although outer space has colossal dimensions with potentially immense resources, our present technological know-how makes it relatively inefficient to engage in extensive exploitation of extraterrestrial mineral resources.

b. **Economic realities:** The problems caused by technological constraints result in economic inefficiencies. Therefore, when present cost-benefit ratios of some space options are examined vis-à-vis terrestrial considerations, the first are found to be mostly unjustified in monetary terms. Again, such might be considered the case of extraterrestrial mineral resources. Nevertheless, I want to make it clear that these statements fit present conditions and are appropriate only for short-term analyses. Technological advancements will change the scenario drastically. Moreover, emphasis on intangibles, such as longer-term security considerations, might motivate an actor to undertake projects that when viewed in strictly monetary terms might appear noneconomic.

c. **Realistic extrapolations:** My concern also involves the elimination of far-fetched speculation. Consequently, I believe that attention should be paid to shorter-term extrapolations, because these can be predicted more realistically. For example, we cannot state that "since the resources of space are colossal we should not worry about a resource, scarcity problem." The resources of space are "colossal," but presently, for all intents and purposes, this characterization is useless because we are not in the position to take full advantage of that. Therefore, when examining resource alternatives, it is recommended that we proceed cautiously and that our extrapolations focus on projects having more realistic bases.

d. **Methodological considerations:** These involve ways of dealing with a problem or a situation. Thus, it is my objective to look at some of the technological, resource, economic, security, and political problems of our time and then make extrapolations concerning their future status.

In conjunction with item (c), I would like to express my belief that a problem will usually remain in a recognizable form for the shorter and medium term only.[3] In the longer-term, though, we either overcome the old, problematic conditions or these acquire such drastic changes that they might be perceived as representing problems of a totally different nature. Therefore, I thought it was overtly important to concentrate on the space resources whose utilization will be considered realistically feasible within the time horizon examined in this work so as to make sure, among other things, that the methodological approach is sound and reflects pragmatic conditions. Here, I should make it clear that all the manufacturing and industrial experiments and activities that will take place in SPARE I will utilize exclusively raw materials carried there from Earth.

For reference purposes only, I would like to state the following: One can conceptualize of a *Second Space Regime (SPARE II)*, that extents from GSO to the lunar orbit. Therefore, in this case, one might conceivably utilize the materials of the lunar surface itself for manufacturing and industrial activities, and the Lagrangian Points[4] as security objectives.

The *Third Space Regime (SPARE III)* contains the planetary bodies, their satellites, and the asteroids.[5] The exploitation of this regime will open the road for a real planetary civilization. With this regime, the bounty of the raw material resources at human disposal will be mind boggling (Dyson, 1999; Ellis, 2002; Forward, 1988; Pournelle, 1979; Schmidt & Zubrin, 1996) Finally, one might conceive of the space engulfing all the above, and extending, for all intents and purposes, to infinity. This will be called the *deep outer space regime* (Lewis, 1997).

A MACROPROJECT ILLUSTRATION

The following is an approach that can be used to think about macroprojects.

Macroproject Assignment

Establish a self-sustaining industrial facility on the lunar, Martian, or asteroidal surface(s).

This macroproject will involve an international group of actors, institutions (agencies) of a private, quasi-public, and public nature.

- **Develop an executive summary:** Describe the project in a general outline. Highlight its major aspects.
- **Write a feasibility study:** Examine all pertinent variables involving the planning, deployment, and successful operation of this macroproject. The planning horizon (i.e., the time frame of your feasibility study) will be of, at least, 20 years duration. Identify the resources of the new realms. Discuss their utilization aspects. Highlight appropriate scientific and technological ideas.
- **Finances:** How do you pull the capital resources together to finance the project?
  Cost-benefit analysis: Calculate the streams of revenues and costs involving the deployment and implementation of this project.
- **Strategic management**: Identify the mission statement, objectives, and goals of the project.
  - How do you manage a multicultural, multinational, and technocratic group of people?
  - How do you interface with the other *stakeholders* involved with the project?
  - What is the decision-making *structure* of this *consortium*(?) like? (Identify the organizational structure of the *consortium*).

- **Marketing and public relations:** How do you promote and eventually sell the idea to the public and corporate sectors?
- **Implications:** Consider the consequences of the macroproject. Analyze the national/global, economic, cultural, managerial-organizational-legal, societal, and political impacts of such projects. Examples:

  a. Discuss (speculate about) the organizational characteristics of the corporate actors involved with the macroprojects (Will these be the *mega corporations* of the future)?
  b. Discuss (speculate about) the nature of the transnational institutions—*INGOs*—involved with the macroprojects. Will these be actors of a *civil society*?
  c. How will extraterrestrial metallic resources brought to Earth impact the price(s) of metals?
  d. What will the impact of (c) be on the economies of primary natural resource exporters that are primarily LDCs and NICs?
  e. Will the accumulation of new wealth skew the economic/business arenas?
  f. Will (a-e) impact global socioeconomic and political stability?
  g. An *alternative futures* section may also represent an integral part of your paper.

The above is not meant to be an all inclusive outline. Moreover, it is not meant to suggest only one way of approaching a macroproject.

EXTRATERRESTRIAL UTILIZATION PROJECTS: A SYNOPSIS

The following presents an array of space endeavors and resource utilization issues that unfolded historically. Similarly, it contains a futuristic projection of activities as these may occur in the near and more distant futures.

A. Space Endeavors Stage I: 1980s-1990s Domain: 100-22,500 miles from Earth Resources: Stage I
   a. The vacuum
   b. Gravitational forces
   c. Solar energy
   d. Low Earth orbit (LEO)
   e. Geosynchronous Earth orbit (GEO) or (GSO)

   Activities: Microgravity Experiments:
   - Biochemical
   - Pharmaceutical—metallurgical—electronic
   - Scientific studies
   - Survey satellites
   - Solar power security systems
   - Scientific studies
   - Communication satellites

B. Space Endeavors Stage II: Circa 2000 Domain: From GEO to lunar surface Resources: a., b., and c. as above
   a. Chemicals and material of lunar surface
   b. Apollo, Amor asteroids
   c. Lagrangian points

   Activities: Manufacturing: Solar cells—metals
   - Oxygen
   - Diamonds
   - Security
   - Scientific—studies

C. Space Endeavors Stage III: Circa 2018 A.D.
   Domain: From the Martian surface to the limit of the solar system
   Resources:
   a., b., and c. as above
   d. Permanent Martian bases
   e. Permanent Asteroidal habitats

   Activities:
   - Fuel refineries (oxygen)
   - Metals (Martian asteroidal surfaces)

D. Space Endeavors Stage IV: End of 21st Century and After Domain: Planetary Bodies to Cosmic Spaces Resources:
   a. Chemicals
   b. Minerals
   c. Harnessing of gravitational waves, black holes, or exobiological organisms?

   Activities:
   - Mining
   - Terraforming
   - Planetary restructuring: Dyson (spheres), Kardashev (civilizations)
   - Security
   - Scientific (natural curiosity)

or

   - Cosmic post-industrialism
   - Eschatological (teleological pursuits)
   - Entropic future

CONCLUSION

The space regimes that I visualize here are meant to be used primarily and mostly for economic utilization projects. These regimes will be supported by institutions and organizations so as to promote a *global civil society*. Such arrangements are meant to take advantage of the immense possibilities provided to us by the resource reservoir of the extraterrestrial realm.

Deploying permanent human facilities in space, mining planetary surfaces, asteroids, and a host of other activities will require the use of *macroprojects*. These macroprojects will be complex by nature. They will require the use of human and technical networks for their completion, and all that can be done. It can be accomplished by using the skills and talents of people coming from a variety of ethnic, racial, and cultural backgrounds. These macroprojects will require the collective intellect of multicultural teams, of *epistemic communities* that will work in *synergy* so as to realize grandiose goals of a very demanding magnitude and scope.

It is expected that such activities will promote humanity's future. If such macroprojects are successfully implemented, they will propel humanity to heights never imagined before. The possibilities are there, and the final outcome will depend on the human actions to be used in space and time in the years to come. I hope that our decision processes are rational and effective enough so as to attain the goal of harnessing the new extraterrestrial realms.

ACKNOWLEDGMENT

I would like to acknowledge the tremendous assistance and efforts provided by my very efficient Graduate Research Assistants, Mr. Vincent K. Mutai, MBA 2006, and Ms. Esther M. Ngare, MBA 2007, in editing this manuscript.

REFERENCES

Agreement Governing the Activities of States on the Moon and Other Celestial Bodies (The Moon Treaty). Retrieved March 2, 2007, from http://www.unoosa.org/oosa/SpaceLaw/moon.html

Coase, R. H. (1959). The Federal Communications Commission. *Journal of Law and Economics*, 1-40.

Dyson, F. (1999). *The Sun, the Genome, the Internet*. New York: Oxford University Press.

Ellis, G. F. R. (Ed.). (2002). *The far future universe*. Philadelphia: Templeton Foundation Press.

Forward, R. L. (1988). *Future magic*. New York: Avon Books.

Hori, H. (1990). Macro-Engineering: A view from Japan. *Technology in Society, 12*(1), 45-63.

Horwitch, M. (1990). From unitary to distributed objectives. *Technology in Society, 12*(2), 173-195.

Johnson, R. D., & Hobrow, C. (Eds.). (1977). *Space settlement: A design study*. Washington, DC: NASA.

Lainas, Thedoris (2005, September 18). Mines on Mars. *To Vima, H07.*

Levin, H. J. (1971). *The invisible resource: Use and regulation of the radio spectrum*. Baltimore: John Hopkins Press for Resources for the Future.

Lewis, J. S. (1997). *Mining the sky: The untold riches from the asteroids, comets and planets*. New York: Basic Books.

Meckling, W. H. (1968). Management of the frequency spectrum. *Washington University Law Quarterly*, 26-34.

Nelson, K. (2001, February). NASA Ames Center looks at problem of drilling on Mars. *Petroleum News*. Retrieved March 2, 2007, from http://www.petroleumnews.com/pnarchpop/010228-49.html

O'Neill, G. K. (Ed.) et. al. (1979). *Space resources and space settlements*. Washington, DC: NASA.

Pournelle, J. (Ed.). (1979). *The endless frontier* (p. 14). New York: Ace Books.

Schmidt, S., & Zubrin, R. (Eds.). (1996). *Islands in the sky: Bold new ideas for colonizing space*. Wiley Popular Science.

Scientific Committee on Antarctic Research. *The Antarctic Treaty System: An introduction*. Retrieved March 2, 2007, from http://www.scar.org/treaty

Sykes, A. (1990). Macro projects. *Technology in Society, 12*(2), 157-172.

United Nations Committee on the Peaceful Uses of Outer Space. Retrieved March 2, 2007, from http://www.unoosa.org/oosa/COPUOS/copuos.html

Waldrop, M. M. (1983). The commercialization of space. *Science, 221:* 1353-1354.

Weiss, C., Jr. (1988). Sustainable macro projects. *Technology in Society, 10*(1), 375-384.

Young, O. R. (1986). International regimes: Toward a new theory of institutions. *World Politics, 39*(1), 105.

Young, O. R. (1989). *International cooperation: Building regimes for natural resources and the environment* (p. 32). New York: Cornell University Press.

ENDNOTES

1 At a geosynchronous or geostationary orbit, a satellite will remain stationary or fixed vis-à-vis a terrestrial point.

2 By regime, I will mean the existing procedures, rules, or institutions regulating and controlling certain kinds of activities. This is discussed more extensively in the "International Regimes" section of this document.

3 Exact timetable in years for the statements "short, medium, and long-terms are impossible. These depend on the ever shifting socioeconomic and political circumstances and represent a matter of interpretation. As an approximation, though, we might say that "short-term" is equal to 5-10 years, "medium-term," 10-20 years, and "long-term" is more than that.

4 Five points of the earth-moon system where a third body, if placed there, could retain its position with respect to both the Earth and Moon because the gravitational fields of the three bodies would be in balance. Of special importance are the stable Fourth and Fifth Lagrangian Points (i.e., L-4, and L-5), located in the lunar orbit 60 degrees ahead and 60 degrees behind the Moon, correspondingly. Their significance as security objectives will be obvious with the extensive utilization of space.

5 A special class of asteroids though, called Apollo and Amor asteroids, might be said to belong, partially, to SPARE I and SPARE II because of their orbits which cross the Earth's orbit. It is expected that in the future they could be exploited efficiently in our relatively immediate vicinity.

APPENDIX: INTERNATIONAL MILESTONES-AGREEMENTS AS REFERENCE POINTS FOR SPACE UTILIZATION PROJECTS

- The Conference on Peaceful Uses of Outer Space (COPUOUS, 1957)
- The Antarctic Treaty (1959)
- Outer Space Treaty (1967)
- Moon Treaty (1979)
- Law of the Sea Negotiations (1974-1982)
- The Bogota Declaration (1976)

These treaties and international agreements provided the framework for work relating to resource utilization and other issues. Such work involves both terrestrial and extraterrestrial realms. It pertains to mineral resources, the planetary and exoplanetary environments, satellite orbits, pollution, and the fauna and flora of Antarctica.

A more analytical approach of these concepts follows.

Outer Space Law

In space law we have the Treaty on Principles Governing the Activities of States in the Exploration and Use of Outer Space, including the Moon and Other Celestial Bodies (1967).

This treaty that is also known as the *Outer Space Treaty*, was adopted by the General Assembly in its resolution 2222 (XXI). It was opened for signature on January 27, 1967, and was entered into force on October 10, 1967. As of January 1, 2005, it was ratified by 98 countries.

The other interesting development in space law is the Agreement Governing the Activities of States on the Moon and other Celestial Bodies (1979). This agreement is also known as the *Moon Treaty* or *Moon Agreement*.

The agreement was adopted by the General Assembly in 1979 in resolution 34/68. It was not until June 1984, however, that the fifth country, Austria, ratified the agreement, allowing it to enter into force in July 1984. As of January 1, 2006, 12 states have ratified it, and an additional four have signed the Moon Agreement (Moon Treaty, http://www.unoosa.org/oosa/SpaceLaw/moon.html, 2006).

The above documents represent the legal framework that attempts to regulate human activities in space. Nevertheless, specific provisions of the aforesaid documents are very controversial. That is true especially with Article 1 of the Space Treaty (1967) which states that the exploration and use of outer space, including the moon and other

celestial bodies, shall be carried out for the benefit and in the interests of all countries, irrespective of their degree of economic or scientific development, and shall be the province of all mankind.

Article 11 (1) of the Moon Agreement (1979) also provides that "The Moon and its natural resources are the common heritage of mankind."

The two articles have proven to be very contentious when outer space resource issues are debated between or among economically and technologically developed and underdeveloped or developing countries.

United Nations Committee on the Peaceful Uses of Outer Space

The Committee On The Peaceful Uses Of Outer Space was set up by the General Assembly in 1959 (Resolution 1472 (XIV)) to review the scope of international cooperation in peaceful uses of outer space, to devise programs in this field to be undertaken under United Nations auspices, to encourage continued research and the dissemination of information on outer space matters, and to study legal problems arising from the exploration of outer space.

Number of member states in the committee: 67.

The committee has two standing subcommittees of the whole:

- The scientific and technical subcommittee
- The legal subcommittee

The committee and its two subcommittees meet annually to consider questions put before them by the General Assembly, reports submitted to them, and issues raised by the member states. The committee and the subcommittees, working on the basis of consensus, make recommendations to the General Assembly. Detailed information on the work of the committee and the subcommittees are contained in their annual reports.

The 49th session of the committee on the peaceful uses of outer space will be held from June 7-16, 2006, at the United Nation office at Vienna, Vienna International Center, Vienna, Austria (http://www.unoosa.org, 2006).

The Bogotá Declaration

On December 3, 1976, the equatorial states of Ecuador, Colombia, Brazil, Congo, Zaire, Uganda, Kenya, and Indonesia met in Bogotá, Colombia. (Gabon and Somalia, also equatorial states, were not present). These countries adopted what is known today as the Bogotá Declaration. The declaration claimed the right of equatorial states to exercise national sovereignty over the arcs of the geostationary orbit (GSO) that are directly over their territories. The GSO is at a distance of 22,300 miles (i.e., 36,000km) above earth.

Apparently, the Bogotá Declaration was an attempt to alter the international legal status of outer space in favor of those underlying equatorial states. The legal status of the GSO is tied to the controversy over a legal definition of outer space. Both have been debated in the United Nations Committee on the Peaceful Uses of Outer Space (COPUOS) or its legal subcommittee for four decades, and they still remain on the agenda.

The Law of the Sea

The Law of the Sea or the United Nations Convention on the Law of the Sea (UNCLOS), as it is being referred to, is an accumulation of the laws and regulations, including international agreements and treaties, which govern activities at sea or in any navigable waters. The Law of the Sea is a codification of human activities from antiquity to modern times. Rhodian, Byzantine, and Roman laws have been incorporated in our modern documents dealing with the sea. The *mare liberum* doctrine formulated by Grotius, and John Selden's *mare clausum* concept became very important ideas in today's world. Moreover, one can argue that these concepts also impacted the legal treaties dealing with outer space.

The Law of the Sea has been in force since 1984. It has been ratified by a substantial number of countries.

The Antarctic Treaty

The Antarctic Treaty system is the whole complex of arrangements made for the purpose of regulating relations among states in the Antarctic. At its heart is the Antarctic Treaty itself.

The treaty states that "in the interests of all mankind that Antarctica shall continue forever to be used exclusively for peaceful purposes and shall not become the scene or object of international discord." To this end, it prohibits military activity, except in support of science, prohibits nuclear explosions and the disposal of nuclear waste, promotes scientific research and the exchange of data, and holds all territorial claims in abeyance (http://www.scar.org/treaty, 2006). Both the Antarctic and the seas represent an interesting laboratory of human activities. Because of these areas, humans have amassed and continue to accumulate new experiences that will be used in outer space when we finally deploy permanent human facilities there.

Chapter XII
Commercialisation of Space Technology for Tomorrow's Space Missions

Stella Tkatchova
TUDelft University, Spain

Michel van Pelt
ESTEC Centre of the European Space Agency, The Netherlands

ABSTRACT

This chapter presents an initial identification of direct and indirect benefits for space agencies and space and nonspace companies from new markets development, creation of new collaborations, and an analysis of the costs and financing of future human interplanetary exploration. Commercialization of space technology is the process by which private companies commercially exploit space technology, without being its owners. Commercialization of space technology for future interplanetary missions is considered as a primary focus and principle benefit in this vision. Before private companies invest in commercial projects for interplanetary missions, they will have to perform cost benefit analysis for their commercial projects for future interplanetary missions.

INTRODUCTION

This chapter discusses the direct and indirect benefits for the different stakeholders of commercialization of space technologies for future Moon and Mars missions. These include NASA's Moon and Mars vision and the European Space Agency's (ESA) Aurora program. Commercialization of space technology is the process by which private companies commercially exploit space technology, without being its owners. Commercialization of space technology for future interplanetary missions is considered by Aldrige (2004) as a primary focus and principle benefit in this vision. Furthermore, as also concluded by Tkatchova and van Pelt (2004), there are several reasons for considering commercialization, such as new markets development, the high costs for vision implementation, and the need for technology innovation.

Before private companies invest in commercial projects for interplanetary missions, they will have to perform a cost benefit analysis[1] for their commercial projects for future interplanetary missions. Therefore, this chapter presents an initial identification of direct and indirect benefits for space agencies and space and nonspace companies from new markets development, creation of new collaborations, and an analysis of the costs and financing of future human interplanetary exploration.

At present, in 2006, there are discussions on the immense energy benefits to Earth from the exploitation of helium-3, and the development and demonstration of new propulsion, life-support and other technologies, as stated by Safronov and Jakimenko (2006). At the same time, at this early stage of the future programs, the estimation of the costs of large, possibly crewed missions is very difficult.

However, initial assumptions on the possible direct and indirect benefits can be formulated. Direct and indirect benefits from the International Space Station (ISS) program in the context of commercialization of space technology have not yet been defined, due to initiation of ISS commercialization only in the final stages of the ISS program. Therefore, examples of direct and indirect benefits will be taken from the aviation industry. The aviation industry has been quoted by various authors, such as Collins (2002) as a model for "carrying passengers safely, conveniently, and cheaply without depending on taxpayers." Therefore, as a basis for identifying direct and indirect benefits, a number of types of benefits[2] as defined by Eurocontrol (2000) for the aviation industry will be taken.

Direct benefits are revenues from sales, new markets development, cost savings, and reliability[3] of the space technology relevant to these private companies. Direct benefits are revenues generated from the sales of:

- Technology developed and tested for Moon and Mars missions
- Spin-off technology and scientific breakthroughs
- Images, movies, brands from the Moon and Mars space exploration
- Use of intellectual property rights (IPR) and marketing rights from the technology developed and tested for Moon and Mars missions

The benefits to space agencies, customers, and the public are harder to measure and referred to as indirect benefits. These are, for instance, increased international cooperation, technology innovation, improved public image, and emergence of space brands, and thereby increased support to space exploration. The direct and indirect benefits will be different for the various stakeholders in the future Moon and Mars space exploration visions and may change after their actual implementation.

THE PRESENT SITUATION

At present, human space exploration is almost exclusively the domain of government space agencies. Commercial companies are operating some Earth observation satellites and most of the world's telecommunications satellites, but have few commercial projects on board the ISS.

The first commercial activities on-board space stations started in the early 1990s, with the launch of the first Japanese reporter on board Mir. He made daily TV reports, and the Russians were paid $28 million for a 1-week flight on board the station

(Space, 2005). Pizza Hut, for instance, developed a vacuum packed space pizza which can be kept for long times without freezing. It was eaten by Russian cosmonauts onboard the International Space Station; footage of the event was later used in a TV commercial. In Japan, Nissan Food Products, the maker of Cup Noodle, is collaborating with the Japanese Aerospace Exploration Agency (JAXA) to develop instant space noodles for the astronauts onboard the International Space Station. Commercial customers can already use the existing ISS as a technology platform for performing their commercial projects.

However, till now the vast majority of missions to the Moon and further have been strictly government-funded and government-operated. The main reason for this is that investments in, for instance, a telecommunications satellite, usually result in significant profits within the span of a couple of years. In contrast, human space exploration missions require large investments with often high cost risks, and long development durations. The results of such missions are usually strictly scientific, and as a consequence can only be sold to a relatively small group of scientists with very limited budgets.

Even if an exploration mission would manage to find a part of the Moon covered with gold, it would not be worth the investment of developing and operating a spacecraft to mine it.

Investments in space exploration are high, risky, and have low, uncertain, and late profits. In contrast, a typical telecommunication satellite project requires a relatively modest and low-risk investment, while the profit is almost certain, high, and quick.

Nevertheless, there may be some near-term niche-markets for innovative space missions beyond Earth orbit. An example is the California-based private company TransOrbital Inc., which is currently developing the Trailblazer lunar probe without government funding. Instead, the mission is sponsored by a number of companies that are mostly in it for the advertisement possibilities. Computer manufacturer Hewlett Packard has offered to supply the onboard computer for the Trailblazer spacecraft. It envisions the project as an advertising opportunity for its new wireless computer technology, which will enable Trailblazer's computer to be in contact with its onboard equipment via radio-links. It will also make it possible for owners of the right hand-held computers to interact with the spacecraft via e-mail (apart from confirming receipt, it is not clear whether the spacecraft will do anything with these messages, however).

Private persons are involved with the financing of the mission by paying money to put all kinds of small personal items onboard the spacecraft. Sending a small message to the moon on Trailblazer is, for instance, priced at $17, launching a business card costs $2,500, and other items can be flown for $2,500 per gram. There is room for some 10 kilograms (22 pounds) of personal items onboard Trailblazer. Thousands of people have already taken advantage of this unique possibility, making the initiative an important part of the financing of the mission. The whole project is estimated to cost about $20 million, including launch.

Trailblazer's camera will make high-resolution pictures of the lunar surface that TransOrbital intends to sell to scientific institutes. Additional revenue could come from a new lunar map, based on the Trailblazer images. Apart from stills, Trailblazer will make high definition TV-quality videos to be sold for use in documentaries and television commercials.

Eventually, Trailblazer will crash on the Moon. All personal items onboard will be packed inside a crash-resistant capsule that is expected to bury itself 4 to 5 meters deep inside the lunar soil upon impact. There the messages, business cards, letters, hairs, and other items will remain until found by some future lunar explorers.

Apart from the plans of TransOrbital, not much seems to be happening concerning (partially) privately funded interplanetary missions. In 2000, the LunaCorp company received $1 million in backing from the Radio Shack chain of electronics shops. LunaCorp is developing a robotic spacecraft that would be assembled at the International Space Station and from there launched to the Moon. Like Trailblazer, it would generate high-resolution pictures and video and involve public participation. However, the project is still looking for further funding to continue the development. Another plan of LunaCorp is the development of a small rover that the general public could drive by remote control.

In the 1990s, Dutch astronaut Wubbo Ockels was leading an ESA moonlander project named Euromoon that would have been partly funded by private sponsors and the general public. However, budget constraints put an end to that innovative plan.

As long as spaceflight equipment, and especially launches, are prohibitively expensive, the efforts of companies striving to organize privately funded space exploration missions are likely to have limited results. Complicated, expensive projects with long development schedules are likely to stay the domain of government space agencies for the near future.

THE COST OF EXPLORATION

Interplanetary exploration, especially when involving crews, is costly. Spacecraft are usually specifically designed for a certain mission and therefore normally only one is built of any type. Even equipment that can be used on more than one space probe, such as standardized solar arrays or antennas, is only built in small numbers. As a result, space probes do not benefit from the cost reducing economies of scale, as a result of mass production, such as in the computers and car industry.

Moreover, it is impossible to repair robotic interplanetary spacecraft once they are launched into the harsh space environment. The hostility of space means that only components of very high quality can be used. Astronauts onboard crewed spacecraft can sometimes make external repairs, that is, extravehicular activities (EVA), but the need for high safety margins nevertheless results in relatively higher costs[4] than for robotic missions. Moreover, crewed spacecraft require life support, return, landing systems, and so forth, which are not required on robotic missions. The high cost for crewed space missions also results in a high price for commercial users. For example, NASA's price to launch 1 kg of payload to the ISS is $22,000 and to return is also a similar price, as presented by ESA (2001). One hour of astronaut work on an experiment inside the ISS, that is, intravehicular activity (IVA), is priced by NASA at $15,000.

A typical, relatively small Mars orbiter with a dry mass (i.e., without propellant) of about 500 kg, partly based on existing and partly on modified equipment, may cost in the order of $100 million for the so-called spacecraft platform and another $50 million for the scientific instruments onboard. A launch with a soyuz fregat launcher has a price of about $40 million, and another $20 million would be required for tracking and control of the spacecraft during flight. This makes a total of $210 million.

A much more complex project such as the Cassini-Huygens mission to Saturn, requiring a large amount of newly developed equipment and carrying an extensive array of sophisticated instruments, has a much higher price tag. This mission consists of the NASA Cassini orbiter (dry mass 2581 kg) currently investigating the ringed planet and its various moons, and ESA's Huygens (mass 348 kg) probe that in January 2005 successfully descended through the thick atmosphere of Saturn's largest moon, Titan. According to NASA, the total cost of the Cassini-Huygens mission is about $3.26 billion, including $1.4 billion for prelaunch development, $704 million for mission operations, $54 million for tracking, and $422 million for the launch vehicle. The United States contributed $2.6 billion, ESA $500 million, and the Italian Space Agency $160 million.

The total cost of crewed space exploration is much higher. Converted into today's dollars, the Apollo moon program, with its 11 crewed missions, cost around $130 billion, not including the development of the Saturn 1 and 1B launchers that led to the Saturn V launcher used for Moon missions. Even with the technology and experience available today, for an extremely efficient program of crewed Mars missions such as proposed in the Mars Direct plan of Zubrin, Baker, and Gwynne (1991) would cost at least $80 billion for 10 missions performed over 10 years, as concluded by Hunt and van Pelt (2004). In September 2005, NASA's administrator, Griffin, announced that the estimated costs for astronauts to return to the Moon will correspond to $104 billion through 2018, as discussed by Smith (2005). This cost estimate does not include sending astronauts to Mars. The estimate will most likely change, considering that the cost for building and operating the ISS for 10 years is about $100 billion, as mentioned by ESA (2005), paid by all the ISS partners[5]. The assumed high costs for Moon and Mars crewed missions will most likely result in international cooperation (indirect benefits) for the implementation of these missions between space agencies and space companies.

However, compared to the direct and indirect benefits gained, the financial investments required for space exploration are relatively small. For instance, the missions of the two voyager spacecraft that flew by Jupiter, Saturn, Uranus, and Neptune and its moons have cost the U.S. taxpayers less then a simple lunch per person. And for the cost of a couple of beers per year, the European citizens have been able to explore the Sun, the Moon, Venus, Mars, Saturn's moon Titan, and Halley's Comet. Furthermore, technology developed for space exploration missions and paid for by space budgets has found its way to applications on Earth. The ancestors of the compact computers developed for use onboard the cramped crewed spacecraft of the Apollo program are the home and office computers of today. Technology originally developed for communication with spacecraft is now widely used in telecommunication satellites for intercontinental telephone, television, and e-mail traffic. Earth observation satellites are now an indispensable means for providing information on the weather, climate, state of the ozone layer, pollution, and so forth. Processes defined for space exploration, with its need for extremely reliable equipment and failure-tolerant mission scenarios, have been successfully applied in other industries as well.

FUTURE STAKEHOLDERS

To identify the direct and indirect benefits for the different stakeholders from the Moon and Mars visions, an overview of these stakeholders is required.

The successful development and implementation of the future space exploration visions will influenced by the industry, collaborations customers, and general public. Their roles and activities in the future space exploration vision will depend on the benefits that they will foresee from the commercialization of space technology for future Moon and Mars missions. At present in 2006, the main initiators of the future space exploration visions are the space agencies, and in particular NASA, followed by ESA.

Figure 1 shows an overview of the stakeholders. At present, NASA and ESA are the space agencies with official space exploration visions, while the Russian Space Agency (FSA) is performing various technical and scientific feasibility studies for the implementation of human Mars missions. There has been unconfirmed information that China is also looking into the development of Moon and Mars programs. Eventually, other space agencies may join NASA, ESA, or FSA in the implementation of the space exploration visions.

Space industry, as defined by P. Lionnet (personal communication, 2004), is an industry that involves the design, development, and production of space-qualified space hardware and software[6]. Under space industry are also included space industry associations or lobbyists organizations, such as the U.S. Coalition for Space Exploration[7].

As presented in Figure 1, there are various relationships between the U.S., Russian, and European space industries. These relationships illustrate the existence of collaborations, such as joint ventures between United States and Russian space companies or European ones. Many of these collaborations, such as international launch services (ISL), Starsem, and Sea Launch, are for the commercial exploitation of Russian launchers, as discussed by Tkatchova (2005).

NASA's relationship with the private sector, as discussed by Aldridge (2004) "must be decisively transformed to implement the new, multidecadal space exploration vision." NASA Administrator Michael Griffin has announced that he wants NASA in the near future not to by hardware anymore, but rather services: "In other areas the U.S. government bought services—tickets—for cargo space on airplanes. In developing space we at NASA and the Defense Department have largely relied upon buying hardware rather than contracting with industry to provide the services we need," as stated by Morris and Morring (2006).

Collaborations are referring to other organizations as intermediaries between space agencies, industry, and commercial customers and the general public. For example, at present in 2006, for the commercial utilization of the ISS the existing intermediaries are NASA's research space centers (RPC)[8] and ESA's ISS Lab Ruhr. For the future interplanetary missions the collaborations will be responsible for the commercialization of space technologies, attracting technologies and products that can be important for future interplanetary missions (spin-in) and identifying space technologies that can be relevant to nonspace companies (spin-off). As the spin-off of space technologies, as discussed by Aldridge (2004), will lead to a positive economic effect in nonspace industries, such as insulin pumps or forest fire fighting tools derived from space-based infrared camera technology.

Figure 1. Stakeholders in Moon and Mars exploration

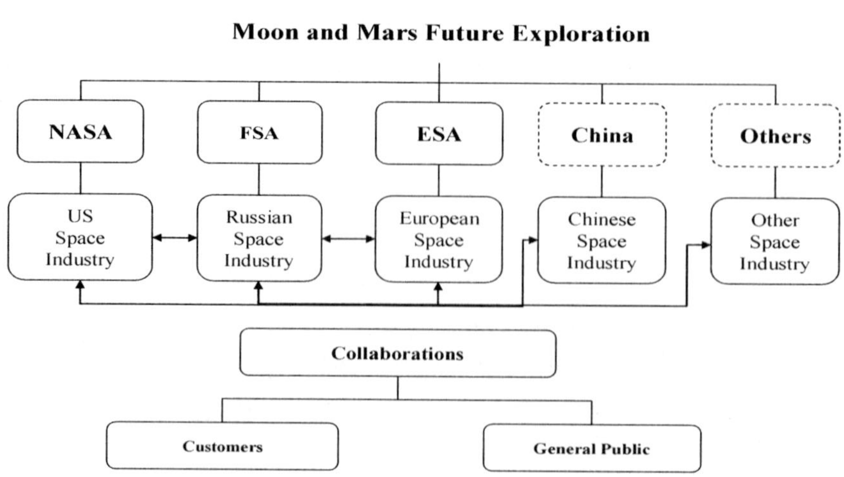

In the future roles and activities of stakeholders in the Moon and Mars exploration visions will change.s

The overview of stakeholders' roles and expected activities from new markets development, creation of collaborations, and financing future Moon and Mars missions will be discussed in the context of the direct and indirect benefits.

MARKET EVOLUTION

In order to identify the direct and indirect benefits for the stakeholders from the new markets, there will be an overview of the expected markets evolution, followed by a detailed description of the various markets.

The markets for space technology of future interplanetary missions are expected to pass through nascent, frenzied, turbulent, and mature phases of market development. As presented in Figure 2, as discussed by Tkatchova (2005), the ISS markets are emerging, in nascent stage of market development and at the end of the lifetime of the International Space Station (ISS) they could be entering their frenzied stage.

In time, a transformation of ISS markets toward the market development for technology for interplanetary missions may be observed during the frenzied and turbulent stages. This market development may be followed by the creation of self-sustainable markets[9]. The stakeholders investment in the markets development during the four phases of market evolution will depend on their roles in the market evolution, and naturally will change during the various phases of development. In the frenzied stage, profits and market demand could be rising (Tkatchova & van Pelt, 2004), and therefore new markets can be developed because of the spin-off technology developed for future interplanetary missions. Space agencies, space industry, collaborations, and customers can use the already identified industrial applications from ISS markets and set commercialization policies (i.e., property rights (IPR) marketing rights) for commercialization of space technology for interplanetary missions. Space agencies can generate revenues from the sales IPR and marketing rights

Figure 2. Phases of market evolution

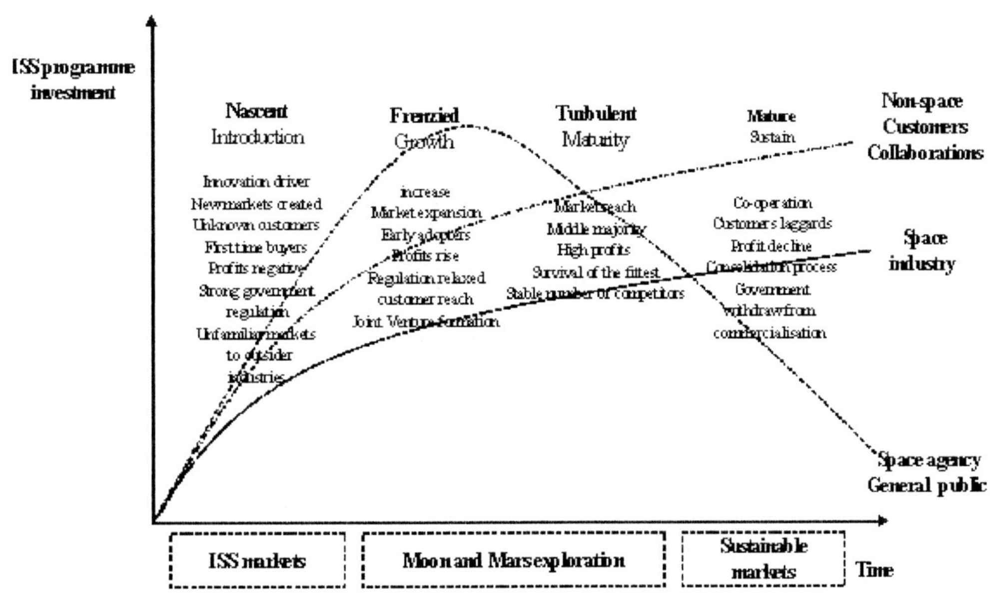

to space companies and collaborations from their commercial projects or technology tested for Moon and Mars missions. Furthermore, the successful implementation of these missions will require international cooperation and the indirect benefits for space agencies will be in international commitments and increased public image.

Space companies can gain direct benefits from proposing and developing technology for space agencies for Moon and Mars missions and identifying potential spin-off technology for collaborations and nonspace companies. The indirect benefits for the space companies will be in technology innovation and the development of space brands. The collaborations will generate revenues from flying and testing new products and later branding them and selling them as space proven and tested products. Also, they will be able to generate revenues[10] from the implementation of spin-off and spin-in projects.

The collaborations can also generate revenues from the sales of images, movies, and royalties from the use of IPR rights from nonspace companies. The indirect benefits of the collaborations will be in increasing their public image and creating and selling space brands.

Figure 2 shows that with the reduction of the space agency's investment in markets development, an increase of space companies' investment is expected. Furthermore, the collaborations will probably invest in market development during the nascent and frenzied stage, as they will have to aim at identifying relevant spin-in and spin-off technologies for and from the interplanetary space missions. Therefore, collaborations investment will be higher than the space companies' ones during the first two phases of market evolution. As the markets further develop and enter the mature, and it is possible that the nonspace customers will also start to invest in the development of projects which use space-based technologies.

The above market evolution will influence space agencies, space companies, and collaborations roles and will result in the development of new market trends in the space industry. The overview of the possible trends is presented in next section. The market evolution presents a general overview of the expected future developments, but not of the potential markets and industrial applications. These will be discussed in the following section.

FUTURE MARKETS

The stakeholders in the future Moon and Mars mission will be able to generate direct and indirect benefits from the developments of the potential nonspace markets. In order to identify the potential markets, first an overview of the selection criteria for their identification is presented, and then an overview of the targeted markets and industrial applications.

The potential markets for space technology of future Moon and Mars missions will differ from the targeted ISS markets, due to the necessity for the development and spin-in of new technologies for implementation in interplanetary missions. The stakeholders will be able to derive direct and indirect benefits from the implementation of the missions, if they consider the processes of commercialization, spin-in, and spin-off of space technologies as common. This means that the presented targeted markets from Figure 3 will have to offer benefits to all stakeholders, and therefore the proposed targeted markets have to be classified based on the criteria below:

- Safety
- Reliability
- Technology innovation

The above criteria will target as potential customers nonspace companies, which can test their products or services and marketing rights, thus generating direct benefits from the sales of their space-tested products and services. However, the same customers can also have technologies that can be useful for future Moon and Mars missions, and therefore the space agencies can incorporate these technologies into their missions (spin-in). At the same time, nonspace companies can be given free marketing rights and offered sponsorship deals. As the space technologies for implementation of these missions further develop, certain units of these technologies can be spin-off into improving the products or processes of these nonspace companies.

Figure 3 illustrates the R&D and novel markets. During space flight, astronauts experience health problems,[11] such as loss of bone and muscle mass, which allows for research in the area of osteoporosis. Astronauts lose around 1% of bone mass per month, and the development of this bone mass loss is the major concern for long-term manned missions. Another health risk is the astronauts' exposure to high radiation levels during space-flight. Therefore, medical devices for ultrasound diagnostics for bone density and radiation diagnostics can be not only tested, but also can be of great benefit for astronauts during long-term missions. Nonspace companies can generate sales (direct benefits) from the marketing rights of these devices and the sales of similar medical devices to nonspace customers.

The future Moon and Mars missions will provide a technology platform for the development of diverse markets.

In the novel markets, IT companies can develop computer games for astronauts, and due to good marketing, can develop new markets, increase their sales (direct benefits), and develop space brands (indirect benefits). Furthermore, with the development of infrastructure services, such as the development of technologies for extraction of helium-3, private companies will achieve technology innovation (indirect benefits). As a result, they will even be able to generate revenues (direct benefits) from the sales of these technologies to oil and gas companies on Earth.

From the development of the markets and industrial applications presented in Figure 3, space agencies can benefit from spin-in technologies and direct benefits from cost saving from not developing a totally new technology, and also benefit from

Figure 3. Potential markets and industrial applications for space-based technology of the future Moon and Mars exploration programmes

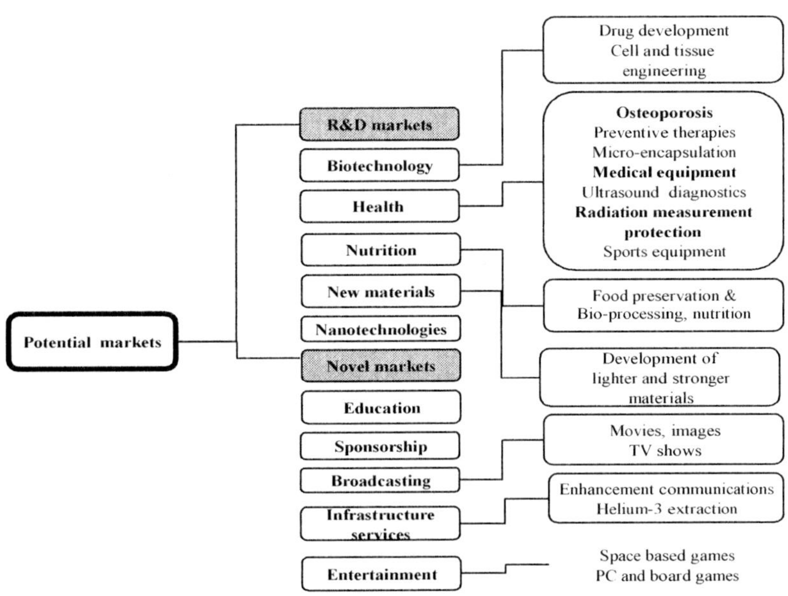

indirect benefits such as technology innovation, international cooperation, and increased public image. Space companies can also increase sales (direct benefits), develop space brands (indirect benefits), and increase the reliability of their space technology (direct benefits). Collaborations can gain direct benefits not only from sales and new markets developments, but also from international cooperation and space brands (indirect benefits). Finally, the general public will gain benefits from the spin-off of technology and scientific breakthroughs. These will be additional indirect benefits of international cooperation, technology innovation, and use of space brands. However, the benefits for the public will be the least clear and transparent, and therefore space agencies, as the initiators of the Moon and Mars visions, will need to clearly define and communicate the benefits of these programs to the public.

As the potential markets develop, the direct and indirect benefits from the Moon and Mars missions for all stakeholders will become clearer. Space agencies need to define and clearly communicate to all the involved stakeholders the benefits from these missions.

PUBLIC BUDGETS AND FINANCING APPROACHES

Space agencies are at present the initiators and investors of the future Moon and Mars space missions. As earlier discussed in the third section, the Apollo program cost corresponded to around $130 billion in today's money, while the estimated cost for returning astronauts to the Moon is $104 billion.

Today, NASA's total space budget represents less than 1% of the total federal expenditures. Even at its peak in 1966, its budget was only 5.5 % of these expenditures. In Europe, the ESA member states spend even less than the United States on space exploration. NASA's civil space budget for 2005 corresponded to $16.2 billion, as stated by Smith (2005). The issue that arises is whether the NASA space budget alone or even in combination with Europe, Canada, Russian, and Japan will be sufficient to finance the development and implementation of such Moon and Mars missions. Figure 4 illustrates the world civil and military space budgets of Europe, the United States, Japan, Russia, China, and Canada from 1999 to 2004.

NASA has the highest space budget, followed by Europe, Japan, and China. However, in 2005, the U.S. Congress allocated to the Department of Defense (DOD) $19.8 billion for military space activities. The military space budget is forecasted by 2008 to reach $28.7 billion per year, as mentioned by Smith (2005). This expected increase of military space budgets can constrain the allocation of U.S. civil space budgets for future Moon and Mars visions. As the need to continue the war on terrorism will most likely require an increase of military space budgets, it may result in reducing the civil budgets. Therefore, in order to secure independent and sufficient financing schemes for the successful implementation of the Moon and Mars visions, new financing approaches will need to be considered. Various authors have come up with different proposals for financing human interplanetary space missions.

Attracting venture capital for financing projects under the programs of future Mars mission has been discussed by Livingston (2002) as a potential approach for securing financing. However, the space agencies, space companies, and collaborations will have to develop business plans and cost benefits analysis including cash flow and net present value (NPV). Space agencies, space companies, and collaborations will not only have to be able to attract funding, but also to achieve cost savings (direct benefits) and identify new industrial applications and new market opportunities for the developed technologies (direct benefits). The above approaches can be considered for securing long-term funding, as the projects under the programs will be valued

Figure 4. World space budgets from 1999 to 2004 (Eurospace, 2004)

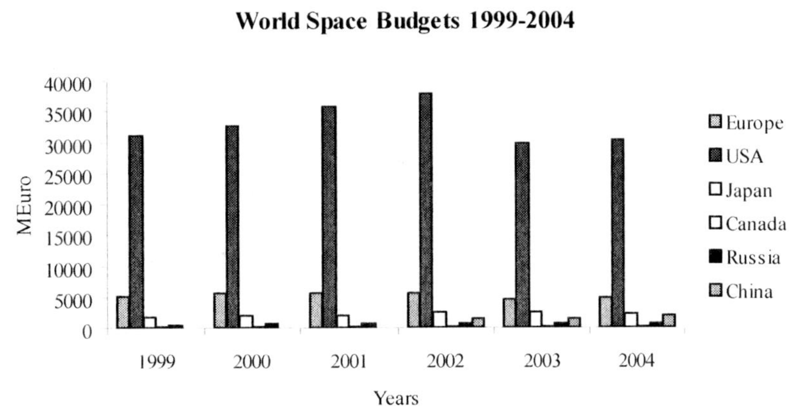

based on market and economic forces instead of political ones.

Private donations, advertising, promotion, and commercialization of space technology are approaches proposed by McCullough (2002). The author proposes to attract funds from the use of donations, sales of television rights, sales of sponsorship to corporations, and sales of merchandising rights. Space agencies will gain revenues (direct benefits) from the sales of the above TV, sponsorship, or merchandising rights and increased positive public image (indirect benefits). Meanwhile, collaborations will be able to develop space brands (indirect benefits) and new markets (direct benefits). Customers and the general public will learn about interplanetary space exploration, international cooperation, technology innovation, and various space brands (indirect benefits). Other ways for encouraging the private investments in the future Moon and Mars missions are the implementation of the future space visions tax incentives to companies, which should encourage investments in space and space technology. Tax law can be changed to make profits from space investment tax free until they reach some predetermined multiple (e.g., five times) of the original amount of the investment (Aldridge, 2004). These approaches can be widely used as complementary sources for attracting short-term funding, and will result in direct and indirect benefits for the stakeholders.

The financing of future Moon and Mars missions has also been investigated by Russian authors such as Safronov and Jakimenko (2006). They propose the creation of a Russian space lottery, which is different and unique by its character in comparison with the other lotteries. In order to attract public attention, the authors propose that the lottery generator is onboard the spacecraft and the cosmonauts participate in the lottery. The space agencies could benefit from revenues from the public for the lottery (direct benefits) and from improved public image (indirect benefit). The collaborations will be able to develop space brands (indirect benefits) and new markets (direct benefits). However, the above funding approach may be able to generate short-term funding for certain projects, but not long-term ones.

The implementation of Moon and Mars visions may require budgets allocation and sufficient financing schemes independent from the U.S. Congress. Attracting venture capitalist can secure long-term funding, but the space agencies, space industry, and collaborations will have to prepare business plans, cost benefit analyses, and NPV analysis. Additional to the venture capital could be revenues from private donations, advertising, promotion or commercialization of space technology, and space-based lotteries. The above approaches are mostly short-term; however, the use of private funding will reduce the impact of the political forces on the programs and increase the importance of the market forces.

FUTURE TRENDS

The analysis of the costs, future markets, public budgets, financing approaches, and benefits for the various stakeholders gives an indication of expected future trends in the implementation the future Moon and Mars interplanetary visions.

Interplanetary space exploration is dangerous and challenging and requires the best technology in order to protect astronauts from the harsh environment of space. Space exploration is costly and often cost estimates of today are at least doubled during the actual implementation of the programs. As discussed in the third section, the cost of the Apollo moon program and 11 crewed missions was around $130 billion of today's money, and the ISS program has cost about $100 billion over 10 years (ESA, 2005). Changes in the cost estimates and possible cost overruns, similar to those experienced by the ISS program, can be expected for future Moon and Mars missions. This may result in higher prices for commercial users willing to use space technologies.

The overview of the future stakeholders in the fourth section showed that space agencies, and in particular NASA and ESA, are officially committed to the implementation of the future Moon and Mars space visions. In the future, NASA is expected to be buying space-based services from the space industry. The political presence of the U.S. space industry will probably increase, as the industry is expected to continue and increase its lobby activities through the U.S. Coalition for Space Exploration. Existing and future collaborations will be involved not only in commercialization of space technologies, but also in spin-in and spin-off of space technologies. Therefore, this involvement will influence and change their objectives, activities, product, services, and types of structure (PPP, Joint Venture).

The analysis of the expected market evolution in the fifth section showed an expected transition of ISS markets toward those of technology for interplanetary missions in the frenzied and turbulent stages. These market developments will result in the creation of self-sustainable markets which will not rely on public investment for their development. Furthermore, they will be characterized by operating in a competitive environment, in which market forces will be dominant. As a result, market entry conditions for new companies in space industry will reduce. Space companies can increase their investment in new market development and space agencies can reduce theirs.

The diverse market sectors presented in the sixth section show that the benefits for the public from these missions of international cooperation, technology innovation, and space brands are the least defined and transparent. However, an expected trend is that space agencies become more transparent in defining and communicating the direct and indirect benefits for the stakeholders from the Moon and Mars space exploration programs.

The allocation of U.S. civil space budgets for the future Moon and Mars visions can be constrained by the expected increase of military space budgets, as discussed in the seventh section. Therefore, space agencies will have to look for alternative and independent financing schemes, such as attracting venture capital, generating revenues from promotions, or implementation of a space lottery. A future trend will be that space agencies, space companies, and collaborations start regularly to apply cost benefit analysis and apply NPV calculations for projects under the interplanetary programs. The use of cost benefit analysis will not only support the identification of the direct and indirect benefits for the various stakeholders, but will also result in an increased public and industry support to the Moon and Mars space exploration visions.

CONCLUSION

Commercialization of space technology of Moon and Mars missions has to be considered not only because of new market opportunities and expected high cost and the need for technology innovation, but also because of the direct and indirect benefits that it can bring to society.

At present, human space exploration equipment, and especially launches, are prohibitively expensive, and companies aiming for organizing privately funded space exploration missions will have limited results. Furthermore, the assumed cost estimates for the implementation of future Moon and Mars missions most likely will change and may even result in cost overruns. The high cost most likely will result in international cooperation (indirect benefit) between space agencies and space companies.

The ISS markets development of nascent stage will probably enter the frenzied and turbulent markets evolution stages for space technology of Moon and Mars missions. In these stages profits will rise, new markets will be developed, and the stakeholders can use the already identified industrial applications from ISS markets and set policies (i.e., property rights (IPR) and marketing rights). This market evolution will result in the creation of self-sustainable markets that will not rely on public investment for market development. Space agencies and collaborations can generate direct benefits from the sales of IPR rights, marketing rights, images, and movies. At the same time, space companies can gain direct benefits from the development of reliable technology. Increased public image and international cooperation are indirect benefits for space agencies and collaborations. Technology innovation and development of space brands are indirect benefits for space companies.

Potential markets for the space technology of Moon and Mars missions will be different from the ISS markets, due to the necessity for the development and spin-in of new technologies. Safety, reliability, and technology innovation are criteria considered when targeting markets and nonspace customers.

Moon and Mars missions can become a platform for the potential development of diverse market sectors, such as osteoporosis prevention and other medical therapies, games, and infrastructure services. Nonspace companies can generate sales (direct benefits) from marketing rights and medical devices (diagnostics, etc.) and computer games, or they can develop space brands. In this way, space companies can also increase their sales, develop space brands, and increase the reliability of their technology. However, the direct benefits for the public will be less clear and transparent, and space agencies will therefore need to clearly define them and communicate them.

The NASA cost estimate of $104 billion for the implementation of part of the Moon and Mars space visions will probably increase and is liable to cost overruns. Furthermore, the expected increase of future military space budgets may constrain the allocation of civil ones. Attracting venture capital is a long-term financing approach that will require from space agencies, space companies, and collaborations to prepare business plans and implement cost-benefits analyses for the various projects under the Moon and Mars programs. Private donations, advertising, promotions, commercialization of space technology, and space-based lotteries are more short-term funding approaches. Space agencies will have to secure independent and sufficient financing for the implementation of the Moon and Mars visions.

Identifying, defining, and analyzing the direct and indirect benefits from commercialization for the future stakeholders in Moon and Mars missions will contribute to public and political support and their successful implementation.

REFERENCES

Aldridge, P. (2004). *A journey to inspire, innovate, and discover.* Report of the President's Commission on Implementation of United States Space Exploration Policy. Washington, DC: U.S. Government Printing Office.

Bush, G. (2004). *Renewed spirit of discovery: The president's vision for U.S. space exploration.*

Сафронов, М. С., & Якименко, И.Р. (2006). Некаторые перспективы космической деятельности на пороге освоения ресурсов Луны и дальнего космоса. Федеральное Космическое Агенство. Retrieved March 3, 2007, from http://www.roscosmos.ru/luna.html

Churchill, S. (1999). Introduction to human space life sciences. In A. Houston & M. Rycroft (Eds.), *Keys to space* (pp. 18-13). USA: McGraw-Hill.

Coalition for Space Exploration. (2004). *CSA joins Coalition for Space Exploration.* Retrieved March 3, 2006, from http://www.californiaspace-authority.org

Collins, P. (2002. June 4-7). Preparing for passenger space travel, the future of human spaceflight. In M. Rycroft (Ed.), *Beyond the ISS: The future of human spaceflight. Proceedings of an International Symposium* (pp.165-172). Strasbourg. Kluwer Academic Publishers.

ESA. (2001). *General policies and procedures, process, pricing and conditions for sale.*

ESA. (2005). *How much does it cost?* Retrieved March 3, 2007, from http://www.esa.int/esaHS/ESAQHA0VMOC_iss_0.html

Eurocontrol. (2000). *Guidelines for the economic appraisal of EATMP projects: The effective use of cost-benefit studies.* PLC.ETL.STO7.1000-GUI-01-00.

Eurospace. (2004). *European space directory* (19th ed.). Eurospace, Paris.

Hunt, C. D., & van Pelt, M. O. (2004). Comparing NASA and ESA cost estimating methods for human missions to Mars. In *Paper presented at the ISPA*, Frascati, Italy.

Livingston, D. (2002). $pace: The final financial frontier? In R. Zubrin & F. Crossman (Eds.), *On to Mars: Colonizing a New World* (pp.59-63). Apogee Books Space Series.

McCullough, S. (2002). Financing a manned mission to Mars? In R. Zubrin & F. Grossman (Eds.), *On to Mars, Colonizing a New World.* Apogee Books Space Series.

Morris, J., & Morring, F. (2006, June). Forewarned: Griffin cautions that industry inefficiency could damage U.S. space program. *Aviation Week and Space Technology.*

NASA. (2001). *General policies and procedures, process, pricing and conditions for sale.*

NASA. (2003). *Taking business to the stars: Nasa's commercial space centers.* Retrieved March 3, 2007, from http://spd.nasa.gov/csc.html

NASA. (2004). *The vision of space exploration.* Retrieved March 3, 2007, from www.nasa.gov

Rosaviakosmos. (2001). *ISS commercial program rosaviakosmos policies and procedures.*

Smith, M. (2005). *U.S. space programs: Civilian, military and commercial* (CRS Issue Brief for Congress, Code IB92011). The Library of Congress: Congressional Research Service (CRS).

Space. (2005). Mir milestones. Retrieved March 3, 2007, from http://www.space.com/missionlaunches/missions/mir/_landmarks/_010307.html

Space Daily. (2003). *Government space budgets to continue growth.* Retrieved March 3, 2007, from http://www.spacedaily.com/news/satellite-biz-03zzzl.html

Tkatchova, S. (2005). *From space exploration to commercialization.* Unpublished doctoral dissertation, TU Delft, Delft.

Tkatchova, S., & van Pelt, M. (2004). ISS commercialization paths the way for tomorrow's future commercial missions. In *Final Papers 55th IAC 2004* (p. 567), Vancouver, Canada.

Zubrin, R. M. (1997). *The case for Mars.* USA: Free Press.

Zubrin, R. M., Baker, D. A., & Gwynne, O. (1991). *Mars direct: A Simple, Robust, and cost effective architecture for the Space Exploration Initiative* (AIAA-91-0328).

ENDNOTES

1 Cost benefit analysis is widely used by private and public organisations for appraisal, planning, and decision support of an investment project. It is an analysis that aims at quantifying the costs and benefits of a project.

2 In the cost benefit analysis, Eurocontrol divides the benefits into quantitative and qualitative benefits. Quantitative are cost savings, capacity, reliability, and delays, while qualitative benefits are safety, environmental, international commitments, contingency, and upgradeability.

3 Reliability of space technology is of crucial importance for preserving the lives of astronauts and cosmonauts during their journeys to the Moon and Mars. Furthermore, technology failures problems as, observed by Livingstone (2002), can result in a venture's capitalist lack of interest in the commercial space projects.

4 In 2001, as presented by Rosaviakosmos (2001), the Russians set a price from $2 up to $4 million for an EVA exit of a Russian cosmonaut onboard the ISS. Later in 2002, they withdrew their ISS prices and, at present in 2006, official information on the prices of ISS products and services was not found.

5 The ISS partners building the space station are NASA, FSA, ESA, JAXA, and CSA.

6 This includes ground segment equipment and related services such as satellite tracking, station keeping, and launcher tracking.

7 The Coalition is an ad-hoc alliance created by U.S. space companies (i.e., Aerojet, Boeing, Lockheed Martin, etc.) and its mission is to ensure that the United States remains the leader in space, science, and technology by ensuring that the NASA Moon and Mars initiative is understood and supported both by the public and the U.S. Congress. The Coalition is compromised of three parts; a Public Affairs/Public Outreach team, an Action/Implementation team, and a legislative team, as presented by Coalition for Space exploration (2004).

8 RPC are partnerships between NASA, industrial partners, and universities. The activities of these centres are in space research and product development in the areas of material science, biotechnology, combustion, and agribusiness (NASA, 2003).

9 Self-sustainable markets are markets that do not need to rely on public investment for their development and are dominated by market forces of supply and demand. These markets will continue to develop after the end of the life-time of the ISS, will be able to meet the needs of the future commercial customers for space-based products and services.

10 In the case of spin-in projects in which nonspace companies propose and provide technologies and products to space agencies for future interplanetary missions, the collaborations will be intermediaries between the agencies and the nonspace companies. This means that they can prepare the proposals to ESA and also support the companies in winning contracts for becoming suppli-

ers of certain technologies and products to the space agencies. In return the nonspace companies can repay to the collaborations a certain percentage of the funding of the project, which has been won with the help of the collaborations.

11 Microgravity affects various systems of human physiology, including cardiovascular, respiratory, nervous, human sensory, and balance, resulting in bone mass loss, muscle atrophy, changes in the metabolism, body posture, and others (Churchill, 1999).

Chapter XIII
Challenges in Knowledge Management:
Maintaining Capabilities Through Innovative Space Missions

Larry J. Paxton
The Johns Hopkins University, USA

ABSTRACT

One of the key problems faced by organizations is that of managing knowledge. How does an organization improve and maintain performance by generating, maintaining, and sharing knowledge? High tech organizations are much more dependent on knowledge as a commodity than those in the manufacturing sector. NASA certainly is the epitome of a high tech organization. It faces complex and deep challenges, not the least of which is how to address the loss of knowledge as the workforce ages and retires. In addition, NASA faces the consequences of a program that, in the face of programmatic constraints, subsumes the process of generating knowledge to the demands of maintaining commitments. Those commitments may not provide the optimal path for generating knowledge relevant to the future success of the organization. For a space-faring organization, mission cadence is one of the key determinants of cost and risk. Mission cadence is also important, as it determines the number of people in the organization with direct and relevant experience with space missions. Under a constrained budget, mission cadence can be increased by reducing the size and scope of the missions. Small spacecraft missions can afford to be innovative and thus create a culture in which new ideas are welcomed and sought. These smaller missions can preserve and generate knowledge by training the next generation of scientists, engineers, and program managers.

INTRODUCTION

Knowledge management consists of the tools, techniques, and practices that enable an organization to improve and maintain its performance by generating, sharing, maintaining, and distributing knowledge. NASA is faced with a series of critical knowledge management challenges in the next decade. As the NASA work force ages[1,2] the expertise that enabled robotic and manned missions to the near Earth environment and to the other members of the solar system could be lost unless steps are taken to ensure continuity of experience. This paper will explore a path that is both programmatically and scientifically valid for managing that critical space-faring knowledge.

In addition to the highly visible human exploration program, NASA operates a less visible program of unmanned space science that generally offers a high science return on investment. These missions range from small rockets designed for suborbital flight (a few million dollars each) to the elements of the great observatory program at more than $1 billion. These robotic space missions are key to generating and maintaining core capabilities, not only by providing the training ground for the next generation of scientists, engineers, and managers, but also by providing a proving ground for testing and validation of new technologies. In this important role, they serve as reservoirs and proving grounds for the human exploration program.

In the near term, NASA faces a transforming challenge in order to address President Bush's *Vision for Space Exploration*: the organization must solve the budgetary and technical problem of the shuttle, come to closure on its international obligations to support the International Space Station, repair the Hubble Space Telescope, rebuild infrastructure damaged during Hurricanes Katrina and Rita, develop new vehicles to go to the Moon and Mars, develop the infrastructure needed to support these missions, and accomplish this mission within its current budget while still carrying out its scientific research and unmanned exploration missions. The goals of the vision for space exploration (VSE) places all current and future NASA science programs under tremendous budget pressure (Committee on an Assessment of Balance in NASA's Science Programs, 2006). The primary, long-term, nonbudgetary challenge, then, is that of knowledge management, that is, maintaining the core capabilities required to carry out the broader agency role in space exploration.

At NASA science missions in space are, in many ways, carried out in a more entrepreneurial setting than the human exploration program. A satellite mission to the near-Earth space environment or another planet or solar system object is essentially a very sophisticated robot, though hardly an autonomous robot. The development and operation of these missions are managed either as *facility-class* missions by NASA centers, or their surrogates (e.g., the jet propulsion laboratory (JPL) or The Johns Hopkins University's applied physics laboratory (APL)), or are lead by a principal investigator (PI) and that PI's organization, typically a university or other academic institution. To the average citizen, the rigorous and highly competitive manner in which these robotic space missions come to be may not be obvious. The science community participates in NASA-directed "roadmap" activities[3] that provide a rough cut at a timeline for the development of facility-class missions. A facility-class mission is, typically, managed by a NASA center, such as TERRA or AQUA for Earth Science, or the Mars Rovers for planetary exploration. That mission is designed to provide data to meet broad, programmatically-driven science and measurement requirements. The priority of missions within a discipline is determined through a series of meetings, trade studies, and internal and external assessments. The timeline for these facility class missions, for example, the *Living with a Star Program*, may change in response to budgetary pressure or changes in priorities. In addition to these facility-class missions, which are ideally protected by the identification of a "program line" which appears in the NASA budget, ad-hoc science teams are spontaneously formed among members of the research community to address high-value science problems that are broadly recognized as being urgent and relevant to NASA-wide goals. These ad-hoc teams are lead by a PI—frequently, the PI is not from a NASA center and may be from a university—and respond to broad announcements of opportunity (AO) to propose a science mission in a particular discipline area within a particular cost range. The AOs are released at somewhat regular intervals, although the gap between AOs will respond to budgetary pressure[4]. The selection process involves the review of the science proposed and is conducted by a peer review committee and an assessment of the programmatic and technical risk of the proposed mission. The term that has the greatest currency among the science community and denotes the cachet that these missions must possess is that these proposals must address *compelling science* in order to be selected. An overview of this process with a focus on PI-led missions is to be found in the national research council report (2005).

This system of doing science-driven research has been largely separate from agency-wide policy and the activities of other NASA directorates until recently and that change has in and of itself created a significant difference of perception between NASA and the science community[5]. Under NASA Administrator Mike Griffin, there has been a stronger coupling between the NASA science missions (in particular, the activities in the exploration systems and science directorates) and the crewed missions as these basic research directorates have taken on responsibilities under the vision for space exploration[6] to address the needs of future missions to the Moon and Mars. Administrator Griffin[7] has said "his biggest challenge has been to find a way to fund missions under the vision for space exploration while maintaining stable workforces at each of the agency's 10 field centers." Note that many of the field centers carry out a significant amount of basic research and that reductions in force have already taken place.

During this period of intense pressure on the science budget, NASA managers and members of the science community should ask: How can the core knowledge and capabilities be preserved? One thing seems clear, that in order to maintain the vigor of the robotic science program, mission costs will have to be reduced to accommodate the constraints of contracting budgets.

THE FOUR GOALS OF INNOVATIVE SPACE MISSION DESIGN: DOING THINGS FASTER, BETTER, CHEAPER AND WITH LOWER RISK

One often hears "You want faster, better, and cheaper? Choose any two!" This has certainly been true of the human spaceflight program. Implicit in this articulation of the mission design trade set is the notion that risk is to be avoided; in particular, risk that would lead to mission loss (especially loss of life). The desire of an organization—and the people that comprise or identify with that organization—to avoid failure can lead to decisions that may not be the soundest from the standpoint of developing new capabilities. As a heuristic example, consider a purely hypothetical case in which the probability of flying a \$250 million is 90%. If the cost of increasing the probability of success to 98% is \$100 million, is it worth it? From a cost viewpoint it is not, but the benefit of avoiding a failure may be worth far more than the \$100 million cost if the consequences of a failure are deemed too dire.

The cost of failure is readily appreciated: the cost of failing to innovate is much less readily apparent—particularly in a culture where "failure is not an option" (e.g., Kranz, 2000). Innovation in the aviation industry was fostered by a willingness to take risks and the early, pioneering days of aviation saw a very high accident rate, as evidenced by personal memoirs (see e.g., Doolittle and Glines

(1991) for a particularly engaging look at the risks of the early days of aviation) and official histories. In aviation, that innovation was driven largely by national interests and implemented by commercial ventures (often seeking a competitive advantage in the attempt to acquire government contracts). The impetus for innovative robotic science missions is very different and the argument to be made for accepting risk in these missions is both substantively different and more complex.

Most of us would accept the argument that, all too often, doing things faster means taking bigger risks. One might also argue that doing things faster is a key component of doing things cheaper due to labor costs. Doing things faster, or increasing the mission cadence, has another important and beneficial effect. Mission cadence is a major enabler of innovation and the driver for the training and evaluation of the next generation of managers, engineers, and scientists. A high mission cadence is required to maintain and develop competence in mission design, management, and execution, and for an exploration-driven organization, to develop and train the next generation of leaders. The time between missions must be short enough that careers span the complete life of more than a few missions so that expertise can be acquired and applied to key phases of mission design. A high mission cadence reduces risk because the *lessons learned* are current, applicable, and widely held. Knowledge can be widely disseminated, as the groups involved in mission design are not kept in stasis for years as they work to complete their missions, and their knowledge is more readily disseminated when teams are not held in what amounts to isolation by the long time-scales of the mission life cycle.

This balancing act is conceptualized in Figures 1 and 2. In Figure 1, we consider the case where we are assessing the risk that a mission will be successfully executed for a range of program development durations. Space missions typically are described as having *phases.*[8] These phases range from A (definition), phase B (risk reduction), and phase C/D (implementation through launch and early operations), to phase E (routine mission operations). Experience shows that programs that have a moderate mission development lifetime (phase C/D of 3-5 years) tend to be viewed as more successful (in terms of an assessment of relevance, overall cost, and ratio of return to the cost of operating the satellite, which tends to increase as the technology ages) than programs that have a very long development period. In fact, programs that suffer very long development phases are often seen as having been overcome by events and may have been marginalized in addition to the costs and risks associated with using outdated hardware and ground systems. Figure 1 actually can be applied to any phase or to the entire program; the time axis need only be adjusted accordingly. For the sake of this discussion, we will consider just phase C/D, the phase where active development and construction occurs and where most of the money is spent. Figure 1 is intended to convey a few ideas based on experience in robotic space missions: (1) space missions can be done with a very short turn around and still be successful, (2) to some degree, increasing the mission duration can increase the probability of success as more time for design and test activities can be incorporated, (3) there is a broad middle range where the mission has enough time to be fully developed without overstressing the team members, while still maintaining the community support and relevance that brought it into being, and (4) as the mission duration increases, the program experiences the loss of the original plankholders from the team as they move on to other activities, the technology incorporated in the ground and space segments ages, and the community and agency support that caused the mission to come into being wanes as it is seen as less relevant (workarounds and other data sources will be developed and exploited if the mission objective is important and urgent).

Figure 2 presents another view of the problem in which the schedule, cost, and technical risk

Figure 1. Risk of failure depends on the duration of the program. Determining the optimal duration of a program from a technological risk standpoint may have a different timescale than that for managerial (especially cost and schedule) risk.

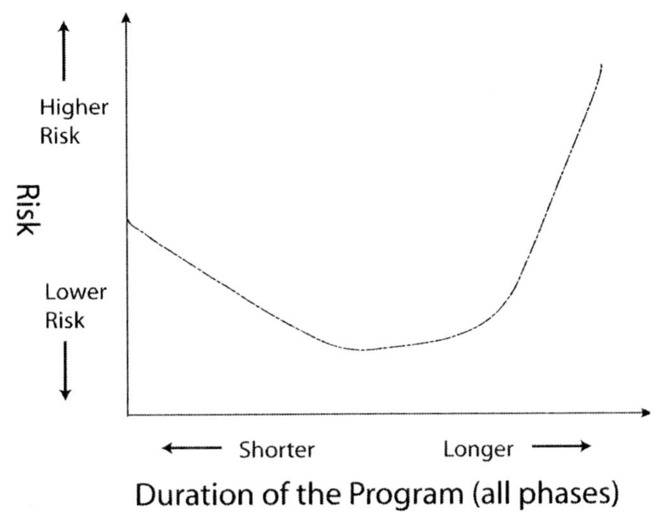

Figure 2. Risk is a function of the time between missions. There exists some optimal frequency for missions. That frequency may vary from institution to institution or even among program elements.

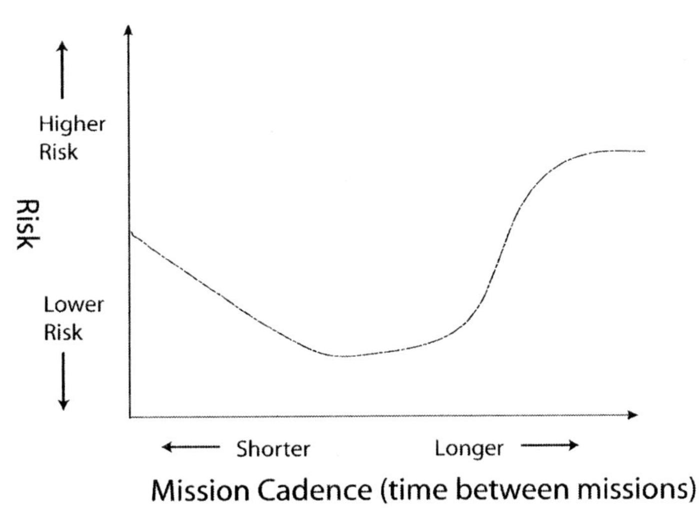

of a mission is viewed in terms of the frequency of missions. If the mission cadence is very high (short duration between missions), risk tends to increase due to scarcity of resources, which lead to resource conflict. This resource may be a launch facility, a launch crew, a thermal-vacuum facility, or any other resource that is used by many different programs. Facilities are a two-edged sword. They reduce costs as long as they are fully utilized, but as the usage rate drops the amortized cost increases. That facility usually represents a large capital investment and may be a rate limiting factor at both ends of the curve. For long durations between missions, critical facilities (or core capabilities) may not be maintained because they are seldom used. This leads to a very high amortized cost. Consider, for example, the U.S. space shuttle program. If the cost of the entire shuttle program is divided by the number of launches and that cost is used to determine the cost per kilogram to orbit, the cost of the shuttle as a payload delivery system is clearly quite high (on the order of $100,000/kg) for access to low Earth orbit. Part of the reason for the high cost has been that expenses continued to accrue during the stand-downs after the loss of the Challenger and Columbia. Figure 2 shows that there is some region where cost, schedule, and technology risks can see significant reductions. As the time between flight opportunities increases, this leads to an increase in risk. At some interval (the approximate value depends on the size of the program, the level of complexity, the stability of funding for the program line, etc.) there will be a fairly rapid increase in risk until the point is reached where each mission is essentially independent of all others, and there is little knowledge transfer between programs as there is little if any concurrency and the mission line stagnates.

Mission design life is also an important cost driver. Mission design life can be driven by a variety of factors. A mission to Pluto, for example, must have a very long design life because it takes many years to reach Pluto. On the other hand, a mission may have a very long design life, or mean time between failure (MTBF) because there must be some assurance that the mission will last for a specific period. There are different ways of achieving a long mission life, particularly if continuity of measurements is the driver for mission life.

The MTBF is typically calculated by determining the probability of failure of a component or subsystem through the use of a reference handbook such as MIL-HDBK-217[9]. Note that the probability of failure of a group of components that must all be functional in order to work is the sum of the probability of failure of each individual component. Thus, complexity reduces lifetime, or to achieve lifetime with complexity a system must be made redundant which increases cost. Redundancy increases reliability because the failure probabilities are additive within a system but multiplicative for parallel systems. In other words, if the probability that a single component will fail before a time t, P(t), is given by:

$$P(t) = 1 - e^{-t/MTB} \tag{1}$$

$$R(t) = e^{-t/MTBF}$$

where *R(t)* is the reliability or probability of lasting to a time t. Note that both *P(t)* and *R(t)* have a range from 0 to 1. Then the probability that a system consisting of *n* dependent or series components will fail after a time t is given by:

$$P(t) = 1 - \prod_{i=1}^{n} R_i(t) \tag{2}$$

If the system is constructed of n components, each of which must fail, then the probability of failure is:

$$P(t) = \prod_{i=1}^{n} \left(1 - R_i(t)\right) \tag{3}$$

One can readily demonstrate that the reliability of two components that must work in series in order for the system to work (i.e., a *single string* system) is

$$R(t) = R_1 R_2 \qquad (4)$$

and that for a *dual string* system, which will work if any one of the components fails, is

$$R(t) = R_1 + R_2 - R_1 R_2 \qquad (5)$$

where $R_1(t)$ is the probability that that component 1 will last a time t and $R_2(t)$ is the probability that component 2 will last a time t. Equation 5, the reliability of a dual string system, is always greater than Equation 4 a single string system, provided that the MTBF for both systems are the same. This expression also shows that breaking a system up into redundant components can reduce the requirement on the MTBF, which may allow one to consider using more cost-effective solutions that have a lower MTBF.

There is an additional benefit of relaxing the requirement on the mission lifetime. Since the MTBF requirement grows exponentially to meet the lifetime requirement, there are fewer parts that meet those requirements (this precludes the use of low cost commercial off the shelf (COTS) technology) and they are much less capable and cost more. Redundancy provides a partial solution, as it does come with a penalty in mass and component expenses and testing. In practice, it is often wise to ensure longevity by making certain components of the system dual-string and leaving others as single-string. Parts grade is another variable. A single string design using class S parts (parts that are designed for high reliability space missions) can actually have a higher reliability than a redundant class B design, though it may be much more expensive. This is all part of the trade-space that must be investigated, especially when attempting an innovative design. One should note that a low cadence of space missions does lead to a reduction in the availability of class S parts, and the lack of impetus for the development of new, better class S parts.

Figure 3 provides a qualitative view of the factors that affect the mission cost as a function of the required lifetime of the system. Note that the placement of the elements along the vertical axis will vary from mission to mission. The graph is, however, intended to convey how these costs change with mission duration. In Figure 3, the spacecraft and payload are combined into a single element labeled *spacecraft* on the graph. Spacecraft costs are often driven by the requirement to meet mission performance requirements at the end of life of the mission. In order to achieve these requirements, the spacecraft design may have to incorporate redundancy or high reliability electronics. This will drive the cost up and it may even increase the launch vehicle cost if the total spacecraft mass grows in response to these performance requirements. Launch costs tend to increase with mission duration rather than staying constant in fixed year dollars because a lower cadence, such as dictated by longer mission durations (under the assumption of fixed budgets), will lead to fewer launches, which in turn leads to higher costs per launch (due to the practice of amortizing the costs of the facilities both at the manufacturer and at the integration and launch site over the number of missions). This last trend has just begun to impose a significant burden on NASA missions through the application of what has come to be called at NASA *full cost accounting*. Full cost accounting (FCA) is a familiar business practice that seeks to assess the true cost (direct and indirect) of an activity. Only in the last few years has NASA begun to address the full cost of a particular project. The impact of FCA on the ability to sustain a mission line with few launches and major requirements for the long term use and availability of unique and underutilized facilities has yet to become fully apparent. Ground segment costs tend to rise with mission duration for two reasons: the first is that the duration of the mission itself leads to an increase in the total cost

Figure 3. Cost as a function of mission design life. As mission design life increases, both total cost and the cost per year of the mission increases for phase C/D. Increasing phase E lifetimes may reduce the average (over all phases) per year cost .

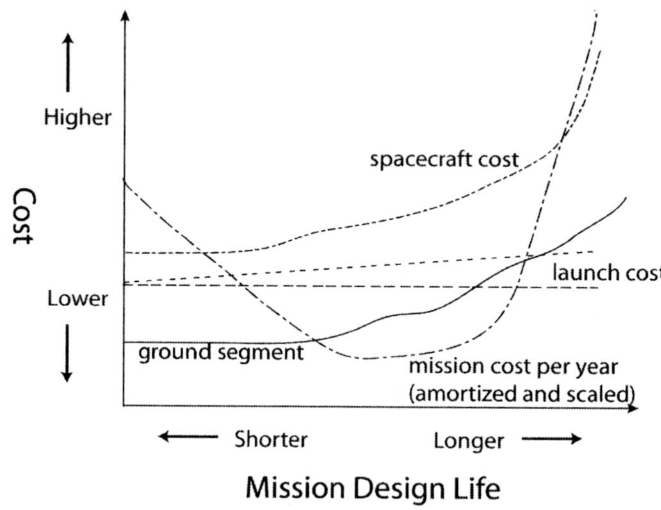

of the mission and the ground segment must be continually staffed (so-called *marching army costs*) though, provided the knowledge base of the mission can be efficiently managed, this can be contained to some degree; the second is that technology and standards evolve and the mission must go through periodic upgrades or be forced to maintain aging and difficult-to-support systems. For the later driver, either course carries a concomitant personnel charge. The per year cost of the mission will then have a minimum that represents the optimal duration for a mission in terms of being the most cost-effective way to maintain a capability. For a "monitoring" mission (one in which certain parameters must be observed for a long term, as in the practice with weather satellites and terrestrial imaging) it may well be more cost effective to launch a series of missions with a more modest lifetime requirement than one "Battlestar Galactica" mission that provides all capabilities for a very long time. This is particularly true if the launch costs represent a small fraction of the total mission cost (perhaps less than 10%). The obvious corollary benefit of this approach is that mission cadence is increased with the concomitant benefits to knowledge management as well as providing the opportunity to refresh the implemented technology.

BALANCING NATIONAL POLICY AND SCIENCE AT NASA

NASA's organizational goals have changed in the last few years, largely in response to President George W. Bush's vision for space exploration (VSE, 2004)[10]. As the document states (p. 5): "The fundamental goal of this vision is to advance U.S. scientific, security, and economic interests through a robust space exploration program. In support of this goal, the United States will:

- Implement a sustained and affordable human and robotic program to explore the solar system and beyond

- Extend human presence across the solar system, starting with a human return to the Moon by the year 2020, in preparation for human exploration of Mars and other destinations
- Develop the innovative technologies, knowledge, and infrastructures both to explore and to support decisions about the destinations for human exploration
- Promote international and commercial participation in exploration to further U.S. scientific, security, and economic interests"

This policy statement was first articulated in 2004 under then NASA administrator Sean O'Keefe. Dr. Mike Griffin became NASA administrator in 2005 and began to reshape NASA programs to bring them into alignment with national policy goals, in particular the VSE. As Dr. Griffin pointed out[11], the NASA budget is not sufficient to address all the programmatic requirements of the VSE and maintain the same level of funding for basic research. The budget limitations require a rebalancing of priorities. The current priorities are to return to flight after the loss of Columbia so as to fulfill the international obligations to the International Space Station (ISS) and to develop the new crew exploration vehicle (CEV). This new venture has been interpreted as requiring a significant long-term reduction in the amount of funding available for basic research and analysis (over $3 billion between now and 2010 in the science mission directorate). Dr. Griffin prefers to take the long view, as articulated in the February 17, 2006, hearings before the House Science Committee[12]. There, Dr. Griffin called the reductions a necessary part of the plan to get the human exploration program back on track. But will there be a loss of the knowledge base that is embodied in the faster, cheaper missions driven by the basic research community?

The Space Studies Board of the National Academy released a report that reviewed the impact of these cuts on the science programs. That study (Committee on an Assessment of Balance in NASA's Science Programs, 2006) pointed out that, while the reduction taken in FY06 is significant, it should be compared to the liens against the system that were put in place as late as 2004 (when the VSE was announced). In 2004, before the VSE, the budget for NASA's science mission directorate (SMD) was to grow from $5.5 billion in 2004 to $7 billion in 2008 to accommodate planned science missions. As pointed out in the NRC report, the FY07 request for SMD is $200 million less than the FY04 request even before taking inflation into account. The following years are planned to show a decrease in available funding as the projected budget growth rate is lower than the estimated inflation rate. The other NASA directorate that deals with robotic space exploration missions is the exploration mission directorate (EMD). There, the cuts are even larger, as the current plans show a drop from $950 million in 2002 to under $300 million in 2007 and subsequent years. The NRC report found that the research programs in SMD and EMD were not sustainable and were "not properly balanced to support a healthy mix of small, moderate-sized and large missions." Administrator Griffin, in testimony before the Senate Commerce Committee[12], made it clear that he views the science programs as being carried out on a "pay as you go basis." This statement and the clear prioritization of science as secondary to the ISS, Shuttle, and CEV means that "we must defer some missions that we would prefer to do now, but simply cannot afford at this time."

THE CHALLENGE

In the face of a declining budget for science missions, the obvious question is what can be done to manage the knowledge held in the collective experience of those currently doing science missions on the ISS, shuttle, or satellites. Small missions hold the key to preserving core competencies as well as maintaining the momentum of existing programs. In the remainder of this chapter, I consider lessons from the past, in particular how things can be done faster, better, and cheaper, and what that means in terms of transforming the culture in at least one part of NASA.

NASA's robotic missions are generally competed. These competitions are held through announcements of opportunity (AO) for different mission classes. Each mission class has a fixed cost cap. For example, the scout-class missions to Mars have a cost-cap of around $450 million, including reserves, launch vehicle and run-out costs. Experience indicates that the vast majority of these PI-led proposals come in right at the cost cap. This brinksmanship arises because the proposals are judged at two levels: the scientific content which is the key enabler and the first step in the selection process, and the second step which assesses the reasonableness of the cost and the technical risk. With that knowledge, program participants may well be tempted to place liens against programmatic reserves even during the proposal process because they know that unless they can achieve a stunning science goal, they are not going to be selected. There is a temptation to feel that this is a reasonable approach because "we'll figure this out in phase A (risk reduction)." Reserves are often viewed by proposers as an impediment to pricing a winning proposal, rather than as a necessary part of the management process because the margins are large, typically 20% to 30% of the total mission cost (including the launcher). To understand why this might be so, the margin should be viewed in contrast to the direct salary support for science support, which often amounts to less than 5% of the total budget.

Facility-class missions also have issues with cost management. All too often, NASA, in response to external factors (ranging from Congressional earmarks to budget cuts in program elements), is unable to meet its commitments for funding and unable to control costs due to a lack of sufficient programmatic insight and an unwillingness to prevent changes in scope in the mission design. These factors lead to cost and schedule growth and tend to have a snowball effect. One mission's cost growth in a fixed budget program (e.g., the solar dynamics observatory in the Living With a Star line) will lead to delays in future elements in the program and cause increases in the total out-year costs because, in order to keep those delayed programs alive, additional funds are required to keep the teams together. This pushes the start of the next mission in the line further back and further erodes the science return because many missions rely on the availability of other information from existing, finite-lifetime, space assets to leverage their science.

Clearly, NASA needs to turn its attention, once again, to doing things "faster, better, and cheaper." But that experience seems to have been a failure, at least in the minds of some. What are the lessons to be learned from the past, then?

Under then administrator Goldin, NASA was challenged to do things "faster, better, and cheaper."[13] Mr. Goldin came to NASA with a background in small satellite development and saw the need to transform the mission timescale from decades to years. FBC did revitalize space science at NASA. The FBC approach led to some failures but, more significantly, an increase in the cadence of missions. There were far more missions, large and small, than there had been in the previous decade, and confidence was restored in the agency to such an extent that Congress supported substantial increases in the science and exploration programs.

Under the FBC impetus, planetary exploration was reorganized and transformed from the huge probes of the 1970s and 1980s (such as Voyager and Galileo) to the new Discovery program with the low cost NEAR mission (http://near.jhuapl.edu/) as the first example. Space science was also ripe for transformation and was able to start two important mission lines within the Sun Earth Connections program element: the solar terrestrial probe (STP) and living with a star (LWS). These lines have embodied much of the spirit of FBC, especially the first of the STP line, the TIMED mission (http://www.timed.jhuapl.edu/).

During Goldin's term as administrator, the Earth Science Enterprise (now part of the Science Mission Directorate) was obligated to carry out the very ambitious Earth observing system (EOS) program. The program was restructured during the early 1990s as the result of cost overruns and budget pressure. The scope of the EOS coverage was reduced from a series of large platforms to three missions. With the recent launch of the last of these, AURA, the legacy of the mid 1980s is closed for Earth sciences. It remains to be seen what will replace those programs. During the FBC era, new approaches were attempted, such as the successful SeaWifs and QuickSCAT programs, and the creation of a series of smaller satellite missions, including the ESSP program's CALIPSO flight. As an example of an innovative solution to the vexing problem of requiring data continuity, enhanced capabilities, and simultaneous coverage, the EO-1 mission, launched in November 2000, was designed to fly in constellation with Landsat 7 satellite. EO-1 demonstrated that a small satellite mission could fill a gap and provide new measurement capabilities.

Programs conducted under FBC experienced failures and FBC, as a management tool, suffered from a number of well-documented problems (see e.g., McCurdy, 2003; OIG, 2001; Spear, 2000). The core weakness of the NASA FBC implementation was that managers were encouraged to change the way they thought and the way they did business without establishing FBC as a management policy. There was no defined policy or guidance for FBC, and without that NASA could not communicate its principles to the staff in an effective way. In addition, FBC goals, objectives, and metrics could not be part of the NASA strategic management process. What, then, did people think "faster, better, cheaper" meant?

DOING THINGS FASTER, BETTER, CHEAPER

Much of the space community would point to Kelly Johnson, director of Lockheed Martin Skunk Works, as someone who was able to successfully implement "faster, better, cheaper" as a management approach. While Mr. Johnson's record of accomplishment is astonishing (e.g., Janos & Rich, 1996; Johnson & Smith, 1990), NASA operates in a very different environment. The Skunk Works projects were executed in an environment notably devoid of public or congressional scrutiny and isolated from internal Lockheed oversight. A single sponsor existed, and that sponsor was able to operate with near total autonomy and so could afford to take risks and sustain funding in the face of *challenges*. In addition, the Skunk Works was doing work that was universally seen, during its heyday, as being vital to the security of the United States. The key principles that Mr. Johnson based his organization on are found on many Web sites, including the Lockheed Martin site and in various books about the Skunk Works and Kelly Johnson. Some of the major principles are that:

- The manager has to be in complete control and be guaranteed continuity and adequacy of funding.
- The team should be small and devoted to the project and should internalize as much of the design process as possible and minimize documentation.
- The core competency has to be maintained where the work is done and by the group responsible for the work (managing the work of others is not enough).
- Status and pay for managers should be derived from success of the project, not the number of people supervised.

These guiding principles minimize the number of people involved, as well as formal interfaces and documentation. In practice, the Skunk Works was almost completely isolated from the parent corporation and was relegated to a rather Spartan setting—nothing like the research campus settings commonly sought for R&D activities today—and Johnson strove to preserve that isolation and embodied that in his operating tenets in several forms. One point that is often overlooked in reading these tenets is that the entire organization was located at one facility, all the functions were integrated, the work was carried out with a *hands on* approach, and that organization was tight-knit and shared a common vision and a sense of the importance and priority of their work and mission. These factors established a group identity and a unifying sense of purpose. The Skunk Works experience does hold some truths that can be incorporated into individual projects, and PI-led robotic space missions are particularly well-suited to the application of these principles.

While the Apollo program demonstrated that even an extremely large program can succeed provided there is a shared sense of vision and purpose and adequate, continued funding, NASA had begun to sow the seeds for the dissolution of the group identity even as it developed the Apollo program. In order to meet the ambitious development schedule, the Apollo program had to make extensive use of contractors. NASA's experienced technical people became responsible for oversight and contract management and, as they became more divorced from the day-to-day technical activities, instituted a series of formal systems for managing the project that further eroded group cohesiveness. At the conclusion of the Apollo program, the space program faced a fiscal and ideological crisis yet many of the management structures and practices, developed for a very different environment, survived. One of the key cultural artifacts was that risk should be avoided and that failure was not acceptable. This management system, developed for manned spaceflight and generally instituted, carries a large cost burden.

Execution of the vision for space exploration will be played out over decades and with hundreds of billions dollars. Risk and the management of that risk will be a key factor in that vision. The VSE has the potential to provide a renewed sense of purpose and a new organizing principle for NASA. In order to meet VSE's major objectives and still address the needs of the science community, Dr. Griffin must establish a culture at NASA that handles science missions in a different fashion from the human exploration program. The key to success is going to be how risk is managed and how an appreciation of that risk is managed within the organization.

NASA still bears the burden of the loss of the Challenger and Columbia shuttles, and that burden will shape any crewed exploration concept. In addition, there have been the highly publicized losses of robotic missions, including the Mars missions (Mars Climate Orbiter, Mars Polar Lander, and the DS-2 probes) as well as the CONTOUR loss and GENESIS failure. The approach to managing risk should, in principle, be tailored to the scope of that particular activity. For example, a university-class explorer ($10 million or less) can not afford to be held to the same risk management standards that the crewed exploration vehicle must be. Risk management and its incarnation (testing, documentation, reviews, *quality assurance*) can have an enervating and demoralizing effect on the staff if not managed carefully. In fact, the burden to prove that risks have been identified and *managed* can be costly and onerous, affect team productivity and cohesiveness, and lead to schedule and cost risk. A key question is: where does the balance lie?

NASA has experimented with space science missions lead by PI. The PI is, generally, the one person totally responsible for a focused mission. These people have successfully managed to lead a team that wrote a science, technical, and management proposal that has been selected from a peer-review in an open competition as being the very best (subject to programmatic considerations) answer to a scientific question that can only be addressed by a space mission. Many of these PIs are from universities and are used to working in an environment very much like that specified by Kelly Johnson. However, as one can readily appreciate, the drive and focus that has lead to the development of the PI's career as a world-recognized leader in a particular research area has not presented the PI with many opportunities to develop management skills. In addition, many PIs hold academic positions.

Academic institutions may, or may not, recognize the value of having a faculty member lead a space mission. In addition, few academic institutions have the facilities and contractual mechanisms to manage such an activity. PI-led missions have come under scrutiny because of failures (e.g., Kerr, 2004; Space Science Board, 2004). PI-led missions are, however, essential, as they address grass-roots sciences goals not met by facility-class programs and, by being freed from the constraints of a large process-oriented organization, can achieve faster and better missions, which will cost less in the long run.

WHY IS DOING SPACE MISSIONS FASTER, BETTER, AND CHEAPER SO HARD?

Risk, whether it is in our individual lives or in a commercial or technical endeavor, is often deemed acceptable until a failure occurs. The determining factor is whether, in hindsight, that risk is viewed as arising from *weakness* or as a natural consequence of the creative and boundary-defining experience. The fate of any organism is determined by its response to risk and failure. When NASA's articulated approach to doing space missions changed to FBC, successes and failures occurred. We can learn from how those events were handled.

In order to change the way things are done, the NASA administrator and staff must send a clear and unequivocal message that new and innovative solutions to specific kinds of problems are sought and that risk (and failure) must be accepted in certain areas or the organization will not change. Accidents occur in all systems, simple or complex (see e.g., Dorner, 1996; Perrow, 1999), and yet we tend to view accidents as the result of a lack of planning or foresight when, from a purely practical sense, the planning and foresight required to prevent the accident would have been so expensive as to have rendered the mission *not selectable*. Managers must accept the fact that accidents may happen. If accidents are seen as a failure, there comes a time when the sponsors or the managers lose faith in their ability to proceed. With a loss of faith in the ability to get past that roadblock come reviews, failure boards, reorganizations, loss of personnel and morale, and eventually the erosion of the organic capability for innovation. In addition, it should be recognized that not everyone wants to see innovative programs succeed, Innovation can be very threatening to some individuals and elements of an organization. So standards for failure must be established. In other words, just as we have standards of excellence and *zero defect campaigns*, one must establish the appropriate level of documentation, testing, and review that is cost-effective and a self-consistent metric that acknowledges that for a certain level of quality assurance (and cost), a certain failure rate will occur.

Management Challenges, Structure, and Lessons to be Learned

Much of the shape of things at NASA is driven by the manned space flight program, past and present. That management model is applied to many levels of the organization, particularly with respect to risk assessment. The application of large-scale management techniques, developed where loss of human life was considered unacceptable, was not appropriate to smaller scale, research-driven missions. Nonetheless, this dichotomy remains.

The operating principles of the Skunk Works provide an informative contrast to those of NASA. For example, the personnel at the Skunk Works were involved in performing all tasks from design to manufacturing and testing, not solely monitoring contracts. Inspection work was also done, at a higher level, by the people that would actually be responsible for the success of the project. The individual subcontract deliverables were inspected by the subcontractor. This indicates that there were sufficient in-house personnel available to meet the project requirements without subcontracting out key tasks, thus maintaining in-house expertise. The essential management challenge is to resist the temptation to outsource technical capability in order to reduce the quarterly bottom line. Management experience, judgment, and integrity are required in order to preserve enabling technical capabilities, even if they are not *cost effective*.

While not explicitly called out, the Skunk Works and many other innovators have relied on a young workforce or a youth ethos. A young staff, under effective management, offers lower costs, greater stamina (fewer health issues, fewer outside commitments, and fewer vacation days) and the key lack of experience of failure (they believe that they can do great things and change the way things are done). Note that there are many older professionals that exhibit the *youth ethos*, and age is not the sole determining factor. With that in mind, one should note that NASA's over 60 workforce outnumbers the under 30 workforce, in the science and technical areas, by a factor of 3. About 15% of the science and engineering employees are eligible for retirement, and another 25% will be eligible within the next 5 years. This is not readily remedied, as experience and a GAO report (2003) to Congress show, because new hires needed considerable training and lack the experience of the workers they replace. This question of knowledge management is a key one that can be addressed by maintaining a commitment to small, low-cost missions, developed on short time scales, to address focused problems. These focused, streamlined, short duration projects would involve new hires in a setting that would make the most of their strengths while incorporating the experience of the organization's key asset, its people.

The other Skunk Works guidelines also make it clear that project management lines of authority and responsibility must be clear, understood by all involved, and not be compromised. When NASA implements a satellite program, there are many *bosses* within the organization and within each subcontracting organization, each of whom has responsibilities other than the successful completion of that particular project. The generation of multiple lines of authority and reporting arises, in large part, from the reduction in the federal workforce because those lines of authority are now held in many places outside of the federal workforce. For more than two decades, the administrations have been pushing to privatize NASA to ever greater extents. In 1984, Congress amended NASA's charter to include "to seek and encourage to the maximum extent possible the fullest commercial use of space activities." Under former NASA Administrator Dan Goldin, NASA moved to privatize manned space programs, the space shuttle and the International Space Station (ISS). In 1995, the Kraft report[14] urged the amalgamation of shuttle operations under a single contract saying that the shuttle should be thought of as operational rather than experimental. In 1998, the Commercial Space Act forbade NASA from building space launch vehicles and directed NASA to plan for the privatization of the shuttle and the ISS while encouraging private sector development and operation of future reusable launch vehicles. In 2001 and 2002, two studies were released recommending that a private contractor handle the safety and technical oversight requirements for shuttle safety. In 2003, the *operational* Shuttle Columbia was destroyed upon atmospheric reentry. The Columbia Accident Investigation Board (CAIB) report (Gehrman, 2003) "discuss[ed] the attributes of an organization that could more safely and reliably operate the inherently risky Space Shuttle, but does not provide a detailed organizational prescription. Among those attributes are: a robust and in-dependent program technical authority that has complete control over specifications and requirements, and waivers to them; an independent safety assurance organization with line authority over all levels of safety oversight; and an organizational culture that reflects the best characteristics of a learning organization."

The reduction in the federal workforce at NASA has lead to an increased reliance on contractors to perform the actual implementation. In many cases NASA employees provide technical or managerial oversight to the contractors. This relationship brings with it advantages, but also can be a source of conflict, competition, and miscommunication. Contractors, while recognizing the need to satisfy their customer, are subject to many pressures. At all levels in the project, their performance is evaluated in terms of their ability to satisfy their external managers and their own, internal, managers as well as to yield a profit for their corporation. Even when the various parties maintain the best of relationships, miscommunications, or the absence of a complete transfer of information (and assumptions) can lead to disaster. In September 1999, the Mars Climate Orbiter was lost due to a failure to communicate assumptions (Mars Climate Orbiter Mishap Investigation Board Phase I Report, November 1, 1999): the thruster data used in the trajectory software was in English units, while the software itself used metric units. The undocumented assumption of *standard units* is a product of the culture of the originators. There is some evidence that operators recognized that there were anomalies in the flight data before the final burn, but there were not enough people (insufficient funds) to investigate what were, at the time, seen as relatively minor anomalies. The Mars Polar Lander was lost in December 1999 when the braking thrusters failed to fire properly due to a transient vibration caused during the deployment of the vehicle's landing legs. NASA's internal investigation found that no end-to-end tests had been performed on the Mars Polar Lander in its final flight configuration[15]. This may have been the reason but it wasn't the cause. The project management team must have realized that there was risk associated with not performing the end-to-end tests on the *as-flown* configuration but addressing this might have lead to delays to repair the problem and increased the pressure on the schedule. The managers may have weighed this against the cost and schedule risk and the concomitant possibility of cancellation, and found that it was best to proceed. In a similar vein, two Deep Probe 2 microprobes accompanying the Polar Lander were also lost. From the vantage point of hindsight we can see that JPL and Lockheed Martin tried to perform a very complex mission with very limited resources: these costs were small compared to a typical NASA Earth mission, much less a typical planetary mission. These projects were understaffed to begin with, and then new personnel were added and overtime was required. This drove costs up and resulted in an overworked team. This situation arose, in part, from launch dates (set by the planetary alignment) and a science plan that dictated that the spacecraft rendezvous at Mars. Part of the problem may well have been that, in order to remain competitive and appear compliant with their customer's needs, the team may have underestimated the mission resource requirements by adopting the most optimistic scenario.

Improving the national technical abilities and, in particular, the capabilities of the scientific missions flown by NASA, involve political and managerial challenges. The major technological challenge is that driven by the perception of inherent capabilities of modern technology. The common availability of cheap, high speed microprocessors has given rise to the myth that "it can be handled in software" and the "if Microsoft or Intel or Google or... can do it, why can't NASA?" challenge. In reality, the proliferation of processing power has greatly increased the complexity of spacecraft operations. In addition, the reliance on software and the processors that run this software (at 1/100 the speed of a home PC) has increased the complexity of the design and test effort. Software testing and code design are the two areas where most space mission managers have the least experience. Ideally, a full fidelity simulation of the complete system and its interactions during critical transient operations (thruster firings, spacecraft maneuvers, solar array deployments, instrument cover openings, etc.) would be conducted. The *test as you fly, fly as you test* approach is generally followed, but not every condition of spaceflight is always known before launch. Failure review boards have the remarkable ability to locate software and design issues after a failure occurs. One might naively ask why this can't be done before flight but the answer is immediately obvious: something went wrong and the failure review board knew where to begin looking. The study of the FBC failures does indicate that software failures, in particular, the flight software's ability to tolerate *faults* during transient phenomena, are one of the root causes of mission loss. Additionally, testing of software components is not sufficient, and there must be an end-to-end test capability.

MANAGEMENT IMPERATIVES

The single most important driver in a group activity, and among the leaders of that group, is a sense of mission. The VSE has that sense, yet it remains to be seen if that vision alone can transform the organization and enable new approaches to unmanned missions. In order for NASA to maintain the core capabilities that have been built up over half a century of robotic space missions, the mission cadence must be increased. In order for the mission cadence to be increased, under a fixed budget, the way projects are managed and the management practices that are used or are implicit in the way the project is carried out must be changed. Each aspect of the way the project is carried out needs to be examined in terms of whether that practice is cost-effective rather than just "business as usual."

In order to be cost effective the mission must meet the needs of the customer. Those needs are not just the measurement objective (this may, in fact, be the least important and most widely hedged need) but must include a tangible sense of accomplishment which comes from completion of the project. This means that, in order to insure a cost-effective mission, the end-to-end design must be addressed: costs must be managed and products delivered including on-orbit performance.

NASA, like any organization, needs to examine its past and current practices and change them. This is hard work and may require the next generation of managers to accomplish. Dörner (1996) identifies four thought patterns that lead to repeated failures:

1. We reduce the complexity of the problem because it takes more energy to constantly remain aware of that complexity.
2. Because we need to feel confident, we tend to try to apply an approach we are familiar with to the same problem.
3. Because we can not readily capture and assimilate large amounts of data, we prefer unchanging mental models.
4. We tend to focus on immediate, pressing problems and ignore the problems our solutions will create.

The current management practice for space missions includes what are termed *design reviews*. These design reviews can often embody the patterns expressed by Dörner above. Current *best practice* insists upon a series of design reviews as the project goes from one mission phase to the next. These design reviews are highly ritualized. A design review typically consists of a review panel which may or may not include outside reviewers. These reviewers are charged with providing a critical review of the material to be presented. Preparation for the review is a time-consuming process. Even for a relatively simple instrument with a high degree of heritage, it is not uncommon for the instrument team to prepare literally hundreds of charts. Typically, the speakers review the system, subsystem, or component in a series of viewgraphs: even during software walkthroughs the audience is largely passive. For a large program, a PDR may take a week or more and contain thousands of charts. Yet, even then, the level of detail in such a review is often not sufficient to provide a critical insight into the design. The difficulties with the approach may be uncovered during test and the design may be changed (given sufficient time and money). We hold design reviews because they serve two important functions: 1) they demonstrate proper management techniques (a check has been placed in the box labeled *hold CDR*, for example), 2) they provide the only opportunity for the sponsor and other members of the team to get *the big picture*. With respect to point 2, this point is not made lightly: there is, generally, no ready reference for the mission, its systems, or subsystems, other than the PDR or CDR package. Design reviews are not without value: questions and action items do come up at design reviews, and often they are based on the reviewers' own experience or mental model of the difficulties of a particular approach and can bring important insights into the process. Issues are documented in the form of action items. An action item is a requirement for further study that is generated during the review process and usually focuses on a concern about some element of the mission. All too often the action item is viewed as a criticism, where any criticism (particularly a valid one) is viewed as bad and an impediment to progress. Program managers strive to hold reviews where the minimum number of action items are generated. The downside of this is, of course, that there may be times when significant issues or concerns are not surfaced as this would provoke an action item. The people presenting at the review may be tempted to focus on the pressing problems of the design rather than taking the longer view. Once the review is over, there is a collective sigh of relief and the team *gets back to work* as though the review were an obstacle to progress. Time spent on preparing for the review, completing the action items, and sharing lessons learned from the review must be viewed as time well spent by the team members, as well as management. In point of fact, the most important reason for a team to hold a review is not so that they can get comments from the review board, but so they are forced to take time out from their "work" to think about what their assumptions mean in terms of the other elements of the project and to communicate those assumptions to the other team members. These reviews have the potential to enable a reduction in the amount of formal documentation by encouraging small-group, informal interactions rather than a process of designing through interface requirements.

The project review process should include a more rigorous review of software. Since so much of the capability of modern space missions depends on software, a new approach is indicated in order to review, test, and certify spaceflight software. As a robotic space mission transitions from launch (during the end of phase C/D) to routine mission operations (phase E), the most expensive part of the program can become mission operations. Today's spacecraft make use of the advances in microprocessors and software design to implement autonomous operations on the spacecraft. A key use of this is to transition the spacecraft from its normal operating mode to a *safe mode*. In safe mode, one typically sees an orderly shutdown of instruments and other nonessential systems to reduce power requirements and the execution of an autonomous maneuver that is intended to maximize power output from the solar arrays while minimizing other nonessential power requirements. This autonomy software may have hundreds of rules and must operate to secure the spacecraft until ground teams can interpret the problem and provide corrective action. Autonomy software is enmeshed in a system of complex interactions that can be difficult if not impossible to test on the ground. In fact, the Near Earth Asteroid Rendezous mission experienced an anomaly during a thruster firing that had a significant consequence for the mission[16]. In this case, the autonomy software acted as expected, but lateral transients during thruster firing were higher than expected and errors were missed during testing.

The hardware driven didactic approach does not seem to be adequate for a complex, nonlinear, interacting system such as autonomy software. Project reviews should institute *code breakers* in their hardware/software design review process. This is more easily said than done, as the "code breakers" must be a trusted and respected part of the team, and few of the current generation of program managers have significant software design experience. This bond of trust and respect will present a significant managerial challenge because it goes against the current culture of the review process: no one likes to have their *mistakes* pointed out to them. The code breakers would be tasked with developing the test sets to perform the independent validation and verification of the functionality of the flight systems. They would have a charter broad enough to encompass the interaction between many systems and insure that all interfaces are correctly handled. This may well be an effective means of reducing the cost of mission operations for long duration space missions with a long time lag or limited connectivity to the Earth, such as missions to Mars or rovers operating on the surface of Mars.

CHANGING UNMANNED SPACE MISSION PROGRAMS AT NASA

The reorganized NASA presents opportunities for a revitalized and redirected unmanned program. The new NASA Science Mission Directorate should institute a budget line that has three elements: 1) a cross disciplinary technology development and demonstration program, 2) a unified *explorer* office responsible for small to medium missions, and 3) a larger *Discovery-class* office for larger, more complex missions. This division of satellite programs would be patterned after that already in place in what had been the office of space science. The first step for the Earth Sciences activity must be to replace the Earth observing system (EOS) program with a Discovery-class line that would explore more complex measurement requirements that were set by the over-arching goals of the agency. The EOS program laid the groundwork for the observations that can be converted to the category of *monitoring*. Ideally, these monitoring activities should be carried out by an agency with an operational mandate such as the National Oceanic and Atmospheric Administration (NOAA), though we cannot fail to note that the NPOESS program and its problems seems to demonstrate the difficulties of implementing a large requirement-driven mission[17].

The current Earth observation budget, including the near-Earth environment, is of the order of \$1.5 billion. This program could be reconstructed to consist of a line of small university explorers with a mission cost of \$30 million, small explorers at \$250 million, and medium explorers at \$400 million. The cadence to be achieved would be two university explorers per year, one small explorer per year and a medium explorer every other year. The Discovery-class line would be planned to achieve a cadence of a mission every 3 years. The reorganized NASA Science Mission Directorate will bring the experience of the Earth system science pathfinder (ESSP) program into contact with the more successful and mature Discovery and Explorer programs that were managed by the office of space sciences. As an interim solution to the follow-on to the EOS program, NASA should consider a series of small satellites, in the mold of EO-1, that will rendezvous and fly in constellation with the EOS platforms TERRA, AQUA, and AURA to augment or restore capabilities that were inherent in the original EOS platform. This will be necessary and may be the only viable option because there is still a need to baseline the Earth biogeochemical system and this requires at least a solar cycle of observations.

The increased cadence of missions will require major management changes in order to be effective. The teams picked to implement these missions can not be run in the same way that they were in the past. The project teams must be integrated and collocated. NASA personnel may not be the leaders of these teams: other organizational models (FFRDCs, UARCs, or innovative partnerships, as suggested by Aldridge, Fiorina, Jackson, Leshin, Lyles, Spudis et al. (2004)) may be required to make the changes that will keep the Earth-observing program alive. The immediate requirement will be to rebuild the program management structure within NASA to be more responsive to opportunities for the incorporation of new technologies. Risk management must be achieved: failures will occur but they cannot be allowed to become mission killers. Failure-tolerant mission design must be stressed. This means that mission design must be rethought: the functional requirements to survive launch and orbital operation are the key focus and this must be aggressively red-teamed.

Unforeseen problems must be accepted as a fact of programmatic life. This change in attitude is probably the most difficult to impose upon any technical organization. This cultural change means that the success of the project must be identified as a team success rather than the result of the contribution of independent contractors doing the best they can within a limited budget. The freedom to incentivize contracts and structure award-fees based on performance must be put in place.[18] Contract procurement processes need to be revamped to include past performance and transition costs in the assessment of the bottom line. Program management staffs at NASA have to be expanded to include a strong risk assessment component that holds the corporate memory. Not enough is being done to disseminate the lessons learned from NASA successes and failures. As an example, NASA maintains a "lessons learned" database[19] that allows access to the public, NASA, and contractors. An unfiltered and unrestricted database of *lessons learned* sends a powerful message about the corporate culture: it acknowledges that mistakes will be made and that the community needs to communicate to reduce the number of mistakes made. The Earth observing system lessons learned[20] is particularly interesting, as the first page of the site actually makes mention of the issues that are faced in knowledge sharing, and posits an approach and a way to measure success.

Small programs, with their inherently smaller base of support within the scientific community and more focused appeal, are vulnerable to budget cuts. In the budget planning process, a key stabilizing factor is the establishment of program lines. A program line appears as a "line item" in the NASA budget, and has a funding profile that covers at least 5 years in the projections and contains accepted placeholders (established via the roadmaps) for activities beyond that point. A program line provides a unifying theme that can be sold to Congress. Once that theme has been established, the selection of the missions follows. The program line has, by definition, a finite lifetime, thus ensuring the opportunity for new science. In this particular case, NASA needs to establish a line, with clearly stated metrics and objectives, that enables new, innovative approaches to space missions.

In addition to having more flights, new technology must be infused into the flight programs. From a management viewpoint, this is a problem. There is at NASA a *tyranny of the TRL*. The TRL or technology readiness level is used to indicate the maturity of a technology and the level of risk associated with its use. The highest level is for components or systems that have flown in space. Clearly, if it has flown in space it is not innovative.

The JPL deep space program line is an example of a technology infusion program line that has had significant longevity and success. A comprehensive plan to reinvigorate proven space technologies relies upon a higher launch cadence.

In order to effectively manage smaller, cost-effective programs, the vision of project management must be transformed. In a typical work environment, a manager really interacts with relatively few people above and below them, and this interaction realm effectively restricts the flow of knowledge in the project. To effectively manage the flow of knowledge within the project, the manager must:

1. Provide direction and remind the group of what is important and why their work makes a difference
2. Provide support for new ideas, provide the impetus for action, and manage risk taking so that the group's capabilities can grow and their internal biases can be tested
3. Generate a climate of trust and work to sustain that trust within the group so that unproductive paths (commonly known as *mistakes*) can be redirected and so that dissent or criticism can be accepted

These ideas can arise spontaneously within small areas in an organization like NASA, but to achieve agency-wide acceptance will mean that a larger paradigm shift must be accomplished. The NASA study of FBC (Spear, 2000) said "There is an intangible element there is a team spirit associated with doing FBC, and people are the most important ingredient." A management initiative whose goal it is to make missions happen faster and do things better for less money, while managing risk effectively, could prove valuable as a means of transforming the broader NASA culture, but that change must be directed and supported from above.

SUMMARY

In the years ahead, NASA will face significant knowledge management issues that will arise from budget contraction in some areas, reordered priorities, the loss of a significant fraction of its workforce to retirement, and a reduced mission cadence for science missions. This must be addressed as part of NASA's strategic plan. NASA's success with FBC shows that even large organizations can change their direction and adapt to new circumstances. A reinvented FBC approach holds the key to effectively managing knowledge at NASA and in the greater space-using community. In order to be sustainable and effective, these changes depend on a buy-in from the organization's internal and external constituency. For NASA, that constituency includes the employees, contractors, and their customers, Congress and the public. To achieve that buy-in the organization's managers must be aware of the risks involved and cultivate reasonable expectations. A consistent set of expectations, practices, and procedures must be established that offers a tangible reward for positive change for the internal and external constituency. An organization undertaking a task driven by innovation, with its concomitant element of risk, must be able to demonstrate via established metrics that the transformation is successful even if there appear to be occasional failures within the system.

The space business is still a contact sport: accidents happen and things don't go as planned. This is a direct outcome of doing that which has not been done before, and doing it on a limited budget and schedule. Mission participants should continually make the customer aware of the risks inherent in the process, and management would be well-advised not to forget those risks either. Achieving the balance between that which is bold and innovative and that which is proven and known is, in and of itself, part of a voyage of discovery and enrichment that we should all participate in.

REFERENCES

Aldridge, E. C., Jr., Fiorina, C. S., Jackson, M. P., Leshin, L. A., Lyles, L. L., Spudis, P. D. et al. (2004). Report of the president's commission on implementation of United States exploration policy. Retrieved March 3, 2007, from http://www.nasa.gov/mission_pages/exploration/main/index.html

Bilstein, R. E. (2003a). *Testing aircraft, exploring space: An illustrated history of NACA and NASA*. Baltimore: Johns Hopkins University Press.

Bilstein, R. E. (2003b). *Stages to Saturn: A technological history of the Apollo/Saturn launch vehicles*. Melbourne, FL: University Press of Florida.

Committee on an Assessment of Balance in NASA's Science Programs. (2006). *An assessment of balance in NASA's science programs*. National Research Council. National Academy Press.

Diaz, A. (2004). Renewed commitment to excellence: An assessment of the NASA agency-wide applicability of the Columbia accident investigation board report.

Doolittle, J., & Glines, C. V. (1991). *I could never be so lucky again: An autobiography*. New York: Bantam Publishing.

Dorner, D. (1996). *The logic of failure* (R. Kimber & R. Kimber, Trans.). New York: Metropolitan Books.

Gehrman, H. W., Jr. (2003). Columbia accident investigation board report. Washington, DC: Government Printing Office.

Government Accounting Office. (2002). *Major management challenges and program risks: National Aeronautics and Space Administration* (No. GAO-03-114). Washington, DC: General Accounting Office.

Government Accounting Office. (2003). *Space shuttle—Actions needed to better position NASA to sustain its workforce through retiremen*t (No. GAO-05-230) Washington, D.C.: General Accounting Office.

Inspector General's Office. (2001). *Faster, better, cheaper: Policy, strategic planning, and human resource alignment* (Audit Rep. No. IG-01-009).

Janos, L., & Rich, B. R. (1996), *Skunk works: A personal memoir of my years at Lockheed*. Back Bay Books.

Johnson, C., & Smith, M. (1990). *Kelly: More than my share of it all*. Smithsonian Institute Press.

Kerr, R. A. (2004). NASA reigning in PI-led planetary missions. *Science, 305*(5683), 467.

Kranz, G. (2000). *Failure is not an option: Mission control from Mercury to Apollo 13 and beyond*. New York: Simon and Shuster.

McCurdy, H. E. (2003). *Faster, better, cheaper: Low-cost innovation in the U.S. space program*. Baltimore: The Johns Hopkins University Press (See also the other books in The Johns Hopkins University Press Series on the History of NASA).

National Research Council. (2005). *Principal-investigator-led mission in the space sciences*. Washington, DC: National Academies Press.

Office of the Inspector General. (2001, March 31). *Faster, better, cheaper: Policy, strategic planning, and human resource alignment* (No. IG-01-009).

Space Studies Board. (2004). *Steps to facilitate principal-investigator-led Earth science missions*. Washington, DC: National Academies Press.

Space Studies Board. (2005). *Earth science and applications from space: Urgent needs and opportunities to serve the nation*. Washington, DC: National Academies Press.

Spear, T. (2000). *NASA faster, better, cheaper task final report*. Washington, DC: Government Printing Office.

ENDNOTES

1 *"Aging workforce issues program*: Section 404 would express a Sense of Congress that the Administrator of NASA should implement a program to address aging workforce issues in aerospace that would: (1) document technical and management experiences of senior NASA employees before they leave NASA, (2) provide incentives for retirees to return to NASA to teach new NASA employees about their lesions and experiences, and (3) provide for the development of an award to recognize and reward senior NASA employees for their contributions to knowledge sharing. Several reports have identified the aging of NASA's workforce as one of the challenges to maintaining an appropriately skilled cadre of NASA employees." from Senate Report 109-285 - The American Innovation And Competitiveness Act Of 2006.

2 "Issues Affecting the Future of U.S. Space Science and Engineering Workforce: Interim Report,' National Research Council, 2006, *http://fermat.nap.edu/catalog/11642.html*; "Status of NASA's Efforts to Address Workforce Issues Related to the Space Shuttle Retirement," GAO testimony to the Committee, May 18, 2005, *http://www.gao.gov/new.items/d05718t.pdf*; and "Space Shuttle: Actions Needed to Better Position NASA to Sustain Its Workforce through Retirement," GAO, March 2005 (see SR 109-285).

3 See e.g., http://images.spaceref.com/news/2005/earth_roadmap.pdf and http://images.spaceref.com/news/2005/sssc_10_final.pdf for the Earth sciences and Space Sciences Roadmaps, respectively.

4 See, for example, http://nspires.nasaprs.com/external/ which lists current research opportunities at NASA

5 See transcript of 9/12/2006 speech by Mike Griffin to GSFC employees, available through a link on the page http://www.nasa.gov/lb/audience/formedia/speeches/index.html accessed on September 12, 2006.

6 http://www.nasa.gov/mission_pages/exploration/main/index.html

7 http://www.spaceref.com/news/viewsr.html?pid=20785. Note that in the keynote address to the Utah State University Small Satellite Conference, Mr. Griffin reiterated the need for maintaining the knowledge base at NASA (http://www.sltrib.com/utah/ci_4182555 and http://deseretnews.com/dn/view/0,1249,645193239,00.html) while clearly stating that education programs, as such, would not be explicitly funded by NASA. Education or learning by doing must be accomplished either through industry or in the course of carrying out NASA's main business.

8 See NASA Procedural Requirements http://nodis3.gsfc.nasa.gov/displayDir.cfm?Internal_ID=N_PR_7120_005C_&page_name=main

9 Standard reference handbooks, such as this one, are widely available on the Web. A useful government site is http://assist.daps.dla.mil/quicksearch/quicksearch_query.cfm and a search engine such as Google can readily locate many other instances of the manuals, available from commercial organizations that support the U.S. Government and its agencies.

10 The document can be downloaded from the NASA Exploration Web page (http://www.nasa.gov/mission_pages/exploration/main/index.html)

11 NASA Press Release 06-056, 2006, available at http://www.nasa.gov/lb/home/

hqnews/2006/feb/HQ_06056_Budget_Statement.html

[12] Griffin, M., NASA Administrators Statement About FY 2007 Budget, NASA Press Release 06-056, 6 Feb 2006: see also http://www.nasa.gov/home/hqnews/2006/feb/HQ_06056_Budget_Statement.html. See also Griffin, M., Testimony before the Senate Commerce Committee FY 2007 Budget Hearing, April 25, 2006, available at http://www.nasa.gov/audience/formedia/speeches/index.html

[13] The NASA Advisory Council Minutes of March 16-17, 2000, http://www.hq.nasa.gov/office/oer/nac/mins/00-03mins.html, provide an insight into the issues that still faced NASA near the end of Mr. Goldin's tenure as NASA Administrator and an indication that there was a growing perception that FBC was taking place in an environment rife with mission defining issues.

[14] The Kraft report (http://spaceflight1.nasa.gov/shuttle/reference/green/kraft.pdf) was referred to in that year's budget (http://rs9.loc.gov/cgi-bin/cpquery/?&sid=cp104S6pOz&refer=&r_n=sr155.104&db_id=104&item=&sel=TOC_65745&) as a rationale for reducing the operating cost of the shuttle. In hindsight, the words "Equally significant was the Kraft Report's general theme that safety concerns not be used to avoid consideration of ways to downsize the standing army of NASA personnel and the massive infrastructure that operate and maintain the shuttle. The Kraft Report noted that NASA continues to operate the decades-old shuttle as an experimental vehicle, changing 150 items of shuttle hardware after each flight even though an average of only 10 in-flight (mostly inconsequential) problems per shuttle mission typically occur," are particularly telling.

[15] JPL Special Review Board. *Report on the Loss of the Mars Polar Lander and Deep Space 2 Missions*. Report JPL D-18709. March 22, 2000, and available at http://spaceflight.nasa.gov/spacenews/releases/2000/mpl/mpl_report_1.pdf.

[16] http://klabs.org/mapld04/tutorials/mishaps/near.htm

[17] See, for example, the story in Space News on the down-sizing of NPOESS in response ot the Nunn-McCurdy Act http://www.space.com/spacenews/archive06/Npoess_061206.html, accessed September 12, 2006.

[18] See the testimony of Elon Musk before the House Science Committee for a discussion of the problems facing expendable launch suppliers and the need to both streamline and incentivize the process available at http://www.house.gov/science/hearings/space04/mar18/musk.htm.
See also the National Aeronautics and Space Act Pub. L. No. 85568, 72 Stat. 426438 (July 29, 1958) as amended through Pub. L. 109-155, 119 Stat. 2895, (Dec. 30, 2005) at http://www.nasa.gov/offices/ogc/about/space_act1.html. Accessed September, 11, 2006.

[19] http://llis.nasa.gov/

[20] http://eos.gsfc.nasa.gov/eos-ll/pilot.html

Section IV
Space and Society

Chapter XIV
Towards an Ethical Approach to Commercial Space Activities

Jacques Arnould
Centre National d'études Spatiales, France

ABSTRACT

This chapter introduces the ethical questioning in the field of space activities, especially space commerce. If the 1967 Outer Space Treaty defines space as the "property of all" and its exploration as the "province of all mankind," the future utilization of near-Earth (and tomorrow Greater Earth) space probably needs new ethics (if ethics means not only legal applications, but also and for example the application of the rule of three Ps: protection, promotion, and preparation). Orbital debris mitigation, the International Charter on Space and Major Disasters or, in the future, the safety of private astronauts crews, offer lessons in realism and sources of prospective reflections. Space ethics is still in its infancy.

INTRODUCTION

What have we achieved and where are we going with this remarkable adventure, which has its roots deep in the human imagination, which we began just 50 years ago and which has undoubtedly reached its climax with the 12 men who traveled to the Moon? I am, of course, referring to the space adventure. And what have we achieved and where are we going when we consider the use of space for full-scale commercial enterprise? Yesterday's astronauts had the stuff of heroes, but will they be succeeded by bellboys in space hotels open to wealthy space tourists, while sentinel constellations offer costly services for those using satellite data to support their terrestrial businesses? Business as usual?

The time would seem ripe for public—and private—sector space players to consider how, insofar as we can envision this evolution, to bring some ethics into the equation through codes of conduct, regulations, and treaties. There are many issues to address.

Today, orbits are occupied according to international procedures approved by all, but will the same apply to the disposal of satellites at the end of their life and to orbital debris mitigation? What resources should governments be prepared to commit to tracking such debris, for the benefit of commercial business? And what of the use of satellite-based observation, communication, and positioning systems? Once strictly private and commercial, are they no longer bound by rules of discretion? What ties will they maintain with services like those provided by the International Charter on Space and Major Disasters? And how will the safety of private crews be guaranteed? If they are no longer "envoys of humanity" in the accepted modern legal sense, what duties will governments have toward them on Earth and in space?

Clearly, space ethics is still in its infancy. It is building on the experience gained in other scientific and technical fields, and is adding its own touches. Not the least of these is its readiness to question the ends of astronautic activities before tackling questions so urgent or precise that it might be tempted to put the cart before the horse. The ethical question of the commercialization of space is one we must address without delay.

WHAT DO WE MEAN BY SPACE ETHICS?

Our societies, and we ourselves, consider ethical issues on two occasions.

First, when we need to answer doubts and fears inspired by an urgent decision likely to have a major impact on humans, or by a catastrophic event. For this reason, many governments have set up ethics committees to support the development and application of new medical technologies or genetic engineering techniques. In 1986, the Chernobyl disaster and Challenger shuttle accident led to questions and decisions, and with them a greater awareness of the ethical issues involved (Bell, 1998; Martin & Schinzinger,1996).

The second occasion, which is less dramatic and part of the making of social history, is when we are ready to question rationale and purpose before or after taking an action, as well as the consequences and responsibilities that go with it. This type of questioning precedes or accompanies the drafting of treaties, agreements, legislation, and codes of conduct; indeed, we expect decision-makers, politicians, engineers, and medical experts to consider these issues, without necessarily talking about ethics *per se*. Yet ethics it is. In other words, people often take an ethical approach without even realizing it. And setting up ethics committees or appointing ethics officers within organizations must never prevent or excuse people from addressing these concerns themselves.

It would be wrong to approach ethics in a way that reduces it to a simple authorization/veto procedure applied to certain sensitive activities or, worse still, that leaves the whole issue to the ethics committee. Rather than getting into the details of historical, philosophical, or cultural nuances, my aim is to briefly consider the ethical approach from two perspectives. First, the perspective outlined above, which can be summed up in two questions: What are the rationale and purpose behind an action or activity? And what are the consequences and responsibilities that go with it? The second perspective I call the "rule of three Ps": protection, promotion, and preparation. Protect the past, promote the present, and prepare for the future. In other words, the ethical approach is tied to the notion of heritage, which we will discuss below. Space ethics takes these two perspectives very seriously.

Over the last 50 years, the space endeavour has seen its share of disasters; I have already mentioned the Challenger accident, but we could also cite the Columbia disaster and others that have caused loss of material assets and, in the worst cases, human life. While we have had to wait until the start of the third millennium for space agencies to take an overt interest in ethical issues[1], space players have not waited so long. Working groups have existed since the mid 1980s, dedicated to protecting our planet (such as the COSPAR Scientific Commission that advises the UN on space science and related matters) and mitigating orbital debris (like the Inter-Agency Space Debris Coordination Committee, IADC). There are many issues to address. Some were enshrined in the Space Treaty—or the treaty on principles governing the activities of states in the exploration and use of outer space, including the Moon and other celestial bodies—signed in 1967. Others will need more time before finding their place in space law. The commercialization of space is a case in point.

For all that, we already have plenty to go on as a basis for our discussion on the ethics of space commerce. The ideas of common heritage and province offer a good starting point.

SPACE AS A RES COMMUNIS: "THE PROPERTY OF ALL"...

Article 1 of the Space Treaty lays the foundations of the status of space:

"§ 1. The exploration and use of outer space, including the Moon and other celestial bodies, shall be carried out for the benefit and in the interests of all countries, irrespective of their degree of economic or scientific development, and shall be the province of all mankind."

"§ 2. Outer space, including the Moon and other celestial bodies, shall be free for exploration and use by all states without discrimination of any kind, on a basis of equality and in accordance with international law, and there shall be free access to all areas of celestial bodies."

"§ 3. There shall be freedom of scientific investigation in outer space, including the Moon and other celestial bodies, and states shall facilitate and encourage international cooperation in such investigation."

What does the 1967 treaty say? That space is not a "no man's land" or a *terra nullius*—"territory belonging to no one"—to be occupied and exploited by the first to arrive or by just anyone. Space, says the treaty, is international in nature and cannot therefore be appropriated by any one state. In the mid 1960s, such an affirmation was a bold step: with the United States and the Soviet Union engaged in the Cold War and racing to be the first on the Moon, the treaty set out to protect space from any national claims or military aspirations, and to offer it to humanity as a whole. Its intentions were generous, prophetic, and perhaps even utopian; however, its great merit is that it has been accepted, supported, and confirmed by numerous states. However, it remained—and still remains—to be put it into practice. From a purely legal perspective, the question soon arose as to the precise legal status this article and space law in general would give to space. Two possibilities emerged: the qualification of *res communis* and that of common heritage. They are defined as follows:

that which belongs to no one and may therefore be accessed and used by all; qualified under Roman law as rex communis, or the property of all;
that which belongs to all and must consequently be used with due regard for the common interest and the guarantee that all may share in its resources; this category is what is usually referred to as the common heritage of humankind.

International waters and the Antarctic are two examples of this second qualification[2]. Turning back to space, applicable law does not clearly distinguish between the two possible qualifications. Article 1 of the Space Treaty would tend to infer the status of common heritage of humankind, but Article 2 seems to limit the application of common heritage in favour of *res communis*, where states are under no obligation to impose particular conditions concerning the use of space, nor concerning simple exploration by their citizens:

- "Article 2: Outer space, including the Moon and other celestial bodies, is not subject to national appropriation by claim of sovereignty, by means of use or occupation, or by any other means."

This presents a difficulty, which goes back to the introduction in Article 1 of the notion of province. The exploration and use of space are thus qualified as the province of mankind, but what exactly does that mean?

...OR SPACE AS A PROVINCE?

How do we interpret the notion of "province"? Article 4 of the 1979 Moon Treaty, drawn up to govern the activities of states on the Moon and other celestial bodies, states that: "The exploration and use of the Moon shall be the province of all mankind and shall be carried out for the benefit and in the interests of all countries, irrespective of their degree of economic or scientific development. Due regard shall be paid to interests of present and future generations." In the age of the French monarchy, the French word *apanage* (province) referred to the portion of the royal estate awarded to younger members of the royal family in lieu of accession to the crown. Since then, *apanage* has taken on the broader meaning of property, possession, or inheritance, though still inferring a sense of elitism. The application of the notion of province to space law is interesting. On the one hand, province puts humanity in its rightful place: neither of domination (we are not the rulers of the universe), nor of submission (we nonetheless inherit it). On the other hand, it presents us not with a territory, but a mission: that of using and developing a portion of this universe (here, the Moon) for our own interests and those of future generations. However, we must ask ourselves whether the notion of province is not in conflict with the idea of protecting our space heritage, and whether it reduces space to the status of public property. The example of the exploitation of maritime fisheries could justifiably leave us somewhat perplexed. Indeed, the law seems powerless to resolve one of the fundamental difficulties inherent in space: its very definition. Article 1 of the Space Treaty confronts this problem in its discussion of outer space and its exploration and use.

The International Council of the French Language recognizes this problem in its *Dictionnaire de spatiologie* (Space science dictionary); the term *espace* (space) is defined thus: 1. Usual abbreviated form of *espace extra-atmosphérique* (outer space: literally, space beyond Earth's atmosphere), or 2. Sphere of human activity associated with outer space. Outer space is itself qualified as "the region of the Universe situated beyond the part of Earth's atmosphere in which aircraft can fly." The expression "outer space" is used in space law with no specific definition or delineation[3].

The definition by the International Council of the French Language may seem surprising, but it is perfectly accurate. While it is possible to define the inner limit of space, typically an altitude of 100 kilometres, the outer limit remains a mystery and the subject of much scientific speculation. That being the case, should we not abandon the notion of "solid" space or an ether, as held by the Ancients, in favour of the idea of a collection of bodies in motion, not just celestial bodies, but now also human inventions and even human beings themselves, a set of movements and trajectories, knowledge and technologies, exchanges and relationships[4]? In short, space could be defined as much by what we turn it into as by what we do there. Space law refers to the place itself as property or heritage, and it refers to activities (exploration and use) as province. Are these two notions compatible? And should we not recognize that any usage is a form of appropriation, albeit temporary, from running a software application without a licence to crossing an ocean in a boat? Space is no exception, as illustrated by the case of geostationary orbit and the notion of launching state.

In the early days of the space endeavour, certain nations on the equator—Brazil, Colombia, Congo, Indonesia, Kenya, Uganda, and Zaire—laid claim to the portions of the geostationary orbit above their territories; in 1976, they even signed the so-called Bogotá Declaration to affirm their sovereignty. This orbit, 35,800 kilometres above the equator, has the unique feature of being in phase with the rotation of the Earth. A satellite in this orbit appears to be stationary with respect to the point on the surface directly beneath it. This feature of geostationary orbit (from which it derives its name) is particularly useful for telecommunication satellites, because it allows them to provide uninterrupted transmissions covering a predetermined area of the Earth's surface. Consequently, this orbit is occupied by a host of satellites. The Bogotá Declaration prompted the International Telecommunication Union (ITU) to take charge of distributing and assigning slots in geostationary orbit and, in so doing, block the attempt by equatorial states to appropriate this portion of space on the basis of geographic criteria. But is this a *bona fide* application of the principle of nonappropriation of space? Are satellites themselves not a form of appropriation of a position or orbital slot, even beyond their operational lives and activities, because they can continue to occupy positions as orbital debris?

The notion of the launching state is another illustration of the difficulties with these two definitions of space. It is defined by the Convention on International Liability for Damage Caused by Space Objects, concluded in 1972: "The expression launching state refers to a state that launches or procures the launching of a space object; a state from whose territory or facility a space object is launched." These four subdefinitions of a launching state give us two ways to understand their responsibilities: monitoring and control of their activities (assuming *responsibility*) and financial compensation for damages caused by these activities (accepting *liability*). On this second point, Article 2 of the 1972 convention specifies: "A launching state shall be absolutely liable to pay compensation for damage caused by its space object on the surface of the Earth or to aircraft flight." However, behind the clarity of these definitions lie a number of issues for space law experts. For instance, how does this definition of responsibility apply when ownership of a spacecraft is transferred after launch, or in the case of Pegasus, Sea Launch, or similar scenarios, that is, craft released from aircraft in flight or maritime platforms in international waters? And how does this type of responsibility apply when it is impossible to set limits in time? Space objects can remain in orbit for centuries. And during this time they can change owners (referred to as in-orbit handover) and be maneuvered whenever necessary without the launching state being informed, yet remaining legally responsible. While it is clearly not a question of appropriation or property, it would seem difficult to apply such a principle of responsibility without states feeling they have some share in the "province" of space, and thus in space itself. The Committee on the Peaceful Use of Space (COPUOS) is another illustration of the possible shift toward province as opposed to public property or heritage.

While not attempting an exhaustive discussion, it is important to briefly mention the product (*fructus*), as well as the use (*usus*), of space. The use of space can take the form of business operations. In other words, the extraction or manufacture of material or immaterial products and resources and their subsequent promotion, as well as possible by-products. What status should be attributed to these space products and by-products, drawn from the *res communis* or common heritage? Could they be subject to some sort of private exemption? Undoubtedly, as shown by the example of remote sensing: information resources, gathered by satellites then processed by teams on the ground, become the property of those who have the *ad hoc* technical and financial means. In discussing the future of space, we cannot help but wonder whether market pressure will eventually become the overriding factor in decisions made, including ethical decisions.

BUSINESS AS USUAL?

At the dawn of the 21st century, the state of our planet is stark testimony to the dramatic consequences of the "business as usual" attitude: devastating pollution, toxic waste all too easily ignored, global warming, and so forth. While there are grounds for hope that the players concerned will come together and take steps to curb these phenomena, we must recognize that few are optimistic about the realities of such a response and how long it might take. Will the same attitude necessarily be taken concerning the future of space? The growing emphasis on the utilitarian dimension of space and the role of industrial and financial companies, largely independent from governments, would lead us to think and fear so.

Three attitudes can thus be envisaged. First, that of absolute sanctuary status, which would give space not just complete protection, but also recognition of its transcendent nature, beyond any temporal economic calculations. Although this attitude is somewhat alien to the market-driven and so-called developed civilizations of our modern age, it is no doubt familiar to other peoples, present and future, who have the same rights as us. Second is the attitude of controlled exploitation, in accordance with terms and conditions that would govern all forms of exploitation. Applicable documents would be drafted at the outset and not on-the-fly or as a knee-jerk reaction to some crisis, as is often the case with terrestrial enterprises. Third is the "business as usual" attitude, just as neatly summed up in the expression "first come, first served," where the rules of engagement are designed to facilitate economic efficiency.

The sanctuary attitude, in other words, the complete freeze on all celestial bodies that enter our domestic universe, can only be seen as a default stop-gap solution. Whenever a use with high economic or strategic potential presents itself, the pressure to pursue and exploit it will become irresistible. But market forces on their own are not enough to carry forward humanity's long-term interests. At the risk of drifting from my earthly orbit, let me take the example of celestial bodies. While they form part of our future development plans (despite their limited resources), we must not forget that they are bearers of memories—physical testimony to billions of years of evolution.

Turning to controlled exploitation, what form might this take? Based on the status of *res communis*, the community of nations could concede to a request from the public or private sector for the right to exploit this public resource for a given project and for a given duration; once again, terms and conditions should already be in place to guarantee that planned operations are acceptable. For example, should nearby celestial bodies prove to offer substantial economic prospects, it would be good for the human community as a whole to benefit from this new resource and for it to be exploited in the interests of all.

Such quests, combined with the application of standards to ensure good stewardship here in our earthly home, our *oikos*, are surely just the simple expression of good economics. From that perspective, the space of tomorrow would therefore have to be economically viable, without falling into the hands of the market *per se*. Market dynamics is an important part of planning and managing an economy, but its rationale is too limited for it to be allowed to shape and direct our societies. Concerning the use of our new "community village," the coming decades are certain to see a management approach and policy to monitor and control the market in a way that respects people and human societies, particularly their diversity. In the future, this community village could take another name: Greater Earth.

GREATER EARTH

Greater Earth refers to the sphere or spatial territory around the Earth where most space activities could take place without requiring too much propulsive energy. Around our planet there exists a roughly spherical volume, more than a million kilometres across, in which a space platform could either be held in orbit by natural gravitational pull, or be maintained in Earth's vicinity by modest artificial propulsion. The L3 Lagrange point in the Earth-Sun system is one such example. Rather than distant outposts at the end of a long haul to escape Earth's gravity, bases in this area of space are easily accessible; from Earth, the propulsion required is just 1 to 2 km/s—the same needed to loft a satellite to geostationary orbit. Radio wave transmission time is no more than 3 to 4 seconds, which means two-way communication with Earth could be continuously maintained in near-real time, rather than at intervals. For platforms with people aboard, the time needed to return to Earth is only a few days, on a par with lunar missions and requiring even less fuel.

For many, the logical next step in the space endeavour and the rational use of space is to develop this region. It is a logical progression for three reasons: it is useful, the technologies needed are reasonably conceivable and the constraints imposed by human presence would not dictate the terms of each mission (although humans would eventually play a vital role within these systems, many missions could be performed remotely). Despite the obvious benefits, we must not overlook the need to manage space debris and for competing demand for limited resources.

It would therefore seem wise to devise and implement an economy for this growing sphere of space, this Greater Earth. We do not need to wait for a market to grow up around it; on the contrary, while this region of space offers potential, but is as yet unexploited, now is the time to create and promote the values and forms of expression of an appropriate space economy.

So what would this economy be based on? Some advocate cooperation, out of concern for the most disadvantaged populations. All well and good, but this raises another question common to many areas of the space endeavour. What structure should be responsible for developing, implementing, monitoring ,and even controlling such an economy? Once again, space is confronted with the issue of sovereignty.

What do we mean by sovereignty? Roger Lesgards suggests "the capacity to be truly autonomous, to depend on oneself alone to uphold one's own security, defense, interests and even political system and status as an international power" (1998, p. 89ss). He also underlines the extent to which space activities today contribute to the sovereign status of nations or entities like Europe, particularly in military and economic terms (Lesgards, 1998). Indeed, space is so closely tied to the notion of power and dominance that no nation that has embarked on the space adventure has yet given up on it completely. On the contrary, new players are seeking to get involved, even among the so-called developing nations. Today, even though the idea of power is expressed more through participation than through opposition or conflict, it is nonetheless true that these two fields still hold great appeal. The military field, because space systems offer specific and critical observation, surveillance and transmission capabilities, and the economic, because the rapid growth of new information and communication technologies offers huge opportunities, but is largely reliant on space technologies.

Greater Earth will not be exempt from these constraints, but they will be managed and overcome if the space endeavour continues in the same spirit it has embodied thus far, that of treaties and agreements drawn up out of concern for all humanity and future generations. This is no small feat: despite its lofty achievements, the aura it continues to inspire and the fact it is something we now take almost for granted, space is still seeking to gain credibility.

SPACE DEBRIS: LESSONS IN REALISM

I mentioned space debris in passing when I cited the Inter-Agency Space Debris Coordination Committee (IADC) earlier as one of the groups dedicated to protecting our planet. Space debris raises technical and legal issues that are beyond our scope here, but I would like to highlight a couple of points useful to our discussion.

First, we must not over dramatize the debris issue. Whatever we call it (debris, scrap, waste, etc.) and however we describe its effects (contaminating, polluting, degrading, etc.), we should get one thing straight: pollution is natural. In an age and culture preoccupied with health and hygiene, protection and conservation, we often overlook the fact that all living things move, breath, eat, drink, and produce waste. Most of the time we simply do not think or even care about what happens to that waste, even though we produce some of it. In nature, countless processes are constantly recovering, converting and reusing waste for new purposes. And without these processes, waste would soon build up and inexorably would eventually have a devastating effect on our planet's biology and ecology. Human beings—and by extension many human activities—are also part of these natural cycles. The air polluted in our lungs goes back into the atmosphere, waste water finds its way into rivers and filters down through underground layers, and even feces provide humus for the soil.

But we are more than just natural. We are also cultural beings, and throughout our history we have developed activities that are contrary to nature and its processes. Hunter-gatherer became domesticator and cultivator, adding things to the soil and producing new types of waste. Then walkers became drivers, sailors, and aviators, extending the boundaries of human occupation and taking with them not only what they needed to survive, but surplus as well, not to mention other species, both domesticated and parasite. Cave-dwellers built houses, factories, and office blocks, which can upset the balance and cycles of the natural environment. For many centuries, humans have added to existing natural pollution by creating new types of waste, which are the result of our technical and cultural activities and bring into play our freedom and powers of conscience and reason, foresight, and choice. However, by their very origins, these new types of pollution are not necessarily compatible with the natural cycles of elimination and transformation. Even when they are, they often saturate these cycles, with serious consequences. For this reason, we cannot ignore the pollution we produce, or simply leave it for nature to deal with. Instead, because our culture inevitably creates undesirable by-products, we must find "cultural" ways to process them. This is all the more important because of the inherently artificial nature of the pollution we produce, which means it has the potential to create greater imbalances than the pollution observed in nature. In other words, envisaging pollution (including natural pollution) and understanding, evaluating, limiting, and isolating it is part of what it means to be rational beings. After all, do we not call ourselves *Homo sapiens*; free, yet members of a group, society, and species?

My second point on the issue of space debris is also a lesson in realism. The Committee on the Peaceful Use of Space (COPUOS) defines space debris as:

> *...all manmade objects, including fragments and parts thereof, whether their owners can be identified or not, in Earth orbit or re-entering the dense layers of the atmosphere, that are non-functional with no reasonable expectation of their being able to assume or resume their intended functions or any other functions for which they are or can be authorized.*

But how do we interpret "nonfunctional"?

The KEO project is an interesting case. The project's designer, Jean-Marc Philippe, has proposed the idea of orbiting a small satellite carrying messages for our distant descendants. The satellite would circle the Earth for about 50,000 years, and then fall back to the ground to be discovered by future Earthlings, or whoever is living here then. In the meantime, the KEO satellite would be completely passive, and its mission would not be to function as such, but simply to exist. Should it thus be considered as an operational satellite, or as debris, because it is technically nonfunctional? During its protracted mission, KEO would find itself cruising along all possible orbits below 1,800 kilometres, at one time or another. It would therefore present a very real risk of collision with other objects using those orbits. For a conventional satellite, the probability of collision is between 1 in 100,000 and 1 in 10,000 per year. Multiply that by 50,000 years and an impact is virtually guaranteed. The consequences of a crash are obvious, not just for KEO, but also for the other (more useful?) satellites involved, as well as the space environment. So should a distinction be made between useful and harmful objects? The same question arises with the idea of dispatching capsules into space containing the ashes of departed relatives or as corporate publicity stunts. The time is ripe to address these issues and set some clear guidelines, because although the chances of a collision remain low, the seriousness of an impact with a working satellite or crewed vessel cannot be ignored.

As well as getting these definitions straight, there is work to be done to bring applicable space law up to date. I have already touched on some of the difficulties arising from the notion of launching state as a good example of how complicated it can be to define responsibilities and implement the necessary procedures. The need to protect certain orbits and produce less debris, during reorbiting operations for instance, is well established. However, the question remains about how to translate this need into guidelines and legislation in a commercial context. Should we apply a form of the "polluter-pays" principle to space?

Likewise, the question of how to insure space activities must also be addressed, particularly as they become more commercially oriented. Space insurance primarily covers manufacturers and operators against damage to spacecraft. So far at least, most incidents have been due to factors specific to satellites, launchers, and launch operators, with no reported cases of damage caused by orbiting debris. For insurance companies and underwriters, therefore, space debris is hardly at the top of their list of concerns. And yet they are aware of the current limits of their practices. A typical civil liability policy runs for 12 months, with no provision for accidents caused by debris or other damage to orbiting structures after that. Despite this shortcoming, it seems difficult to persuade insurers to offer policies with unlimited duration and liability. The idea of a fund to cover damage caused by space debris has been put forward as an alternative. But would it be workable in a broader commercial context?

WHERE DO WE FIT INTO THIS SCHEME?

Humankind, let us not forget, is at the heart of both the ethical debate and the space endeavour; we are masters of the house, Earth, or Greater Earth. Over and above the workings of commercialization, space must work for us. In the spirit of the 1967 treaty, agreements have been signed by various states to facilitate the distribution and use of satellite data; the International Charter on Space and Major Disasters, signed in 2000 by France and Europe, was one of the first and most eminent examples. The charter signatories—since joined by Canada, NOAA, Japan, India, Argentina, and others—commit to offer a unified system to acquire and deliver satellite data (through authorized users) to nations that fall victim to natural and technological disasters. In its first 5 years, the charter was activated over 60 times.

And what about the astronauts? Back in 1982, one expert said this: "Today, the conquest of space has become a battle on three fronts: commercial, political and strategic. Explorers have been replaced by soldiers, traders and investors." Gabriel Lafferranderie notes:

Astronauts are no longer just pilots, they are scientists, astronomers, doctors, engineers and journalists. And one day they will be gardeners, miners and shopkeepers. Astronauts are contributing to industrial experiments. They live in a restricted space and are 'odd-job men', constantly observed and monitored by ground teams, who send work instructions and tell them when to go to sleep and when to get up again. They are required to do physical exercises, undergo medical experiments and do movements in space that are opposite to what comes naturally (1993, p. 255).

The odd-job astronaut? The expression may sound surprising, but it is quite accurate. So what becomes of the notion of the "envoy of humanity," as enshrined in space law?

In common usage in diplomatic language, the term "envoy" was applied by the 1968 agreement from this new perspective, because it considered astronauts to be the representatives of all humanity. This choice recognized and gave the space endeavour a singular humanist dimension, the same that appealed to the notion of the province of mankind to describe the exploration and use of space. Admittedly, the expression is somewhat imprecise and even ambiguous. Not only did the 1968 accord do little to define or delimit the rights and obligations of this role, now assigned to astronauts, but it left the very meaning of the term "envoy" as something of a grey area. If someone takes the place of others, does s/he simply represent them and defend their interests? Does s/he not also act as forerunner, messenger, missionary, and even scapegoat? If so, who is at the other end, the opposite number, the person we are trying to reach? In short, the idea of the envoy of humanity calls for a clearer understanding of humanity and our place within the universe. Have we really thought these issues through, or even begun to think about them? Possibly not. But does the concept of the envoy of humanity still make any sense when the space endeavour has become an individual enterprise, generates profits, and has more to do with recreation than exploration? And which of the rights granted to astronauts should we retain for the future? Ultimately, who should venture forth from here, the astronaut or the space tourist?

CONCLUSION

In recognizing that the unfolding of human history has been a gradual realization that we belong to a greater whole, it is difficult to imagine the abandon of the space endeavour, but what will become of it in one or two centuries' time? We will almost certainly have established Greater Earth, but will our envoys be reduced to simple representatives of trade and commerce? Whatever the limits of this new Earth, be they scientific, technological, political, or commercial, humanity will still need to accompany this movement by an active ethical reflection. This one soon exists, through treaties and international agreements, codes of good control, or even groups of studies created by the space agencies. However, the questions and the challenges remain numerous, leaving the future of space endeavour open to extraordinary discoveries and possible failures. Its ambition is deeply human, and even humanistic, to bring the cosmos to the human level and take the humanity to the limits of the cosmos. And the concept of the envoy of humanity must not be trampled in the wake; instead, while the concept must evolve, it must nonetheless be part of the exciting future that lies ahead. This humanistic ambition could and should inspire all space activities, including the most commercial ones.

I pointed out it that the space of tomorrow would have to be economically viable, without falling into the hands of the market *per se*. Thus, the trade is the first applicability of a realistic economy, responsible and respectful of humanity, of its terrestrial environment, and at the same time as concerned of their future to all. The implications of such a perspective are huge, and responding to them is a considerable challenge, and not only from an ethical point of view. In my opinion, it reflects the essence of what it means to be human: our hunger for knowledge, our awareness of risk, and our obligation to choose.

SUMMARY

Over the last 50 years, the space endeavour has experienced its share of disasters, most dramatically when accidents have claimed astronauts' lives, and space players have not sought to hide the underlying ethical issues. However, the main ethical challenge for space today lies in its day-to-day and long-term undertakings: protecting our planet, mitigating orbital debris, and the commercialization of space. While space agencies and international bodies like COPUOS are taking the first two of these very seriously, the third is still largely uncharted territory. What will become of the "province of all mankind" enshrined in the 1967 Outer Space Treaty, or the prerogative given to the exploration and use of outer space, once real commercial enterprises have been developed there? Will yesterday's astronaut heroes be succeeded by space hotel grooms and tourists, with sentinel constellations offering costly services for those using satellite data to support their terrestrial businesses?

The time would seem ripe for public and private space players to consider how to bring some ethics into the equation through codes of conduct, regulations, and treaties. There are many issues to address. The prospect of making near-Earth (and tomorrow Greater Earth) space a *res communis*—"the property of all"—rather than a common heritage calls for a new set of rules. Today, orbits are occupied according to international procedures approved by all, but will the same apply to the disposal of satellites at the end of their life and to orbital debris mitigation? What resources should governments be prepared to commit to tracking such debris, for the benefit of commercial business? And what of the use of satellite-based observation, communication, and positioning systems? What ties will they maintain with services like those provided by the International Charter on Space and Major Disasters? How will the safety of private crews be guaranteed? If they are no longer "envoys of mankind" in the accepted modern legal sense, what duties will governments have toward them on Earth and in space? To all these questions it is perhaps not possible to give a final answer today; at least, it is advisable it to see the human dimension, even humanistic, of the space company. Space ethics is still in its infancy.

ACKNOWLEDGMENT

I would particularly like to thank Jacques Breton and Jean-Pierre Haigneré, with whom I began to draft various parts of this discussion at the International Academy of Astronautics congresses.

REFERENCES

Arnould, J. (2001). *La seconde chance d'Icare. Pour une éthique de l'espace*. Paris: Cerf.

Bell, R.(1998). *Les Péchés capitaux de la haute technologie. Superphénix, Eurotunnel, Ariane 5...*, Paris: Seuil.

Lafferranderie, G. (1993). Espace juridique et juridiction de l'espace. In A. Esterle (Ed.), *L'Homme dans l'espace*. Paris: P.U.F.

Lesgards, R. (1998). *Conquête spatiale et démocratie*. Paris: Presses de Science Po.

Martin, M. W., & Schinzinger, R. (1996). *Ethics in engineering.* New York: McGraw-Hill.

Salomon, J. J. (2000). *Survivre à la science. Une certaine idée du futur.* Paris: Albin Michel.

ENDNOTES

1 UNESCO, supported by ESA (European Space Agency) and ECSL (European Centre for Space Law), requested COMEST (World Commission on the Ethics of Scientific Knowledge and Technology) to set up a Sub-commission on the Ethics of Outer Space. In 2001, the French space agency CNES created the post of ethics officer to provide input on ethical issues.

2 In fact, the maritime environment has both qualifications. Only the sea floor, which conceals polymetallic nodules for possible exploitation, is declared the common heritage of humankind; the waters covering it, that is, international waters, are assigned the status of *res communis*. In that respect, the open sea can be used freely: no state has any form of sovereign claim to it; no national law may apply there, although international agreements make provision for certain national intervention.

3 The definition continues: "The lower limit of outer space cannot be associated with a specific altitude; it is generally accepted to be about 50 km. However, spacecraft are subject to drag and heating caused by the atmosphere at much higher altitudes" (Conseil international de la langue française (2001). *Dictionnaire de spatiologie*, Paris).

4 We could apply to space what Jean-Jacques Salomon said of technical objects: "...industrial mechanization—machines reproduced by other machines—brought us out of the natural history of the tool, whose limits were biological, and into a new era, that of the artificial history of the technical object, whose limits are defined only by the limits of our knowledge..." (Salomon, 2000, p. 43).

Chapter XV
Commerce in Space: Aspects of Space Tourism

Robert A. Goehlich
Humboldt University at Berlin, Germany

ABSTRACT

Space tourism is the term broadly applied to the concept of paying customers traveling beyond Earth's atmosphere. Operating reusable launch vehicles might be a first step to realize mass space tourism. Thus, the aim of this chapter is to investigate the potential hurdles, along with other important aspects, of space tourism flights utilizing reusable launch vehicles. The primary elements are social issues, for example, "Is space tourism acceptable concerning ethical aspects?", institutional issues, for example, "Is environmental pollution caused by space tourism harmful compared to other emission sources?", and financial issues, for example, "Are there any potential investors interested in space tourism?".

INTRODUCTION

Figure 1 shows a selection of key aspects of space tourism, which includes hurdles and opposing forces. Hurdles and opposing forces could be harmful for a successful establishment and enhancement of space tourism activities. These potential hurdles in commercial space travel need to be considered thoroughly, preferably before actual activation of the first regular *space tourist services* (Goehlich, 2003). Those aspects are discussed in this chapter after giving some brief information on space tourism in the background section.

BACKGROUND

For past decades, interest in the possibilities of space tourism has increased among engineers, scientists, entrepreneurs, and the public. A continuously growing collection of papers are being published on space tourism and associated subject areas such as reusable launch vehicles, space habitats, space entertainment, and the corresponding law and regulation. Market research promises sufficient interest in tourist space travel to take off and develop into a multibillion-dollar business. This is understandable, not only because of the

Figure 1. Aspects and induced possible hurdles of space tourism

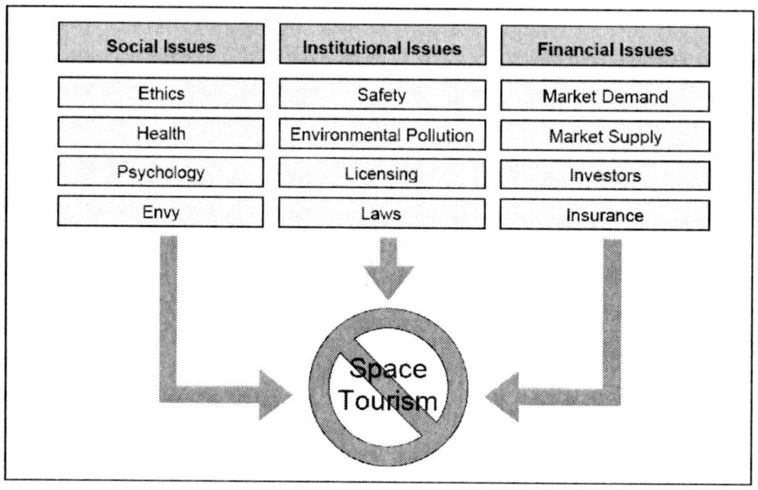

attractiveness of being in space, but also from an exclusivity point of view. In some sense it is equal to the somewhat exclusive luxury cruise ship business, which also requires huge investments. It is all about doing something unusual, similar to an adventure trip or a challenge, like climbing Mount Everest or exploring Antarctica.

Currently, there are only three possibilities for accessing space and orbit Earth as far as human spaceflight is concerned: Russian Soyuz, the USA *Space Shuttle,* and Chinese Shenzhou. For the time being, only Soyuz has been used for space tourism, and there are no indications that the *Space Shuttle* will be used for this purpose in the near future. The Shenzhou is under discussion for tourist transportation in the mid future. The lack of alternative access is a critical factor limiting the supply of space tourism services. A breakthrough in this area, such as the development of a new generation of reusable launchers, could have a significant impact on space tourism. Low-cost and low-risk access to space is assumed to be vital for expansion of the space tourism market. Traditionally, very high levels of public funding and minimal private investment have characterized human spaceflight activities. *Space tourism* flights might have the potential of changing the balance from public to private expenditures in human spaceflight.

Nowadays, the number of commercial space launches in a year is between 70 and 80. Comparing this to 70 or 80 take-offs by commercial aircraft every minute all around the world, it is obvious that the way the space industry works today makes it very difficult for any commercial venture in space to yield a profit. Supporters of space tourism argue that this is where all the excitement about space tourism comes in. Space tourism has the potential to provide a high number of flights on a regular basis. This is essential to reducing launch cost radically. Tourism flights would have virtually no saturation limits: in the early years, though, the focus would be on *suborbital flights*, short trips to orbit, simple orbital hotels, and after that trips to the Moon, perhaps with a stay at a lunar hotel in the far future, or even tours to Mars someday. In other words, people might be the payload of the future.

Today's space tourism flights are in the early so-called pioneer phase, which means approximately one or two tourists per year. For example, Soyuz flights to the International Space Station cost approximately $20 million, and will last approximately 10 days. In April 2001, Dennis Tito was the first space tourist, followed by Mark Shuttleworth in April 2002 and Greg Olsen in October 2005. Tourism related to space, as a commercial activity, is in its infancy compared to other tourist enterprises. However, facts show that the space tourism industry is already larger than most people realize. The true potential of space tourism in coming decades does not rest within one or two flights per year for $20 million per trip, but in providing a wide range of services with different levels of prices. Peak turnovers from ticket sales in the range of $10 billion per year during this century are imaginable, plus additional turnovers from novel secondary markets such as space fashion, space food, space entertainment, space sports, and so forth.

Space tourism is different from any other previous space activity. Like every other form of tourism, space tourism must be customer-focused. This consideration forces engineers and operators to look at manned spaceflights from a slightly different perspective than that used to date for scientific missions. Space tourists need not only to survive and keep healthy in extreme conditions, but also to live comfortably and enjoy their stay as much as possible.

The potential for an introduction of reusable launch vehicles (RLVs) is derived from an assumed increasing demand for the transportation of passengers in the decades to come. Like other professional planned business, once space tourism gets started, it is expected to develop progressively in several phases. Beginning at a relatively small-scale and a high-priced "pioneering phase," the scale of activity might grow and prices might fall as it matures. Finally, it might become a mass-market business, similar to aviation today. The phrase "space adventure or individual travel" is a convenient one to describe the first phase. Customers might be relatively few, from 1 per year to 10 per year; prices would be high, and the service would be more close to "adventure travel" than to luxury hotel-style or first class flight. Orbital accommodation would be safe but "Spartan." The mature phase might see demand increasing from 10 passengers per year to thousands of passengers per year. Tickets to orbit would cost less and flights could depart from many different airports. In the mass phase, ticket prices might fall and customer count might grow from thousands of passengers to hundreds of thousands per year. There would certainly be no limit to possible destinations. Access to space resources that low cost launches might bring about would ensure that economic growth need not end and generate funds needed to open up space to a wide range of human activities.

SOCIAL ISSUES

Ethics

Apart from concerns regarding the feasibility of mass space tourism, there is also a more human ethical issue. This issue has been barely touched upon in literature. Despite the expected progress in safety and reliability of launching rockets, it will still be a risky procedure. Even small errors or faults can result in major dramatic events. The question arises, what level of risks for space travelers would be acceptable for the society?

These days, there are many adventure travelers and extreme competitive athletes climbing Earth's highest mountains, as shown in Figure 2, crossing its largest deserts, exploring the Antarctic, and diving toward the darkest depths of oceans. One by one, dangerous activities with risks of death at every step are done, yet humankind hail them as highly admired heroes. Moreover, if atop of these activities there is a genuine interest for traveling

Figure 2. Mount Everest

to space, potential risks might be certainly no reason to abort.

Conquest of space might contribute to dissemination of a scientific and technical culture. It also could maintain an imaginative horizon and determination to make new discoveries, which might be driving forces of human society. There might evolve many social benefits of space travel in general, and space tourism in particular. Travel does broaden the mind, would give people a better understanding of the complex world they live in, and in this case, would give them a planetary conscience. In all, politicians, scientists, and the public would need a fundamental change from today's views of space activities to enable mass space tourism. It is reasonable to characterize this challenge as a political, social, and economical revolution.

Health

Space tourism flights would be intended for persons in generally good health and without physical disabilities. Topics relevant to medical safety and general well-being of space tourists are accelerations during take-off, reentry, and landing, microgravity in space, cosmic rays, and "jetlag" effect.

Loads (acceleration, noise level, mental stress) are similar to those of military aircraft flights for tourists. A maximum acceleration of 3,5 g should not be exceeded, as shown in Figure 3.

Sudden lack of correspondence between information received from the inner ear and visual cues caused by microgravity provokes disorientation and discomfort and is called space motion sickness. Typical symptoms are pallor, dizziness perspiration, drowsiness, nausea, and psychological stress, and vary individually (ISU, 2001). To protect passengers against space motion sickness, pre-flight training such as biofeedback, preflight prophylactic medication and in-flight medication such as Promethezyne could be used.

Along with space motion sickness, the most immediate effect of microgravity is redout, where blood rushes into the head, as opposed to away from it. Symptoms are puffy heads, "chicken legs," and an increased heart rate to compensate for changes in blood volumes and locations. After returning to normal gravity level, light-headedness and fainting can occur.

Figure 3. Acceleration limits for human body (Source: Lo)

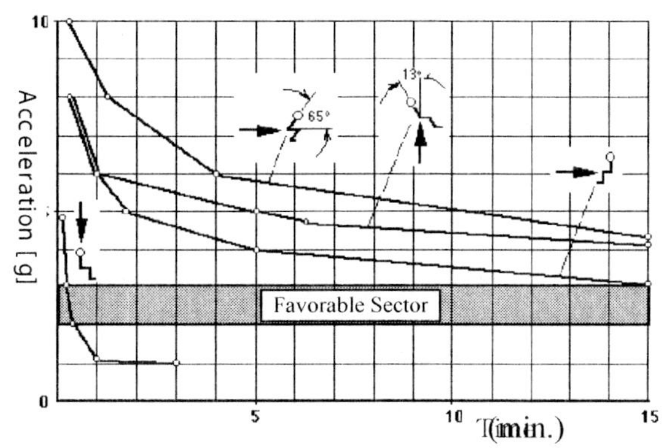

Figure 4. Astronaut exercising on ergometer (Source: NASA)

Unloading of the body in microgravity leads to decomposition of weight-bearing bones and muscles, especially in the legs, hips, and the back. It leads to weakness upon return to normal gravity. An average of 1% to 3% of bone loss per month was observed on Mir space station crews (ISU, 2000). Existing countermeasures consist of several hours of daily exercise on ergometers with bungee cords, as shown in Figure 4. Bone loss is one of the main hurdles for long stays in space, but space tourists spending only a day or a week in space would be not affected.

The average radiation dose to a human on Earth from soil (0,4 millisievert (mSv)), food (0,2-0,5 mSv), and cosmic rays (0,4-1,6 mSv) results in a total of about 1,7 mSv per year (Tascione, 1988). For comparison, each transatlantic flight can account for another 0,04 mSv. Space tourists on a one-day mission would receive a higher dose of about 0,3 mSv (A Med-World AG, 2006) but it is still negligible. Only for very long space flights is space radiation protection needed.

Orbital flights of duration of one day will induce disturbances of circadian rhythms, due to fast night and day cycles. Symptoms are similar to "jetlag" after a long distance flight. Astronauts regularly use medication to assist sleep and tourists would be able to take sleeping pills if necessary.

Psychology

Besides medical standards, it would be necessary to establish psychological standards. Comfort boundaries are strongly affected by physiological issues. Initial difficulties in adapting to space could affect the tourists' enjoyment. Therefore, a parabolic zero gravity flight should be part of preflight training for participants to prepare them for weightlessness. This experience before the spaceflight would help passengers mentally adapt more quickly once in orbit. The preflight training must be organized in such a way that customers consider it part of the adventure and part of what they pay for, not only as preparations for the journey (Abitzsch, 1996).

Overcoming claustrophobia is an area warranting additional study. Simple things such as meals and recreation can influence morale on a large scale and reduce this effect.

A special consideration should be given to religious and cultural ceremonies such as Christmas. Studies on philosophic experience indicate that residency in orbit tends to make individuals more reflective about philosophical questions, such as the meaning of man's existence in the cosmos. Those sessions dedicated to spiritual activities would increase the intensity with which tourists would experience space (ISU, 2000).

Envy

Some persons within the government may view space as an exclusive province of federal national security, and private sector activities of any kind may be considered competition and a threat to their own power base. For instance, the current astronaut corps has many members who have never flown. If capacities of government operated vehicles such as *Space Shuttle* are sold to civilians in lieu of astronauts who have undergone a rigorous selection and training process, it might cause disapproval outside the astronaut corps. Intensified integration of government astronauts into private space business by offering a goal-directed service for government space market in addition to private one could avoid these circumstances.

In case space tourism business would be actually supported by federal funding, some politicians might view it as a taxpayer subsidy to the wealthy, who, barring lotteries or contests, would initially be the only ones who could afford space trips. Even if there is no government funding involved, historically, in the mind of the general public, there is a strong linkage between government's space agencies and space, and therefore any space tourism activity may be perceived as a waste of public funds. Therefore, economic returns, increased tax base, and attractive opportunities should be promoted as byproducts of the space tourism business.

INSTITUTIONAL ISSUES

Safety

Because space tourists are not going to be trained like professional astronauts, a familiarization with emergency procedures would be needed. While a space tourist would require more than the standard 2-minute airline drill on how to fasten seatbelts and use the oxygen system, an intense week of training should be sufficient to learn the basics of how to be a safe passenger. This course may include flight training, medical training, and emergency procedures.

However, it will be more difficult to make vehicles themselves safer. There are two approaches: the first option is to reduce catastrophic failures by redundancy and over-designing of subsystems, to improve maintenance by using an extensive health monitoring system, and to improve operations with many soft abort sequences. The second option is to protect passengers, if a catastrophic failure has occurred, by using safety equipment for passengers and crew such as space suits, ejector seats, emergency shelters, and so forth.

Both options would result in an increased vehicle empty weight and therefore a reduced number of passengers. For vehicles with a large passenger capacity, option one might be more suitable (rescuing the vehicle with passengers as a whole), while for those vehicles with low passenger capacity, option two might be more suitable (rescuing only passengers). A higher safety standard would result in lower economic performance due to less profit, resulting in higher life-cycle costs. A lower safety standard results also in lower economic performance because the higher risk would be unattractive for passengers and ethically unacceptable, resulting in lower demand. More research is needed to find out what the "right" safety standard is for space tourism vehicles.

Note that *Space Shuttle's* statistical vehicle loss rate of less than 0.01 (1 out of 100 flights) is the lowest besides the Soyuz rocket of all space launch vehicles to date. For comparison, the statistical loss rate of civil aircraft is 0.000,001 (1 out of 1 million flights), while 0.0001 (1 out of 10,000 flights) for military aircraft. The author suggests that tourist rockets are required to reach a loss rate of much less than 0.001 (1 out of 1,000 flights) at the start of operations, which needs to be reduced further.

Environmental Pollution

Any chemical propulsion launch system leaves traces of emissions in the atmosphere. Because much more energy is necessary to transport a passenger to suborbit or orbit compared to any other place on Earth by aircraft, there would also be more pollutants generated. For example, Kankoh Maru Plus, as shown in Figure 5, which is a Japanese reusable launch vehicle concept, would need 71 mega grams (Mg) liquid hydrogen as fuel to transport 50 passengers to Low Earth Orbit (LEO) and back to Earth. Thus, each passenger needs 1.4 Mg liquid hydrogen, which is equivalent to an energy of 202 GJ (Giga Joule). A *Boeing B747-400* needs about 150 Mg kerosene as fuel to transport 400 passengers one-way, from one continent to another. Thus, each passenger needs 0.4 Mg kerosene for a one-way flight, which is equivalent to an energy of 17 GJ. Roughly, the energy consumption per passenger of 1 two-way space flight is equivalent to 6 two-way transcontinental flights.

However, from a global scale point, the cumulative energy consumption of space tourism would be relatively small compared to today's annual 1500 million air passengers with expected future annual 0.1 million space passengers. It should be noted that this example is just intended for illus-

Figure 5. Kankoh Maru Plus (left) (Kawasaki) and B747-400 (right) (Source: Rafi)

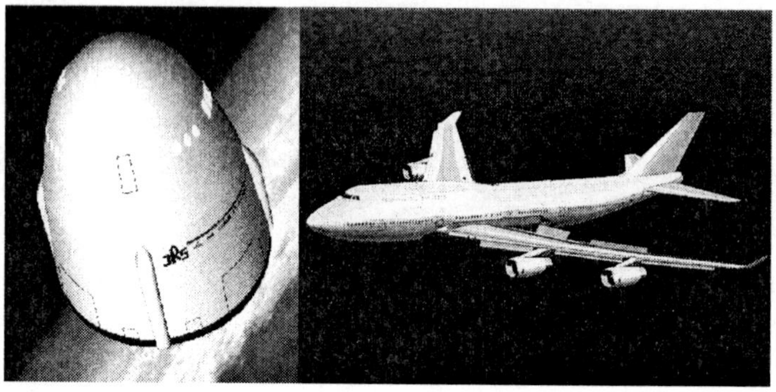

Table 1. Estimated emissions for 2065 (modified from Adirim, Lo, & Paatsch, 1999) Unit: Mg/year

	Sources	H_2O	CO	CO_2	HCl	NO_x
Anthropogenic	Space Transport	< 23	n.a.	< 0.0005	n.a.	< 0.005
	Air Transport	> 436	> 0.26	> 1,070	n.a.	> 5
	Burning of Fossil Fuel	8,300	n.a.	20,350	2	n.a.
	Others	n.a.	1,490	n.a.	n.a.	85
Natural	Volcanoes	n.a.	n.a.	n.a.	5	n.a.
	Oceans	525,000	n.a.	n.a.	330	n.a.

tration of basic numbers and real relations have been simplified. For example, necessary energy consumption to produce kerosene, liquid hydrogen, and liquid oxygen has been neglected.

Table 1 shows a comparison of space transport (excluding the space tourism sector) with other anthropogenic, as well as natural pollution, sources estimated for 2065 (Adirim, Lo, & Paatsch, 1999). It seems that space transport emissions are negligible on a global scale, even if the launch rate would increase by a factor of 100 caused by space tourism. Extensive studies have been investigated on emissions caused by space transport (including space tourism) and air transport in a period from 2010 to 2065 for a scenario with up to 200,000 space passengers per year. It shows that space tourism would only cause between 0.006% to 1.5% of total emissions caused by air transportation (Lo & Paatsch, 1998, 1999).

However, from a local scale point, in altitudes above airline traffic, space vehicles are the only mayor emitters. These emissions along the trajectory in the sensitive upper atmosphere are not negligible; neither is the local pollution at spaceports. It is recommended that operators should be obliged to pay a "keep space clean" fee, depending on the amount of emissions. Unfortunately, effects of some types of emissions, especially in the upper atmosphere, are not well understood. Therefore, ecologically adapted flight profiles cannot be considered, unless emission penalties are quantitatively formulated by atmospheric chemists.

Any additional source of pollution, such as space tourism transportation, should not and cannot be excused by the pre-existence of other pollution sources. This matter is currently not discussed by developers and organizations promoting space tourism flights. However, this is a very sensitive and politically charged issue. Presenting this topic in a wrong manner could possibly lead to a strong rejection of tourist spaceflights.

Licensing

Developing vehicles needed for space tourism is an engineering challenge (low-cost operating procedures, high reliability, safe abort capability at any time, vehicle performance, etc.), but it is also an institutional one (applicable laws and regulations). Currently, there is a deep gap between rocket and aircraft design philosophy, which is illustrated in Figure 6. The mission success of a rocket launch can be merely estimated by a reliability calculation. Thus, the probability of loss

Figure 6. Deterministic (left) vs. probabilistic (right) operation (Source: Yone)

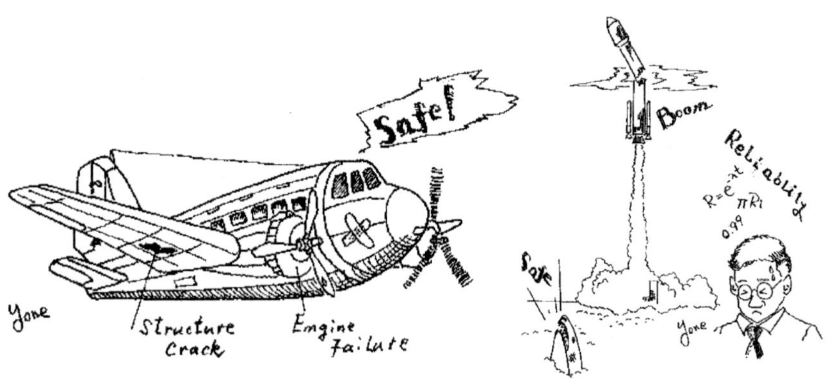

is a figure of the failure rate. This means that the rocket is launched by probabilistic operation for launch success. In contrast, airworthiness requires safe operation even for the case that some subsystems or components of an airplane get out of order during operation. It can be said that aircraft aims at a deterministic operation for safe flights.

Therefore, the safety standard required for certification of space transportation vehicles should not only restrict their design, but should change the fundamental operation process from probabilistic launch to deterministic take-off and landing with enough safety, like aircraft. The existing regulatory and legal environment needs to be reformed to allow for the promotion of commercial passenger flights to and from space. Solving these hurdles is essential in order to allow developers and operating companies to raise the necessary investment from investors. Investors themselves want to understand and control their capital risk, which is only possible in a regulatory market environment. The absence of regulations may make investors afraid that any unknown future regulation may kill the business they are investing in. The following points should be considered:

- Systems will be needed for training, testing and licensing of pilots, cabin staff, and maintenance staff.
- The Federal Aviation Administration (FAA) needs to extend their air traffic control system to include suborbital and orbital surveillance. The FAA is a U.S. governmental institution in charge of regulating and overseeing the aviation industry in the United States.
- Need of certification regulations for passenger RLVs, taking into account vehicle structural integrity and damage tolerance, fire suppression systems, noise-levels, evacuation standards, pollution levels, and maintenance procedures
- Passenger travel services in space need insurance similar to that, which was imposed as a part of the Warsaw Convention for air travel. However, in its early stages, the small scale of the space travel industry and the limited statistical base will not be sufficient to permit an insurance calculation with high confidence. As a result, insurance cost will be high.

For redundancy and safety design requirements and structural verification procedures, the civil aviation model provides a good guideline. Primarily, it is the responsibility of governments to negotiate and ratify such agreements in time to be effective and thus in preventing major accidents or international conflicts. However, regulations mean also a barrier for space tourism. A major issue is how to handle the classification of suborbital and orbital vehicles. Passenger aircraft must go through a certification process, which is handled by the FAA, before they are allowed to carry fare-paying passengers or cargo. More than 1000 test flights are typically needed to gather enough statistical data. The process may run for over 3 years. If the space tourism industry would have to go through the same procedure it would mean great economic difficulties, which would probably put a stop to any start-up company before it even got off the ground.

In March 2004, the House passage of legislation set guidelines for the future space tourism industry. The House bill gives regulatory authority over human flight to the Federal Aviation Administration's Office of Commercial Space Transportation. To make it easier to test new types of reusable suborbital rockets, this bill gives the FAA office the authority to issue experimental permits that can be obtained more quickly and with less bureaucracy than licenses (U.S. Congress, 2006). This approach might be a major step in the further development of commercial human space flight.

In parallel with the U.S. government, private groups have also started to study the regulatory system of the aviation industry as an appropriate model for the passenger space travel industry. For instance, the Japanese Rocket Society (JRS) Transportation Research Committee in 1997 studied the requirements needed for the certification of "Kankoh-Maru" to enable passenger carrying (Collins & Yonemoto, 1998). The Universal Space Clipper Company divided in a study the requirements for passenger space vehicles into categories: type design certificate, production certificate, airworthiness certificate, commercial operator's license, spaceport license, and other approvals, such as component manufacturing and maintenance (Gaubatz, 1998).

Laws

Laws already exist to regulate private sector space endeavors, such as satellite launches. The major space laws treaties are:

- **Outer Space Treaty (1967):** The Outer Space Treaty stipulates the principle of "exploration and use of outer space." It can be considered as the backbone of international space law. However, liability of the launching state for damages caused by a space object is not clear. It is unclear which liability regime would apply in the event that a nongovernmental entity's space mission resulted in harm to a foreign citizen. The Warsaw Convention (1929) provides guidelines regarding monetary compensation. This convention is an international private law treaty and has helped to establish international air travel, by limiting airlines' liability for damages in the event of injuries to passengers or loss of baggage (Collins & Yonemoto, 1998). Nonetheless, the Warsaw Convention is an encouraging precedent for the legal innovation needed to make space activities commercially feasible, and it has been proposed that a space law agreement should be based on this convention (Roberts, 1996).
- **Rescue Agreement (1968):** The agreement does not include passengers, so space tourists may not fall into the scope of the agreement and therefore may not take advantage of the rules stipulated there. On the other hand, it would be a wrongful interpretation to assume the exclusion of passengers, just because they are not mentioned. This gap stems from the period in which the Rescue Agreement was created, when tourist participation was not even considered (Wollersheim, 1999).
- **Liability Convention (1972):** Article II provides the launching state's absolute liability for compensation of damage on the surface of the Earth or to aircraft in flight. Accordingly, states have the right to refuse private enterprises to practice space tourism. A gap of the convention is that "nationals of the launching states are excluded from the scope of the Liability Convention" (Wollersheim, 1999).
- **Registration Convention (1976):** The Registration Convention has on one hand the function to coordinate launches, and on the other hand to ensure identification of the launching state in respect of the Liability Convention. Private enterprises need to comply with the registration procedure. The addition of private registrations will greatly increase with space tourism (Wollersheim, 1999).
- **Moon Agreement (1979):** According to the Moon Agreement, celestial bodies and their resources shall not be subject to sovereignty claims. The Moon Agreement refers to the surface, which could be interpreted in such manner, that buildings or facilities on the Moon's surface remain national property and consequently fall under national sovereignty. Lunar bases are subject to state jurisdiction and are legally treated as space objects. The

Moon Agreement does not have a high practical relevance, because the agreement does not prevent states other than the contracting states from claiming national sovereignty for the respective celestial body (Wollersheim, 1999). This is relevant to space tourist projects, such as a lunar hotel.

Space tourism is a new institutional challenge, because of its yet uncharted territory. There is no legal jurisdiction for regulating commercial human spaceflight. It is likely that analogies will be made to laws applicable to air transportation. However, it will require an extensive innovation in applicable regulations in both national and international law. For example, the question arises of who would have jurisdiction if an international passenger on a space tourism flight commits a crime against another international tourist. In this case, for the International Space Station (ISS), the International Governmental Agreement states that criminal jurisdiction should remain to the state of nationality of the "alleged perpetrator" of a crime, provided that the state is an International Space Station partner state.

Law policy is a fundamental component in space commercial development through space tourism. Economic activity in space must be accompanied by the simultaneous implementation of a law framework in which these activities are going to take place. However, only elementary steps toward this direction have been undertaken. Law policy must be developed by an international organization and gain appropriate endorsement from every state. It is unacceptable to have legal regulation without binding compliance from space-faring nations (Hudgins, 2006).

FINANCIAL ISSUES

Market Demand

Space tourism market surveys have been performed in Japan, Germany, United Kingdom, the United States, and Canada (Geoffrey, 2001). The first market investigation was conducted by the National Aerospace Laboratory (NAL) in Japan in 1993 (Collins, Stockmans, & Maita, 2004). It surveyed 3030 Japanese across all age groups and revealed a significant desire to visit space. More than 70% of those under 60 years old and more than 80% of those under 40 years old said that they would like to visit space at least one time. A number of rudimentary market surveys for space tourist flights have been conducted also by Abitzsch (1998), Bekey (1998), Kelly Space & Technology (2001), and Zogby International (2002).

The different survey data are interpolated, adjusted for inflation to fiscal year 2000, and weighted according to their statement quality by the author for determining a market demand model, with the results in Figure 7. A trend curve for suborbital and *orbital flights* is determined, based on results of market surveys and polls. All dots except Zogby International (orbital flight) estimates are scattered around the two curves, which indicate the relative quality of surveys. The two extreme estimates are neglected. In general, the figure shows that passengers are willing to pay more for orbital flight than for suborbital flight. Additionally, it shows that the bandwidth for ticket prices is immense, but the curves' accuracies are higher for a ticket price from $50,000 to $1 million due to a higher concentration of estimates.

There is a risk that the suborbital *space tourism* market would be almost instantly displaced when a product capable of reaching orbit was

Figure 7. Model of annual passenger rate as function of ticket price

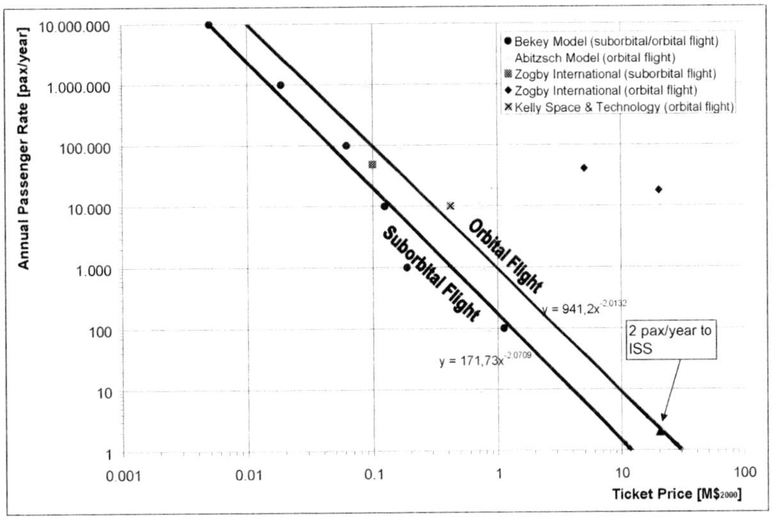

introduced. Therefore, the question arises, which was investigated by Goehlich (2005): "Would a suborbital market last long enough for manufacturers be able to recoup their investments prior to the introduction of a transportation system capable of making orbit?"

However, in the early pioneer phase, it is difficult to forecast the demand/price elasticity correctly because of the following two facts. Firstly, the passengers will be mostly multimillionaires for whom prestige and political causes determine the demand, rather than the ticket price itself. Secondly, there is a difference between saying "I would like to make a trip into space," and the actual payment for a ticket.

Market Supply

Overall, there exist over 300 worldwide proposed vehicle concepts for RLVs, as shown in Table 2, which could be realized by manufacturers from various countries. The majority of these vehicles are proposed for human space flight, but a few are unmanned too. Only 4 RLVs have been realized, namely X-15, Space Shuttle, Buran, and SpaceShipOne. Buran has only been launched one time and therefore reusability has not been verified for this vehicle (Goehlich, 2006)

There are two trends in RLV development. One is developed by the government, like the X-series of United States, and the other one is developed by private companies, partly stimulated by organizations like Ansari X Prize. The Ansari X Prize was a $10 million prize to jumpstart the space tourism industry through competition between entrepreneurs and rocket experts in the world, and was won by Scaled Composites' SpaceShipOne in 2004. The Ansari X Prize competition follows in the footsteps of more than 100 aviation incentive prizes offered between 1905 and 1935 that created today's multibillion-dollar air transport industry (X Prize Foundation, 2006)

Even though the development cost for SpaceShipOne has been much more than the $10 million return from the Ansari X Prize, it will take a lot of flights with test pilots before SpaceShipOne begins transporting paying tourists. In the future, new problems of a technical, political, or economical nature may arise, and the three successful SpaceShipOne's suborbital flights in 2004

Table 2. Worldwide RLV concepts

Country	Suborbital	Orbital	Total
Argentina	1	0	1
Canada	2	1	3
China	0	3	3
France	1	14	15
Germany	3	18	21
India	0	1	1
Israel	1	0	1
Japan	1	8	9
Romania	1	0	1
Russia	3	50	53
United Kingdom	4	10	14
USA	46	132	178
Total	63	237	300

(one test flight and two qualifying flights) have totally changed things. A privately financed rocket is a success story rather than a dream of rocket enthusiasts and science fiction writers.

However, even though the SpaceShipOne's flights have been successful, it might be the case that the "big breakthrough" for space tourism does not really happen. In comparison, after the successful flight of Dennis Tito, the first paying space tourist, in 2001, people thought about a rapid increase in those kinds of flights. Instead, things have happened slower, and none of the ideas such as a space lottery for the general public or TV shows exist on a large scale, as forecasted or planned in 2001. In conclusion, it is difficult to forecast the "ifs" and "woulds," derive from Wright Brothers' powered maiden flight or Dennis Tito's maiden space tourist flight to SpaceShipOne's maiden private financed suborbital flight. But anybody will agree that if we try hard for something we can achieve it.

Investors

By far, the hardest obstacle to any new rocket venture is its being properly financed. Although there is quite a number of start-ups which try to enter the rocket market, only a small fraction of their overall funding requirements were actually supplied by the world's financial markets. Because space tourism is a completely new industry, no data whatsoever on previous experiences are available. The only data available are a few space tourism surveys that have been conducted. No one knows exactly how large the market will be and no reusable rockets today have yet passed safety standards needed to be able to carry passengers. This makes the space tourism industry prospects very speculative. Investors become hesitant, especially considering the large amount of funds needed to develop a completely new vehicle.

Although it is difficult to generalize about high-risk investment and its financing, some broad conclusion can be drawn. New companies are unable to raise risk finance through a stock exchange listing. There are specialized institutions whose main purpose is to provide venture or high-risk capital. Most, however, have maximum limits on the amount of capital they are prepared to provide in any single case. Because they are reluctant to invest in new company ventures, their services appeal mainly to existing companies (Moore, 1983).

Decision making under uncertainty might stay a challenging task because necessary data for a reliable forecast are not available (Bamberg & Coenenberg, 1996). Economics has shown many pitfalls, complications, and even inconsistencies in attempts to measure risk attitudes. Sensitivity to framing, preference reversal, and the gap between willingness-to-accept and willingness-to-pay might well serve to put off any attempt to measure risk attitude (Hartog, Carbonell, & Jonker, 2000). Expressed another way, strategic decisions have a complicated structure, and there are no overall experts. The need to assess alternatives and make significant business decisions with limited information causes many companies to address strategic decisions with models. The necessary requirement is to develop probabilities for the assumptions in the model based upon the uncertainty of each input value (Lorance & Wendling, 1999). A solution might be a cost risk analysis to generate a range of cost and assign a probability level to each cost value in the range. The usual problem is not just to come up with an estimate of the cost of a project, but to predict the range of values into which the cost may fall and with what level of confidence the prediction is made (Dean et al., 1986).

Unless a company is very committed to investing in space tourism, RLV opportunities will most likely have to show that their projected profits must be sufficiently higher than terrestrial alternatives to compensate for the added risk.

Generally, venture capitalists are concerned about the lack of management experience in new space ventures (Livingston, 2000). There seems to be plenty of room for misunderstandings between rocket scientists and finance people, as illustrated in Figure 8. To a rocket scientist, "burning rate" means something totally different from what "burn rate" means to a venture capitalist. Therefore, most commercial rocket ventures have failed to catch the attention of venture capitalists. A professional business approach is needed to make the case for nonspace business and finance communities.

Figure 8. Communication problem (Source: TIM Consulting)

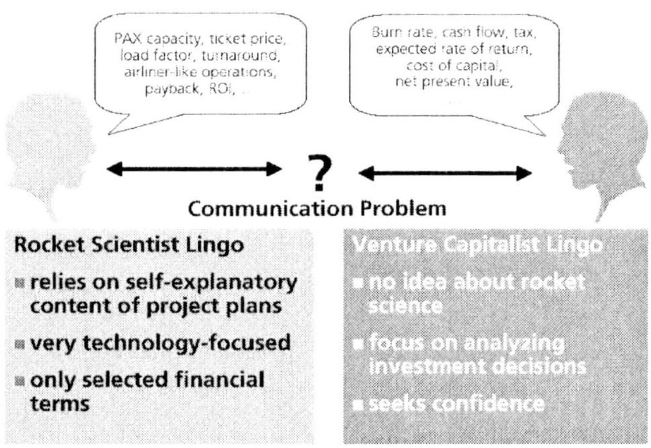

Insurance

Space tourism is a new venture and until it reaches a mature level of development, insurance is going to be a major issue. Both passengers traveling in space and the related equipment and facilities will need insurance. However, the likely small scale of the space travel industry will be insufficient to enable accurate calculation of insurance cost. Accordingly, for tourism to become a vital part of the commercial space equation, limits on the liability of owners and operators of space facilities and vehicles will become necessary (Collins & Yonemoto, 1998).

A high failure rate is not favorable for space tourism, although there are many activities undertaken on ground with even greater risk factors. Due to the assumed high risk of space tourism ventures, insurance premiums are assumed to be very high. As long as insurance companies do not have coherent information about prices generated by implementation of space tourism, this will stay unchanged. Catastrophes like aviation disasters appear to carry more importance in people's minds than would be expected from general considerations of attitudes toward other forms of deaths (Moore, 1983). Even if the level of incident risk is low, the consequences when the risk occurs can be very large, which has implications for any insurance cover sought.

In the case of commercial space launch companies, there exists a law under which these companies are required to carry liability insurance, capped at $500 million, with assurances that the government will compensate for losses above that (U.S. Congress, 2006).

DISCUSSION AND RECOMMENDATION

There is currently no overall framework to deal with hurdles and other aspects influencing space tourism as discussed in this chapter. In particular, for mass space tourism to become a reality, it is important to develop a framework, which needs to be accepted internationally. It should consist of at least ethical, health, psychological, safety, environmental, regulatory, laws, investors, and insurance components, which are listed in the following three main groups.

Social Issues

- An ethical framework is needed to reflect motivations and consequences of public space travel. Ethics might fundamentally influence the development of future space tourism activities.
- A health framework is necessary. Early space tourists such as Dennis Tito were most likely well prepared, ensuring a good level of health and tolerance. In case of mass space tourism, physical and psychological comfort for the average healthy person, as well as for very young or elderly persons, has to be ensured.
- A psychological framework is recommended. Persons with a prior history of personality disorders, claustrophobia, and suicide attempts will have to be excluded.
- Envy cannot be put in a framework. It will depend very much on the way space tourism activities are reported in the media.

Institutional Issues

- The development of a space tourism market might be most sensitive to a safety framework. Studies are needed to find a balance between demanded safety standards for vehicles and the possibility to fulfill these requests by developers and operators using the available technology. Too high safety standards would mean a showstopper for space tourism because it is not technically and economically feasible. However, too low safety standards would mean a showstopper for space tourism too, because ethical aspects render it unfeasible.

- An environmental framework is necessary, similar to the one realized for aviation operations. In particular, issues relating to local spaceport emissions and noise pollution, space debris pollution, and global emission pollutions must be addressed.
- Commercial space activities will require a regulatory framework, as does any commercial endeavor. Major issues to be resolved for space tourism include the training of passengers and crew, the certification of vehicles and launch facilities, and the licensing of space operators.
- A legal framework for space tourism activities is not clearly defined in several fields. In particular, issues relating to jurisdiction, space traffic management, and liability must be addressed.

Financial Issues

- Accurate and accepted market demand research is needed to fulfill the financial requirements of investors, as well as the expected flight program outline from the passengers.
- More space flight demonstrations are needed to motivate investors to invest in those kind of vehicles, to stimulate rocket engineers for thinking of alternative approaches, to increase public and government awareness for space activities, and to create the desire in many of us for "Why not ?" followed by "Why not me ?".
- A financial framework is needed for investors to reduce their risk. This could be realized through the development of an independent consultant agency.
- The topic of insurance as it pertains to space tourism has not been well documented and merits further critical review, because it is a major aspect to reduce financial risk.

CONCLUSION

On the basis of technical, political, and economic investigation, a stable space tourism business for space flights appears to be feasible in the future, but some relevant issues must be taken into account: a start-up market environment with high profits after a few years, as realized in the IT sector, is an illusion for the space tourism sector. The space tourism sector, in particular the *mass space tourism sector* (i.e., aircraft-like operations, using large vehicles, scheduled flights), is assumed to grow slowly and will in one way or another require government support. The reasons are mainly long development periods for new reusable launch systems (around 10 years), high development cost (between \$5 to \$15 billion), and relatively late break-even points for positive cash flow (between 5 to 15 years) (Goehlich, 2005). The *individual space tourism sector* (i.e., adventure type operations, using small vehicles, unscheduled flights) will develop much quicker, as the successful demonstration spaceflights of SpaceShipOne rocket and tourist flights of Soyuz to ISS are already proven.

ACKNOWLEDGMENT

The publication of a space tourism chapter in this book was encouraged by Professor Phillip Olla, which is gratefully acknowledged by the author.

REFERENCES

A Med-World AG. (2006). *Strahlenbelastung in der Raumfahrt.* Retrieved March 3, 2007, from http://www.m-ww.de/enzyklopaedie/strahlenmedizin/strahlenbelastungraumfahrt.html. Berlin, Germany.

Abitzsch, S. (1996). Prospects of space tourism. In *Paper presented at the 9th European Aerospace Congress,* Berlin, Germany.

Abitzsch, S. (1998). *Chancen und Entwicklungsmöglichkeiten der Weltraumtouristik*, Doctoral thesis, Technical University Berlin, Germany.

Adirim, H., Lo, R., & Paatsch, A. (1999). Next generation high performance low pollution launchers: Problems and solutions. In *Paper presented at the 50th International Astronautical Congress, IAA-99-IAA.3.3.01,* Amsterdam, The Netherlands.

Bamberg, G., & Coenenberg, A. G. (1996). *Betriebwirtschaftliche Entscheidungslehre*, ISBN 3-8006-2076-6, pp. 105-113. Munich, Germany: Vahlen.

Bekey, I. (1998). Economically viable public space travel. In *Paper presented at the 49th International Astronautical Congress*, Melbourne, Australia.

Collins, P., Stockmans, R., & Maita, M. (2004). *Demand for space tourism in America and Japan and its implications for future space activities.* Retrieved March 3, 2007, from http://www.spacefuture.com/archive/demand_for_space_tourism_in_america_and_japan.shtml. Tokyo, Japan.

Collins, P., & Yonemoto, K. (1998). Legal and regulatory issues for passenger space travel. In *Paper presented at the 49th IAF Congress,* Melbourne, Autralia. Retrieved March 3, 2007, from http://www.spacefuture.com/archive/legal_and_regulatory_issues_for_passenger_space_travel.shtml

Dean, E. B., Wood, D. A., Moore, A. A., & Bogart, E. H. (1986). Cost risk analysis based on perception of the engineering process. In *Paper presented at the Annual Conference of the International Society of Parametric Analysts*, Kansas City, USA.

Gaubatz, W. (1998). Reusable space transportation —The key infrastructure elements in opening the space frontier to the public. In *Paper presented at the 21st ISTS,* Sonic City, Japan. Retrieved March 3, 2007, from http://www. spacefuture.com/archive/reusable_space_transportation_the_key_infrastructure_element.shtml

Geoffrey, I.C. (2001). Researching the space tourism market. In *Paper presented at the Annual Conference of the Travel and Tourism Research Association.* Retrieved March 3, 2007, from http://www.spacefuture.com/archive/researching_the_space_tourism_market.shtml

Goehlich, R. A. (2003). *A representative program model for developing space tourism* (ISBN 3-936846-29-4). Berlin, Germany: Wissenschaftlicher Verlag Berlin.

Goehlich, R. A. (2005). A ticket pricing strategy for an oligopolistic space tourism market. *Space Policy Journal, 21*(4), 293-306. Amsterdam, The Netherlands: Elsevier Science.

Goehlich, R. A. (2006). *Spaceships—A reference guide to international reusable launch vehicle concepts from 1944 to the present* (ISBN 1-894959-50-7). Burlington, Canada: Apogee Books, CG Publishing.

Hartog, J., Carbonell, A., & Jonker, N. (2000). *On a simple survey measure of individual risk aversion* (pp. 1-6) (CESifo Working Paper No. 363). University of Amsterdam, The Netherlands.

Hudgins, E. L. (2006). *Space policy and space tourism.* Retrieved March 3, 2007, from http://www.cato.org/testimony/ct-eh062601.html. Washington, DC: CATO Institute.

ISU (2000). *Space tourism: From dream to reality.* Chile: ISU SSP.

ISU (2001). *Space tourism: A literature review.* Strasbourg, France: ISU MSS.

Kelly Space and Technology. (2001). *Space transportation market demand 2010-2030.* Retrieved March 3, 2007, from http://www.kellyspace.com. San Bernardino, USA.

Livingston, D. M. (2000). *Space tourism and RLVs: You can't have one without the other!* Tiburon, USA: Livingston Business Solutions.

Lo, R., & Paatsch, A. (1998). *Ermittlung der durch den Weltraumtourismus verursachten Emissionen*. Berlin, Germany: IRTS.

Lo, R., & Paatsch, A. (1999). *Raumfahrtinduzierte Belastungen der Atmosphäre unter besonderer Berücksichtigung des Weltraumtourismus: Berechnung des quasistationären Anfangszustandes der Abgaswolke entlang der Aufstiegsbahn von Raketen*. Berlin, Germany: IRTS.

Lorance, R. B., & Wendling, R. V. (1999). Basic techniques for analyzing and presentation of cost risk analysis. In *Paper presented at the Association for the Advancement of Cost Engineering International Annual Conference*, Denver, USA.

Moore, P. G. (1983). *The business of risk* (ISBN 0-521-28497-X) (pp. 97-115). Cambridge, United Kingdom: Cambridge University Press.

Roberts, L. (1996). Space business incentives: It's time to act. *Ad Astra*, *8*(2).

Tascione, T. (1988). *Introduction to the space environment*. Malabar, USA: Orbit Book Company.

U.S. Congress. (2006). *House passes legislation setting space tourism guidelines*. Retrieved March 3, 2007, from *h*ttp://www.space.com/news/tourism_ congress_040305.html. New York, USA.

Wollersheim, M. (1999). Considerations towards the legal framework of space tourism. In *Paper presented at the 2nd International Symposium on Space Tourism*, Bremen, Germany. Retrieved March 3, 2007, from http://www.spacefuture.com/archive/considerations_towards_the_legal_framework_of_space_tourism.shtml

X Prize Foundation. (2006). *Ansari X Prize*. Retrieved March 3, 2007, from http://www.xprizefoundation.com. St. Louis, USA.

Zogby International. (2002). *Futron/Zogby public space travel poll*. Retrieved March 3, 2007, from http://www.zogby.com. New York, USA.

Chapter XVI
Space Elevator:
Generating Interest in the Future of Space Access

Paul E. Nelson
University of Wisconsin-Madison, USA

ABSTRACT

Currently, transporting cargo into outer space is not only expensive, but a complicated and prolonged process. The space shuttles used today are inadequate, overused, and obsolete. At this time, there are efforts all around the world to make space more accessible. There have been many proposals to solve the space transportation dilemma. One proposal is the creation of a space elevator. The space elevator would provide low-cost, easy access to space by dramatically reducing the cost of sending cargo into space. At \$10-\$100 per pound, the space elevator would provide astounding cost-savings compared to the tens of thousands of dollars per pound it costs today. This low-cost access to space would make it possible to substantially increase the amount of cargo that could be sent into space on a daily basis. The first part of this chapter describes how the space elevator is expected to work, and the advantage of access to space via the SE vs. using primarily rockets. A compendium of information from a variety of sources is included in order to explain how the space elevator would be designed, constructed, and how it could solve the problems of transporting cargo into space easily, cheaply, and frequently. The space elevator is a relatively new topic in the area of realistic science concepts and was merely science fiction not too long ago. The space elevator (SE) concept has only been in the spotlight in the last 5 years due to the work of Dr. Bradley Edwards of Carbon Designs, Inc. Acceptance of the SE will be a difficult task for many reasons. One of these is that most people do not know about the SE concept, and those who do tend to have trouble believing it is possible to build. In order to determine the best way of integrating the SE concept into society, a survey was conducted at Darien High School. The survey included such topics as the naming of "The Space Elevator," and how best to get the younger generation interested in the idea. The second part of this chapter describes how to utilize the survey results to further the SE concept.

BACKGROUND ON THE SPACE ELEVATOR

Since antiquity, people have looked to the sky and wondered what was in outer space. While many ideas have been proposed over the years as to how to actually make space travel accessible, the most common (and currently believed to be the only way to access space) is to use chemical rockets as the main source of energy to exit Earth's atmosphere. However, it is far too expensive to continue using rockets as the primary means of traveling into Space. That is why many people today want to find a cheaper, safer, and more efficient way to access space. One such way is through the use of a space elevator.

The first noted idea of a SE was from a Russian, Konstantin Tsiolkovksy. He envisioned the idea of a link to the stars in the late 1800s after he saw the Eiffel Tower (Clarke, 1978). Time passed—as well as ideas—until a major breakthrough occurred in 1976, when Arthur C. Clarke, a science fiction author, envisioned a SE in his book, *The Fountains of Paradise* (Clarke, 1978; Edwards, 2003). Since 1976, scientists all over the world have been slowly refining the idea of a SE. Most recently, Dr. Bradley Edwards, the past director of research at the Institute for Scientific Research (ISR), has been providing the main source of research and progress for the SE. Initially, he was working under grants from NASA's Institute of Advanced Concepts to research the SE. In fact, he wrote his Phase I and Phase II reports on the SE prior to expiration of his grants from NASA. Recently, Dr. Edwards left ISR to pursue his research on the SE under the auspices of a different group. In addition, he has been chairing International SE Conferences for the past 3 years. As a result of Dr. Edwards's work, the SE is beginning to gain support in both the scientific community and with the general public.

The SE would be a cable going from Earth to space that is capable of moving objects to various levels or distances from Earth. The cable would be connected to a mobile anchor in the ocean a few hundred miles off the Ecuadorian coast. The cable would be roughly 100,000 km long. It would be made out of carbon nanotubes, currently the strongest material known to man (described in more detail later). In Space, the cable would rotate with the Earth's equator, through the geo-

Figure 1. The space elevator base on Earth (European Space Agency)

synchronous orbit location (35,786 km - 22,000 miles above the Earth), with a counterweight at the end of the cable. A climber is the "elevator" that would move payloads up and down on the cable (described in more detail later). Because the cable (or ribbon) would be flat, each climber would use engines to pull itself up the cable. It has been suggested that the climbers, as well as the cable, be made out of carbon nanotubes. A scaled prototype model of a climber has already been designed and constructed by Dr. Edwards (Edwards, 2003).

The SE climbers would receive power from multiple laser-beaming stations on Earth. The laser beams would be received by the SE's photovoltaic arrays. Photovoltaic arrays are panels on the underside of the climbers, which would use the energy received from the laser beams to power the climber during its ascent. The required lasers must be quite powerful. Fortunately, other groups have already constructed a laser (not constructed specifically for the SE, but can be rebuilt for use by the SE) that is already more powerful than currently needed for the creation of a SE. Adaptive optics would be used on the lasers to lower the amount of distortion and losses while transmitting the beam through the Earth's atmosphere.

PREDETERMINED PROBLEMS

It is widely known that traveling to outer space has always been encumbered by huge risks. There are always issues that must be anticipated in order to make sure that everything works correctly. Currently, at ISR, the Space elevator researchers are working on an intricate "map" encompassing each and every area of research, concern, and dilemma. Each one of these areas is called "trade studies." They are made up of each area of expertise, for example, strength of materials, orbital mechanics, and power generation. Each trade is practically a branch of engineering; electrical, mechanical, physics, math, and so forth. These efforts are meant to insure that there are no major errors jeopardizing the space elevator project. This chapter addresses the major problem areas involved in developing and implementing the space elevator.

Even before the "map" has been completed, many problems have already been foreseen, researched, and addressed. The current major problems that have been identified include the following: radiation, heat, large and small asteroids, space debris, meteor showers, weather on earth, water and air traffic, terrorism, explosions, and large and small cable breaks. There have been

Figure 2. The Earth with all known debris that is orbiting around it (ISR)

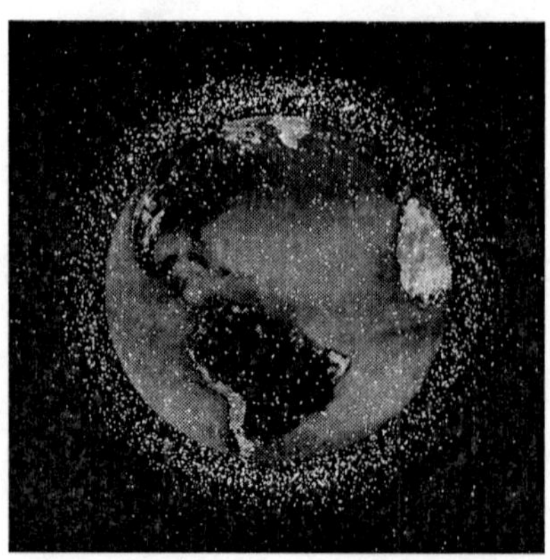

various ideas proposed that would reduce the deterrents for the development of the space elevator. For example, due to the lightness, durability, and strength of the cable, radiation and heat would not be a major concern to the space elevator (Edwards, 2003). While Space is a vast unknown area, and consequentially there might be some problems that cannot be prevented, scientists today are doing the research to prepare for the hostile and changing environment in outer space.

Much planning for safety has already been done in order to make transporting cargo by the space elevator as safe as possible. First of all, to prevent larger objects from interfering with the space elevator, the anchor will be mobile enough to move the cable out of the way of any harmful large objects. In addition, the space elevator would have a built-in system that would be able to plot out potential collisions 2 weeks in advance. This system would direct the anchor to reduce danger to the space elevator.

To protect against collisions with smaller objects, other measures have been taken. Presently, there are minimal countermeasures that can be used to protect against minuscule objects from hitting and even snapping parts of the cable. In that situation some of the individual fibers in the cable would break as a result of the impact with the very small objects. There are studies currently underway to develop connectors and multifilament fibers that may be able to hold the cable together for a period of time. There also is a yet-to-be-researched proposal to develop protection climbers that would block or change the course of debris. Regularly scheduled monthly maintenance by a repair climber has also been planned. A repair climber would mend the cable (Edwards, 2000). Although the proposal for a repair climber has been researched, the research is not yet completed.

There have also been suggestions to use lasers to vaporize debris within a set area around the space elevator. There are also proposals to remove unused satellites and debris from space before the completion of the space elevator. Another idea is to have a safe zone around the space elevator where no satellites may enter and where debris will be vaporized. A flight called STS 75 has already tried using a tether in space. The space tether ended up breaking but this may be attributed to the fact that the tether was not made of CNTs. This kind of experimentation is the first step in testing the possibilities of tethers in space. Other plans have been made for protection from meteor showers. Problematic meteor showers have been studied and a separate plan has been made for this eventuality by making the cable mobile enough hopefully to move out of the way of the larger meteors as much as possible (Edwards, 2000).

One major problem that has become more significant recently is the possibility of a terrorist attack. While there is no way to be completely safe from a terrorist attack, there are many suggestions to minimize the likelihood of a terrorist attack on the space elevator. First of all, the space elevator's platform would be located in the ocean, beyond all air and waterways, so that if any unauthorized planes, boats, and so forth, approached the platform they could be intercepted. Also, the military would protect the space elevator with fighter jets, ships, and submarines. If, however, there was a successful terrorist attack, the only vulnerable parts of the elevator would be the parts within the Earth's atmosphere, that is, the anchor and a minuscule part of the cable, either of which could easily be repaired or replaced. In such a case, the cable could be easily reattached to the platform and back in working order quickly (Edwards, 2003). There has also been a proposal to have an additional platform held up by balloons. In that case, if the cable or main platform were damaged, the balloons could be deflated and the cable could be brought back down to Earth for repair.

After the first space elevator is built, the plan is to build multiple space elevators. That way, in an emergency, any one elevator could supposedly temporarily replace any other elevator. Also, the site chosen for the Earth anchor will be quite far away from any areas densely inhabited by humans. It is unlikely that debris or explosions from the space elevator would cause much damage. In addition, the carbon nanotube cable would be so light that it would cause little damage if it fell to Earth. The probability and damage caused by a climber falling to land would be just as unlikely as and no more damaging than a plane crash. Also, the spot in the Galapagos Islands where the space elevator would be anchored has been shown to be in an area that has minimal problems with hurricanes, currents, storms, and wind.

CONSTRUCTION OF THE FIRST SPACE ELEVATOR

Miniature rockets would be attached to the end of the initial cable stored on the craft. The rockets would give the cable the force to head toward Earth. The natural gravity gradient forces would force the ribbon to lengthen and the ends to pull and unspool. The counterweight would move outward if pushed outward, while the anchor attachment would move toward the Earth. The cable would be pulled down to the anchor and attached to the platform. Every month, for approximately 2 to 3 years, multiple climbers would be sent up the cable attaching another layer to the initial ribbon. Once at the top of the cable the climber would stop and be a part of the counterweight. This would continue for 230 climbers. By that time, the cable would be thick enough to be used for the first space elevator.

The most important of the overall effort to provide easier access to space is the actual construction of the first space elevator. There have been many different proposals from many people about the best way to construct the space elevator. While there is still no consensus as to how to construct the space elevator, the most plausible idea at this time is for four rockets to transport cargo to low Earth orbit with the equipment for the initial large spacecraft. The initial spacecraft would be assembled in outer space and then released. This initial spacecraft would be the beginnings of the space elevator. The spacecraft would use magneto plasma dynamic engines (MPDs) to propel itself to its destination 22,000 miles out in space.

Figure 3. The initial spacecraft for building the first space elevator (ISR)

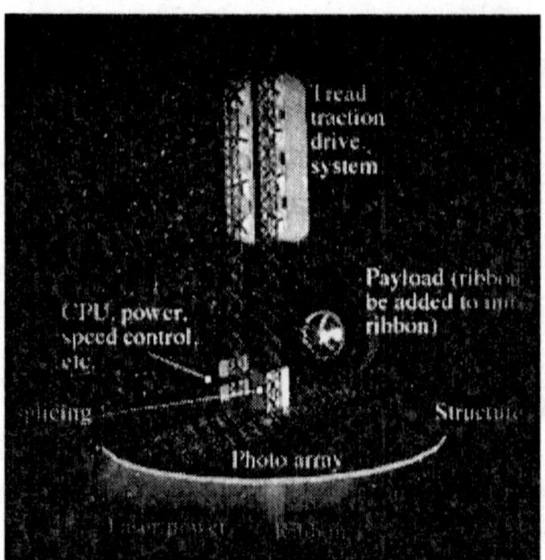

CARBON NANOTUBES

In order to work properly, the carbon nanotubes (CNTs) must be produced at 300 GPa (GPa is gigapascals) tensile strength and 100 composite GPa tensile strength. Once the CNTs are produced at 300 GPa tensile strength, they would need to be mixed with a plastic material to create a 100 composite GPa. In addition, there must be a sufficient mix of CNTs within the plastic material in order for the cable to be strong enough (currently only a 1% CNT composite can be created). Currently, institutes and colleges all over the world are researching and testing Carbon Nanotubes to try to attain the necessary strength. Many of them have been donating free, low-cost samples of the CNTs to the study of a space elevator. Some have said they will do the same for production purposes when the CNTs are at their full strength. One of the major researchers in this area is located at the University of Kentucky. Thanks to Dr. Rodney Andrews, his colleagues and students at the University of Kentucky have been and continue to further the development of the CNTs. The research and support conducted by Dr. Andrews and many others on the CNTs has led to a projection that CNTs will be able to attain their full strength in a few years.

RESOURCES: MONEY AND SUPPORTERS

The proposed budget for the space elevator has been decreasing over time due to the development of new and less costly technology. For example, with the engines that will be used to bring the initial spacecraft into space, it was originally anticipated that it would require large expensive rockets or laser beams, but now the space elevator will be able to use (less expensive and less expansive) MPD engines. The future money and collaboration required for the space elevator to become a reality will likely come from support and donations from groups all over the world. The following are some examples of organizations and institutes that currently provide monetary support for the idea of a space elevator:

- Los Alamos National Laboratory, New Mexico

Table 1. Breakdown of current estimates for the technical budget of space elevator (Dr. Edwards)

Component	Cost Estimate
launch costs to GEO	$1.02B
Cable production	$390M
Spacecraft	$507M
Climbers	$3677M
Power beaming stations	$1.5B
Anchor station	$120M
Tracking facility	$500M
Other	$430M
Contingency (30%)	$1.44B
TOTAL	~$6.2B

- Carbon Nanotechnologies, Inc., Houston, Texas
- University of Kentucky, Lexington, Kentucky
- Institute of Scientific Research, Fairmont, West Virginia
- NASA Institute of Advanced Concepts, Atlanta, Georgia
- National Space Society
- NASA's Centennial Challenge, Washington D.C.
- The Spaceward Foundation, Mountain View, California

In addition, the cost for CNTs is low due to improved technology and monetary support. Currently, it is estimated by Dr. Edwards and his research team that it will cost $6.2 billion for construction of the first space elevator (Edwards, 2003). This is quite inexpensive compared to the money that will be saved transporting cargo to space. Many other space projects cost considerably more, such as the International Space Station, which is costing billions of dollars a year. The breakdown of the estimate is shown in Table 1.

FUTURE POSSIBILITIES

The earliest predicted date for completion of a space elevator is 15 years (Edwards, 2000). According to Dr. Edwards, "In the next 15 years you could have 10 elevators up, you could have large elevators, you could have thrown an elevator to Mars." Dr. Edwards continues to indicate that the first space elevator will be completed before the year 2020 (Edwards, 2003).

Many different types of elevators are possible. Anywhere from a 5-ton to a 100-ton elevator from Earth to space could be built. Dr. Edwards claims that, in the future, a 100-ton elevator could be built if required to carry entire space hotels into outer space (Edwards, 2003). Also it may be possible to use an Earth space elevator to slingshot climbers to different parts of the universe. It may also be possible to create a space elevator on the Moon or Mars. This would enable climbers to travel back and forth from these planets quite easily. The future possibilities of a space elevator are limitless; this will be further explained later.

SUPPORT GENERATING RESEARCH ABOUT THE SPACE ELEVATOR

There have been many suggestions as to how to increase public support for the first space elevator. However, the first obstacle is that many people, both educated and uneducated in this field, do not believe that a space elevator is even possible. From scientists to high school students, professional engineers and college professors, all have shown great doubts. There are reasons these people have doubts.

First of all, as stated by one of the presenters at the Third International Space Elevator Conference, the first thing a person hears about the space elevator is its name. Often when a person first hears the name, space elevator, one thinks of the impossibility of a common building elevator ascending into outer space.

Also, any person who is at all knowledgeable about the space elevator is surprised by the length of the elevator, which would be 100,000 kilometers long. Almost every person that hears about its length questions the space elevator's construction. Once a person does hear about the SE concept, they often are incredulous.

If there is to be general acceptance of the feasibility of a space elevator, there are some steps that must be taken to correct these misconceptions. The first and most important thing that must be done is to change the name. Space elevator is not the right name. It does not send the right message. This was an idea that was discussed at the Third International Space Elevator Conference. One of the names mentioned at the conference, though not actually voted on, was *Sky Line*. This seems to be a better name because the name *Sky Line* brings to mind the sky and a line going through the sky, which is exactly what it would be. There are many other variants to *Sky Line* that would be acceptable, and preferable to space elevator, including *Space Line* and *Sky Way*.

Once the name of the space elevator is changed, the next thing that must be done is to educate the public about how the space elevator would work so that the public can understand why a space elevator is both possible and useful. This can be accomplished in many ways. First of all, it is absolutely necessary to continue publishing articles in publications like Popular Mechanics, the New York Times, and so forth, so that the public will have the opportunity to read about and understand the space elevator. There were many other suggestions presented in Washington D.C. at the space elevator conference, such as constructing a sideways space elevator on Earth to demonstrate to people that it could work. This sideways elevator would be a gondola-like elevator with carbon nanotube cables stretching across a large area. This would demonstrate the strength and the possibilities of carbon nanotubes. Another possibility is the construction of a lunar space elevator in which a mock space elevator would be built on the Moon to test and prepare for the real space elevator. James Pearson suggested this in a paper he wrote in the mid 1970s. Also, it would be helpful to give the space elevator as much publicity as possible by expanding into books, movies, television series, and newscasts. Also, workshops, visits, presentations by scientists at schools, universities, and so forth, are necessary in order to get people responsible for furthering the future of space behind the idea of a space elevator.

WHY THE SE IS NECESSARY: TODAY'S STATUS AND NASA'S ROLE

The space elevator could solve many of today's space access problems, especially the problems that NASA is currently experiencing. NASA is the world's largest space organization and it is in disarray. NASA is using 1970s space shuttles, but, since the crash of the spacecraft Columbia in February 2005, there has been very little done to improve the situation. Also, NASA may be spending too much money on certain projects, especially on the International Space Station.

While NASA is currently unable to successfully get satellites, people, and so forth, to outer space easily or cheaply, other space agencies from Russia, Europe, and China have been developing their space transportation programs. Even so, everyone is spending too much money and working too hard on what are arguably unnecessary goals and equipment. The solution to the problem of accessing space easily and cheaply is to create a space elevator. In previous sections it was explained how the space elevator would work. The space elevator is within technological reach in the near-term in that it is less than two decades away from possible completion. The problem is that not enough people recognize the enormous possibilities of a space elevator or even that it is possible. In that regard, unfortunately, President Bush did not mention the space elevator in his recent speech on the expansion of the space program to the Moon and Mars.

The space elevator would be like what the transcontinental railroad was for opening up the west of the United States for settlement. The space elevator would be the beginning of the colonization of outer space. First, it could be used for payloads such as satellites. It could also retrieve damaged satellites and eventually, to transport humans. Just imagine the rush there was to colonize the west and consider the interest that a space elevator could create in colonizing space. Getting into outer space would be so much cheaper and easier than today that it would not only bring trained astronauts to space but also common citizens. It will put into perspective how important the Earth SE can be.

WHY AND HOW IT WOULD HELP

It seems appropriate to ask why the space elevator would provide such a low-cost easy access into space. The last estimate for the total technical budget for the first space elevator was $ 6.2 billion, which, compared to other comparable projects, is actually quite inexpensive. If it can be completed as proposed, the space elevator would allow constant access to space, and therefore a constant transport of payloads to outer space. One would only have to put the payload on the climber, send it up, and then release it into outer space. It would be that easy. Unlike the shuttles of today, the climbers would not be very expensive. Space shuttles cost millions of dollars, and consequently there are very few and they cannot be used as often as desired. The climbers, on the other hand, would be comparatively cheap and could be used readily and consistently. Payloads could be sent up at low costs, as low as $10 a pound. In contrast, today it requires tens of thousands of dollars per pound to send cargo into space.

Another plus would be that multiple space elevators could be built on Earth immediately after the first one, making it possible for more and more payloads to go to outer space faster and cheaper. The length of the cable would allow for the payloads to be released in many different areas of space, expanding by thousands of miles the area for satellites and materials to orbit around Earth. The space elevator could also be used to pick up space satellite debris, make repairs, and put them back into service. The main advantage, however, is how much money and work the space elevator could save in the future. No matter how we look at it, a space elevator is the best way to solve today's problems of accessibility to space by providing the easiest and most cost-effective means toward expanding throughout outer space, now and in the future.

Table 2. Comparing space access of today to what it could be with the creation of an Earth space elevator

	Now	First Space Elevator
Cost	$50,000/lb	$100/lb ($10/lb)
Capacity	4,000 lb/day ave.	12,000 lb/day
Usefulness	Critical Satellites for gov., large corp., and research	Solar power satellites, manufacturing, Space resources, private environmental monitoring, tourism

EXPLANATION FOR USE OF A SURVEY

In the past few years, research on the SE concept has increased dramatically from science fiction into near-scientific reality due to two major factors. These factors are the discovery of carbon nanotubes and the work of Dr. Bradley Edwards. Even at its current peak of interest, however, the SE concept is still only fully understood by a roomful of people with scientific backgrounds. For this concept to succeed, it is necessary that knowledge and support for the SE concept spread. The most important people that need to be informed are the young people who are currently in high school or college. This is because it is the young people who will be creating future jobs at a time when the SE concept could actually begin construction. Consequently, in order to insure the SE's success, it is necessary at this time to find out how we will be able to gain this support. It seems evident that the most viable way of doing this is to conduct a survey of young people. The purpose of this survey is to gauge the current knowledge and opinions of the future generation on how to gain support for space travel and the SE concept. The purpose of this survey is to learn how to gain the support of young people before they become future leaders.

CREATION OF A SURVEY

This survey is broken into two parts: "knowledge" and "opinion." Together, both parts of the survey total 30 questions. The first part of the survey has 10 multiple-answer questions meant to determine a person's overall knowledge of space and space history. This part is included to find out what people know about space currently, as a first step to determining their knowledge of the SE. Also, at the end of the survey, those completing the survey were asked to list all of the academic science classes they had previously taken or were currently taking in high school. The reason for this is to connect what they knew about space to what classes they had taken. In other words, it is an effort to show how efficient classes are in teaching students about space and space history. This is necessary because if students are not learning basics of space knowledge and history, then they will not be able to comprehend the necessity for space travel, research, and access, which the space elevator concept can provide.

The second part of the survey has 18 multiple-answer questions. The questions cover opinions on NASA, current space access, space exploration, and ideas as to how to gain support for the SE concept. Most of the questions relate to ideas of how to gain support of the SE concept. The survey gathers opinions on when they think the SE will be built, how much they think it will cost, whether they would be willing to go on the SE and other similar questions. Another major area in this part of the survey is questions dealing with opinions about NASA and its current work. It is a very difficult area because people have to mix their patriotism with facts which can cause some surprising results. The last main area in the second part is questions about the survey takers, individual futures, and the future of space explorations. It explores what people think of science in terms of their careers and interests and the future. The survey was completed by a total of 300 students at Darien High School in Darien, Connecticut, and, due to time and monetary constraints, only results from 175 responses were included this chapter.

RESULTS OF SURVEY

ANALYSIS OF RESULTS

- **Question 1:** The correct answer to this question is about 500 million dollars, which corresponds to answer choice b, *hundreds of millions*. Over 95% surveyed believed that it cost hundreds of millions or more dollars. That means that the people who took the survey knew and acknowledged that it costs huge sums of money to use NASA shuttles to space.

- **Question 2:** The correct answer to this question is choice d, *approximately thirty months*. Only 8% of the people surveyed answered the question correctly vs. 25% of those surveyed who answered *around 10 months*. The answers to this question show that the general population of students has limited knowledge or interest in the current space program.
- **Question 3:** The strongest material on Earth is carbon nanotubes, which was the most popular answer with over a 63% response.
- **Question 4:** The correct answer to this question is choice a, *yes*, because it is possible for a material to be 100 or more times stronger than steel. In fact it is true of carbon nanotubes. Over 90% answered this question correctly. This has been interpreted to mean that people believe that super strong materials can be built, which in turn means people should use the carbon nanotube material to construct the SE.
- **Question 5:** The correct answer would probably be billions but no none knows for sure. This was the most popular answer

Question 1: How much do you think it costs to launch a NASA shuttle (in U.S. dollars)?

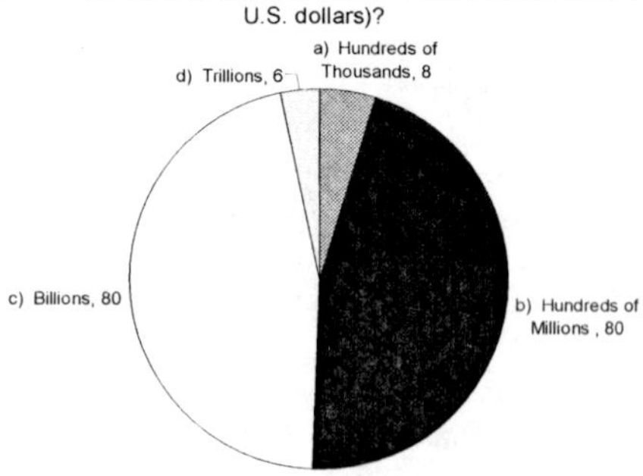

Question 2: How long has it been since a NASA shuttle attempted entering or leaving the Earth's atmosphere (as of 6-1-05)?

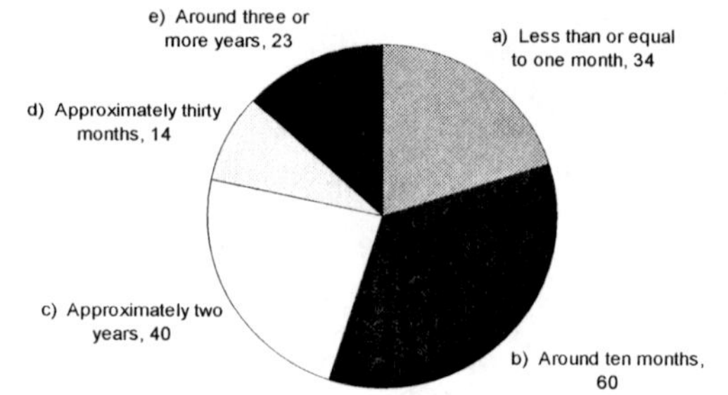

Question 3: To your knowledge what is the strongest material known on Earth?

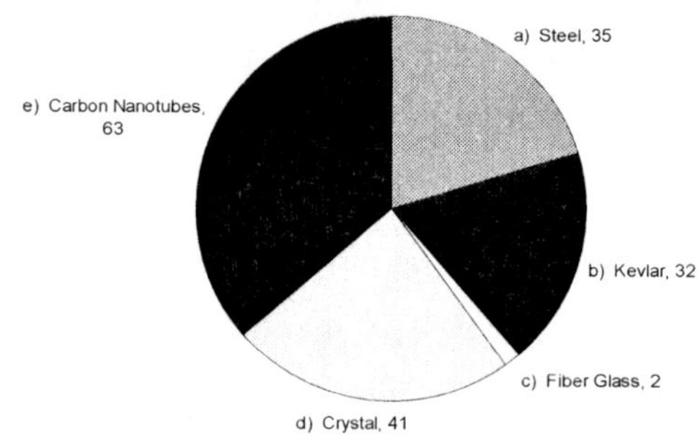

Question 4: Do you think it is possible that a material could be 100 or more times stronger (at full strength) than steel?

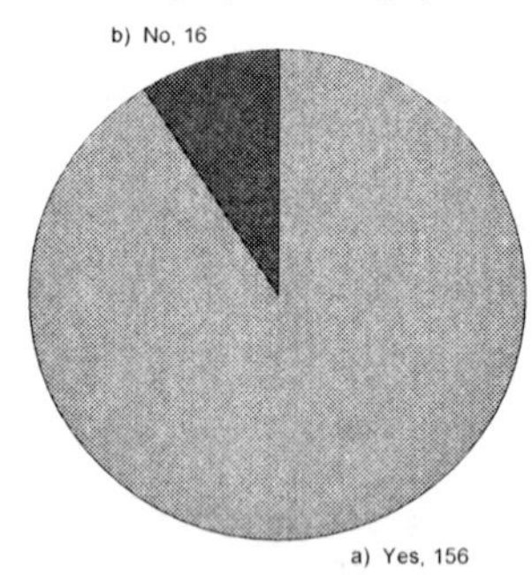

Question 5: How many pieces of debris greater than .5 inches in diameter do you think is currently orbiting within about 1200 miles of Earth's surface?

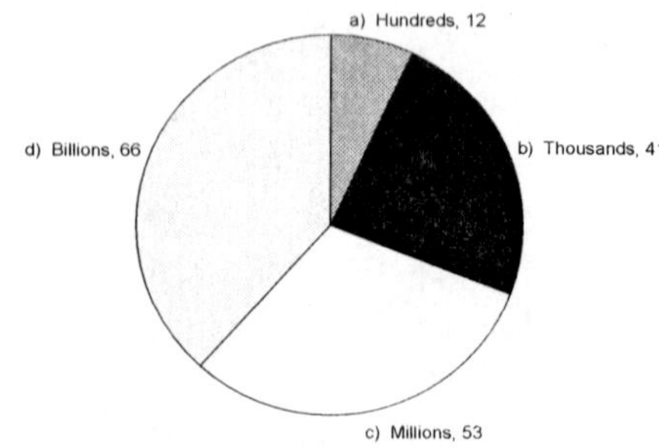

with over 38%. This has been understood to mean that most people are not aware of the enormous amount of manmade debris currently circling the Earth.

- **Question 6:** The answer to this question is *yes*, because orbital debris is of concern to future space explorers. Orbital debris makes the orbital paths around the Earth dangerous. Two-thirds of the people surveyed answered this question correctly. The postulated reason for this high percentage of correct answers is that people mistook *concern* as a danger for ordinary citizens on Earth. If this reason is incorrect, then this would mean people are thinking to the future and do not want debris in space to stop them from visiting and colonizing space.
- **Question 7:** The correct answer to the question is d, *all of the above*. This knowledge question was meant to test historical space-related knowledge, but it was obviously a very easy question as 66% answered correctly. Clearly, people know the answer to the most commonly asked space-knowledge questions.
- **Question 8:** The correct answer to this question is c, *Neil Armstrong*, which received an 83% response. Once again, people know the answer to a commonly asked space question.
- **Question 9:** The correct answer to the question is *Yuri Gagarin*. Only 40% answered correctly. The most popular answer was *John Glenn*. This result was very interesting because the issue of patriotism arises again. The most popular was the first American to orbit the Earth, but not the first in the world to orbit Earth. Rather, it was the Russian Yuri Gagarin who first orbited Earth. It also shows how an important fact about who was first to orbit the Earth is not widely known, once again showing the lack of space knowledge by the public.

Question 6: Do you think that debris orbiting around the Earth is now or will be a major concern?

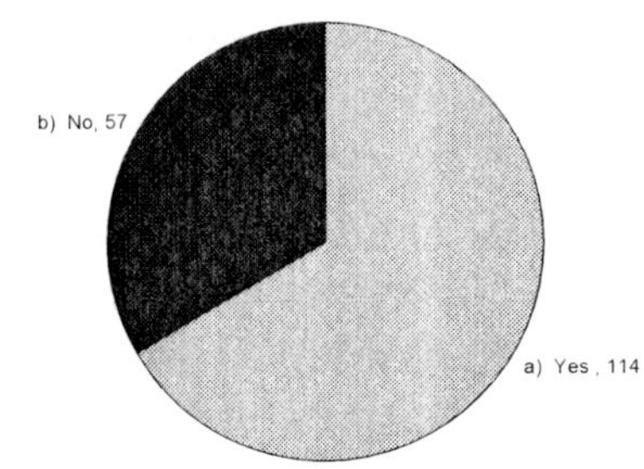

Question 7: Which of the following was invented and/or perfected as a result of the project to land a man on the Moon?

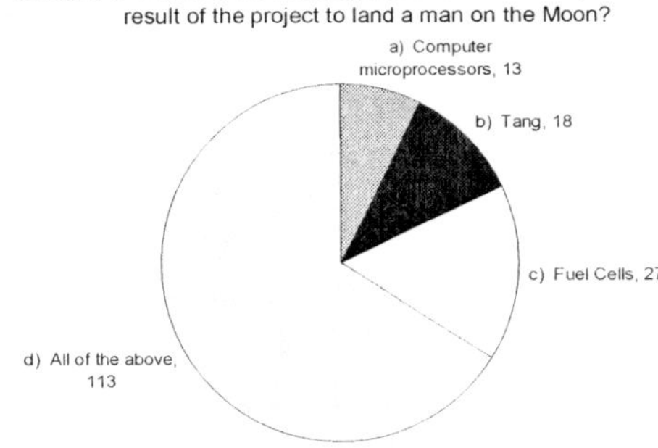

Question 8: Who was the first man to set foot on the Moon?

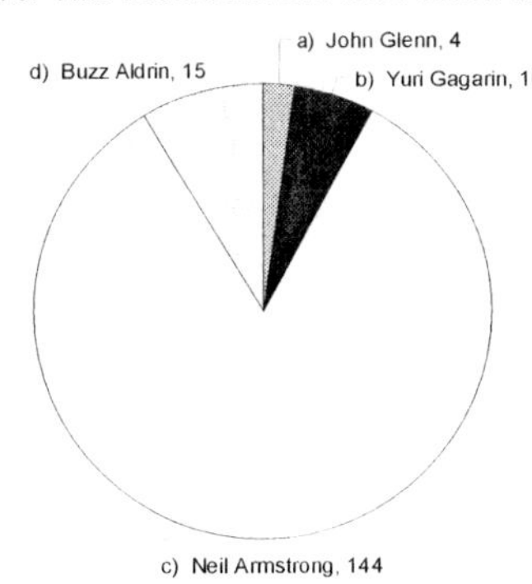

Question 9: Who was the first man to orbit the Earth?

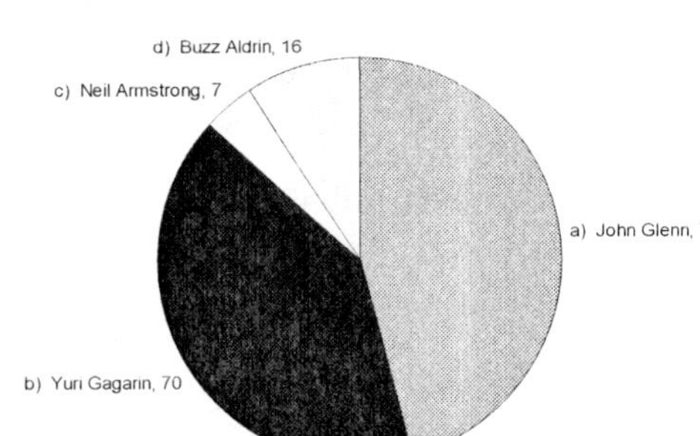

Question 10: What year did the first human land and walk on the Moon?

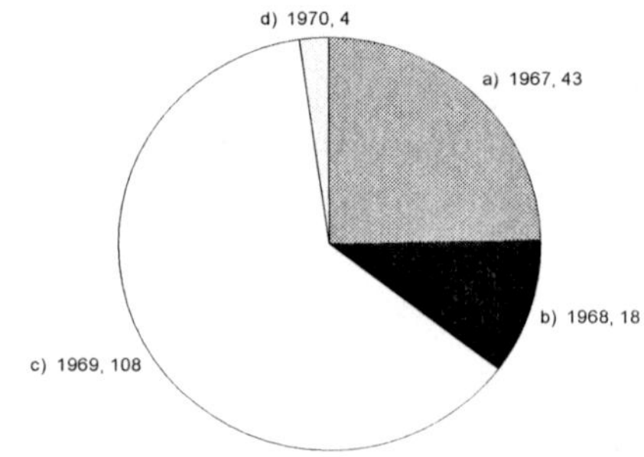

- **Question 10:** This is a follow-up to question 8 by asking when Neil Armstrong landed on the moon. It ended up that the answer to this question was commonly known because the correct answer, c, *1969,* received 62% of the vote. This means people have some information about space knowledge.

Question 11: In your opinion is NASA currently doing a good job in getting astronauts into space?

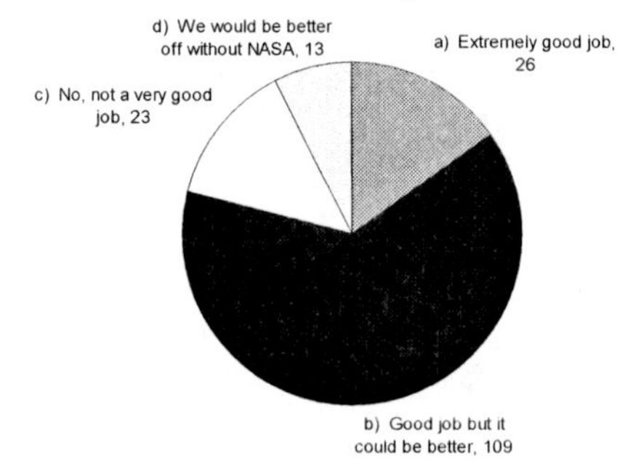

- **Question 11:** The correct answer to this question at the time it was given was answer choice c, n*o, not a very good job*. This is because at the time the survey was given NASA had not sent astronauts in NASA shuttles into space for 30 months, and the last shuttle to be launched exploded during landing, killing all of its passengers. That means that at that time NASA was not doing a very good job of getting astronauts into space. Only about 18% answered the question correctly. Almost 64% said NASA was at least doing a good job. This shows how people will support NASA due to patriotism and naiveté, just because it is NASA, which is run by the U.S. government. It shows that people choose to ignore negative facts about their program in order to say that it is "better" than others in the world. The interesting part of this question is that at the time of the survey other countries in the world were doing a better job than NASA to get people into space, in particular Russia.

Question 12: Who do you think should lead the next push for space travel?

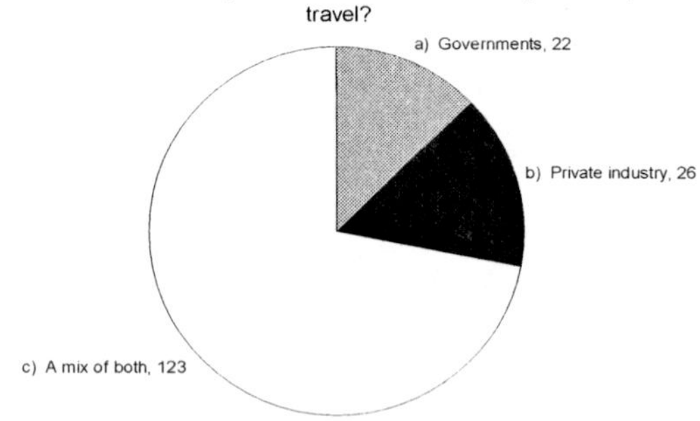

- **Question 12:** The most popular answer to this question is answer choice c, *a mix of both*, with almost 72% choosing that answer. Almost three-fourths of the people believed that future exploration of space should be accomplished by a mix of government and private industry. This result shows how people recognize that governments cannot and should not be solely responsible for exploring space.

Question 13: What do you think is the best cheapest and easiest way for people to go to space currently?

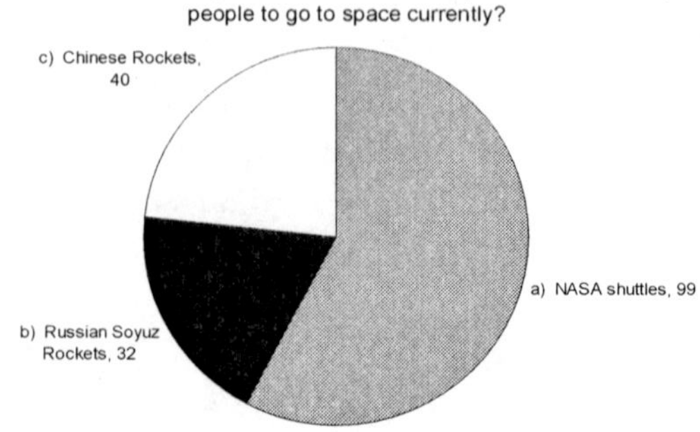

- **Question 13:** The correct answer to this question is probably choice b, *Russian Soyuz rockets*. This is because at the time of the survey, NASA was able to get its astronauts into space on NASA shuttles costing over $500 million per launch. The Russian space program has consistently been getting its cosmonauts into space on their Soyuz Rockets. Only 19% choose the Russian Soyuz rockets. The most popular answer was the NASA shuttles, with 58%. Once again, the issue of patriotism and lack of knowledge comes into play. What is meant by this, in my opinion, is that people chose the incorrect answer of NASA because NASA is considered to be the best because it is run by America, the "superpower."

- **Question 14:** This was meant to find a better name for the SE, but most voted for the SE, so it must mean the name is currently a good one already.
- **Question 15:** SE enthusiasts would hope that most would choose a and *be the first one to go* but, the most popular choice is choice c with 37%. This can be considered a good result though, because over 63% would at least *maybe* go on the SE.
- **Question 16:** It is believed by space elevator enthusiasts that the first space elevator will be built at least within the next 30 years. That would be choice a, which received only 10% of the votes. 90% of those surveyed indicated that in their view it would take 30 or more years to build the first space elevator. The most popular choice was choice d, n*ot until at least 2100 or later* with 39%. This means that people are just not convinced that the space elevator can or will be built anytime soon.
- **Question 17:** Previous questions put a price tag on NASA and other space groups. This question discusses a price for the SE. Dr. Bradley Edwards published a report indicating that the cost of an SE should not go over 10 billion dollars. That would correspond to choices a and b, which received about 30% of the vote. A substantial majority of 66% said at least 20 billion dollars. Yet the most

Question 14: If you had a structure similar to an elevator that could ascend and descend from Earth to space what would you call it?

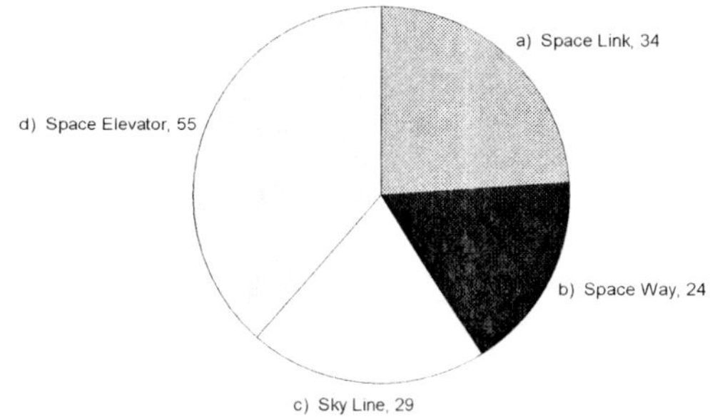

Question 15: If an elevator-like structure existed and it was safe and reliable how willing would you be to go on it?

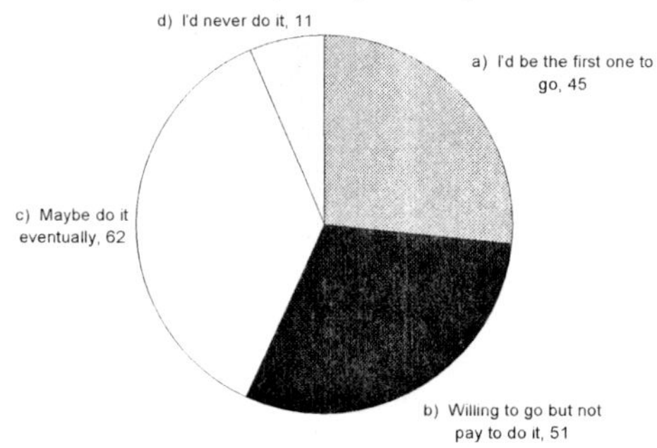

Question 16: With your current knowledge how feasible do you think it would be to build an elevator-like structure?

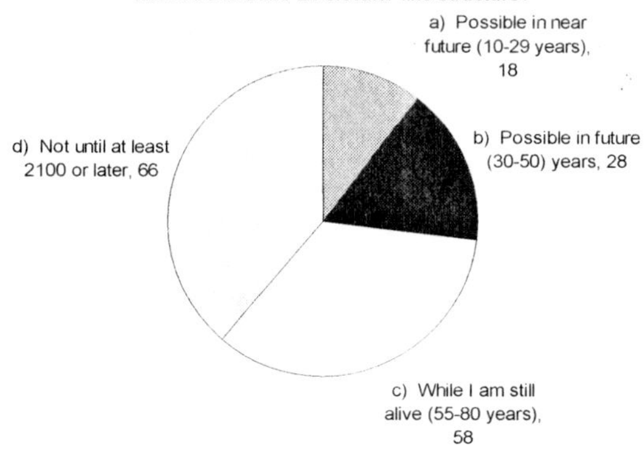

Question 17: How much of an overall budget do you think it would take to build and maintain one elevator-like structure in space in the near future (in U.S. dollars)?

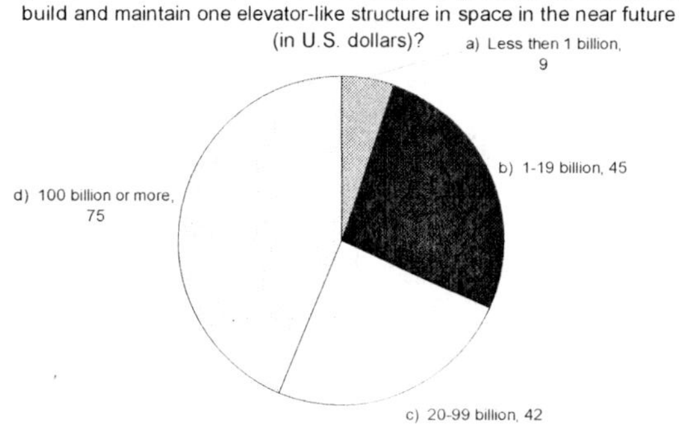

popular answer was choice d with 43% of the votes. This means that people believe the SE will be very expensive, even ridiculously expensive, and helps to explain why people don't support the SE Concept.

- **Question 18:** This question was meant to get a vote on why people would not support the SE. It was later discovered that having all of the above as a choice tainted the results. This is because 67% said *all of the above*. All that can be taken from this is that choices b and c are issues of equal concern to people.
- **Question 19:** There was an idea to protect the SE with the United States Army in the ocean and this question was meant to generate an opinion on this matter. It seems that the question may have been misunderstood. Many survey takers reported misunderstanding the question, as is shown with the result of 53%. That is true unless people really do not believe making it safe is important, which is very hard to believe in America today.
- **Question 20:** Out of 170 people, only about 35% were willing to invest money in the SE Concept. What this means is that currently people are not convinced about either the benefits or the need for the SE. This means that a better job must be done to educate the public about the space elevator.
- **Question 21:** This is a very useful question because it asks people why anyone would want to use a SE. The most popular answer aside from *all of the above* is *future possibilities* with 26%. A probable reason for this is that people know that in the future when humans colonize space the SE will be helpful. The most interesting result from this question is that choice a, *save money accessing space* only got about 4%. This is amazing because the key reason for building the SE is that it would save a great deal of money. This indicates that people are not convinced the SE will be cheaper than NASA's current approach.

Quesiton 18: If you opposed the construction of such an elevator-like structure on Earth what would your reasons include? terrorism concerns

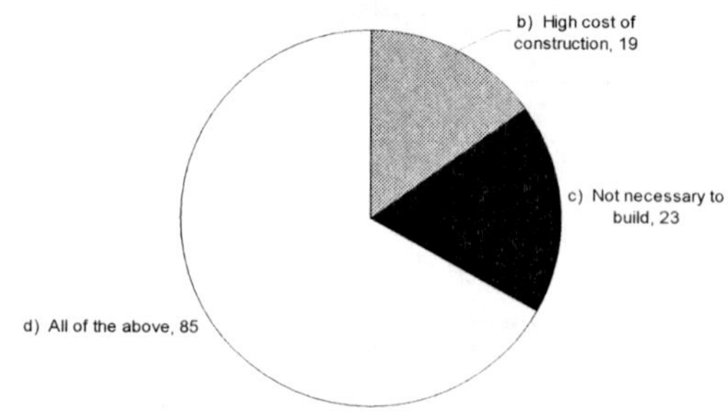

Question 19: If the elevator-like structure were put in a barren secure location protected by the U.S. Military would this affect your decision to support its construction?

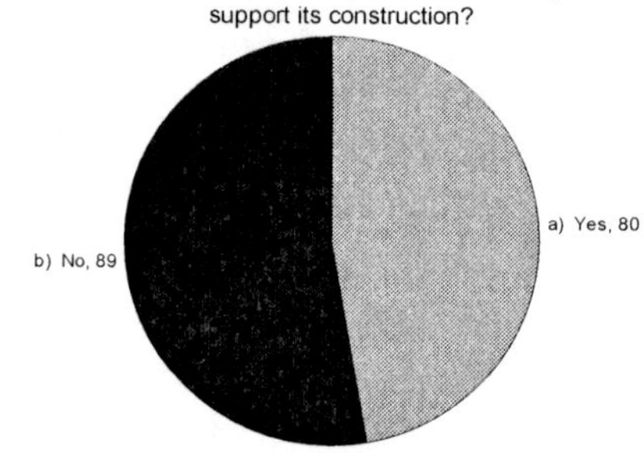

Question 20: If you were a multi millionaire would you invest your money to help build such an elevator-like structure?

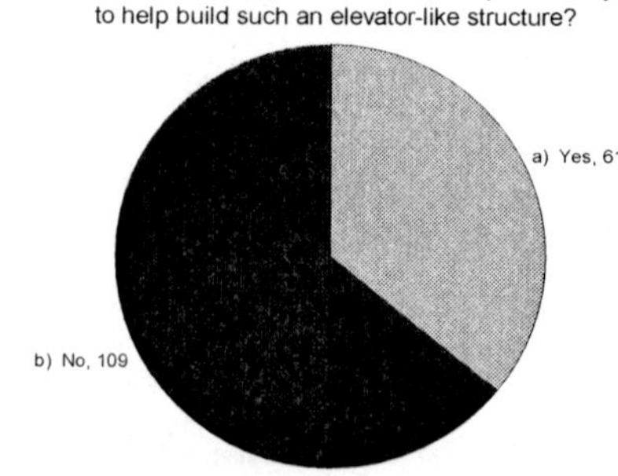

Question 21: What do you think is the main reason to build such an elevator-like structure?

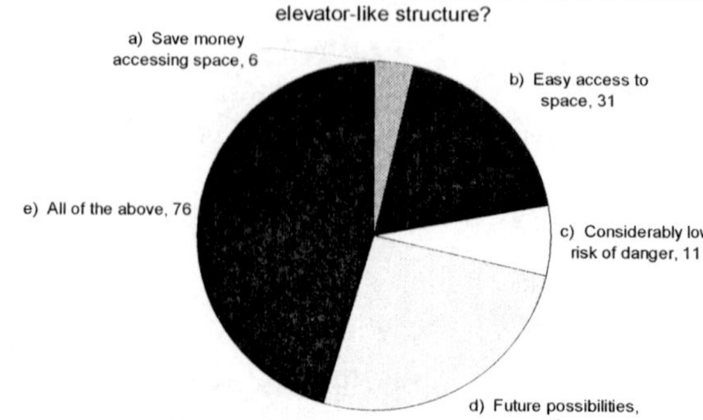

- **Question 22:** This question looks at how to solve the issue of support by looking at different possible solutions. The most popular answer is choice a, *a cool movie* with a 40% response. Second was to use advertisements with 31%. This means that those mediums may help increase support for the SE concept.
- **Question 23:** This question was meant to see if the idea of a Moon SE would gain support for an Earth SE. Over 60% did not think so. This means that the moon SE would not help support for the SE. However, it is possible that, due to the lack of description of a Moon SE, the results were tainted.
- **Question 24:** 56% of the people said that if a *sideways gondola* was build on Earth and was able to successfully carry humans it would improve support for the SE concept.
- **Question 25:** This question is very similar to question 28. This asks who would work in space science. Unfortunately, 61% said *never* while only 5% said *definitely.*
- **Question 26:** Once again, this question asks for an opinion on space in general. It turned out that 60% believed that the exploration of space does not help them in their everyday life. This shows that people are naive and ignorant to the many benefits space exploration creates.

Question 22: In your opinion what would be the best way to convince our young population under 21 years of age to support the construction of such an elevator-like structure?

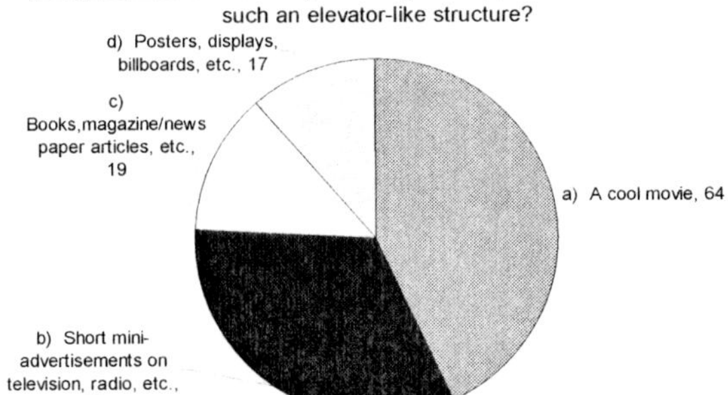

Question 23: If such an elevator-like structure was first built on the Moon would that have an effect on your consideration to support its construction on Earth?

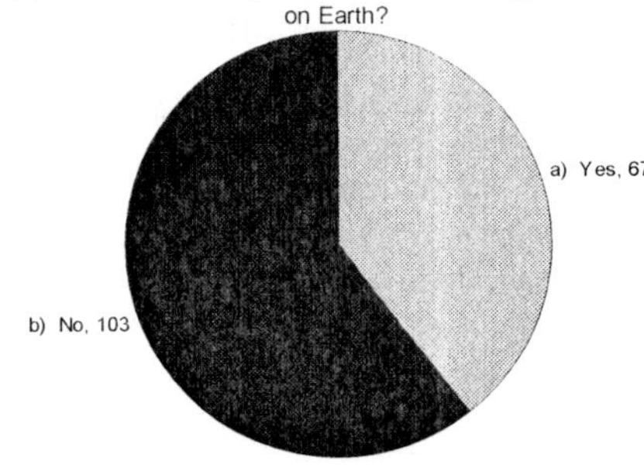

Question 24: If a smaller version of such an elevator-like structure was capable of carrying humans back and forth from places on Earth (eg. Gondola cables) would that have an effect on your decision to support the construction of such a structure?

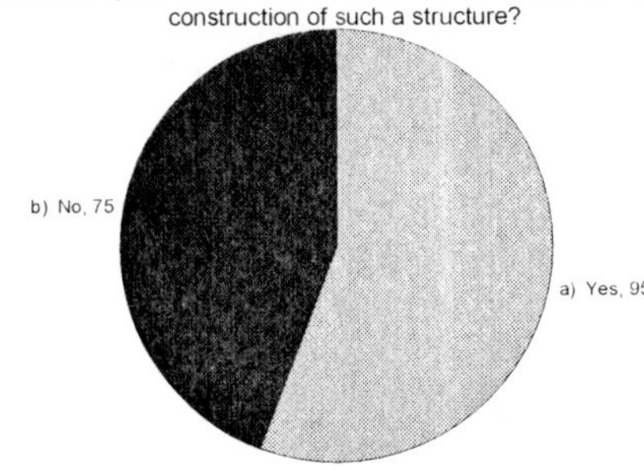

Quesiton 25: Do you expect to work in the space field during your lifetime?

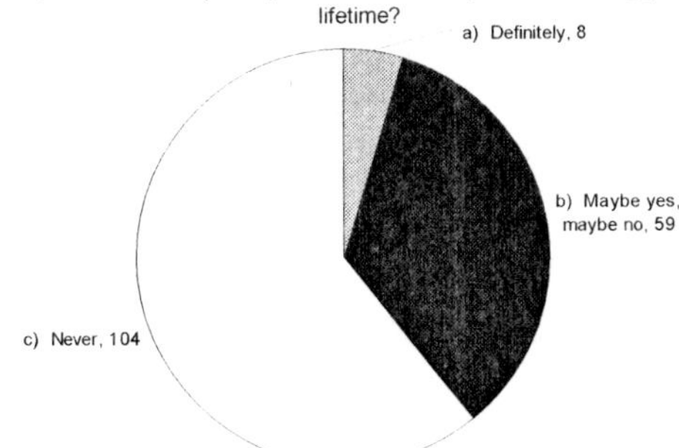

Question 26: Do you believe that the exploration of space helps you in your everyday life?

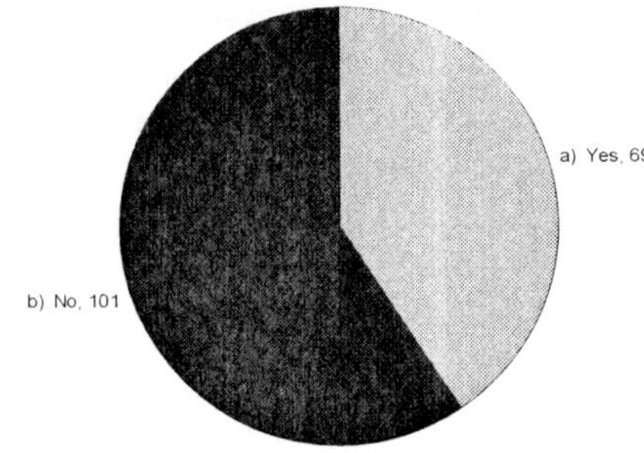

- **Question 27:** This question focuses more on the future of space-exploration ideology than method. It is clear that people prefer sending humans to explore space and planets, as that choice received 60% of the votes. This means that people prefer a cheap and easy way to get into space, therefore preferring the SE.
- **Question 28:** This question focuses on what the people who took the survey want to do in life in the field of science. Over 65% said they did not plan to work in any field of science in the future.
- **Question 29:** A majority of people who took the survey were at the end of 11th grade. Second came 12th graders, followed by 9th, and lastly 10th graders in high school. This means that a majority of the survey takers were soon to graduate from high school and therefore ought to be informed about space and space knowledge.
- **Question 30:** This shows the science classes taken by the surveyed high school students. The top three are biology, chemistry, and physics, respectively. These are the most important classes to the surveyed students.

Queisiton 27: In your opinion what would be better for the future of space exploration?

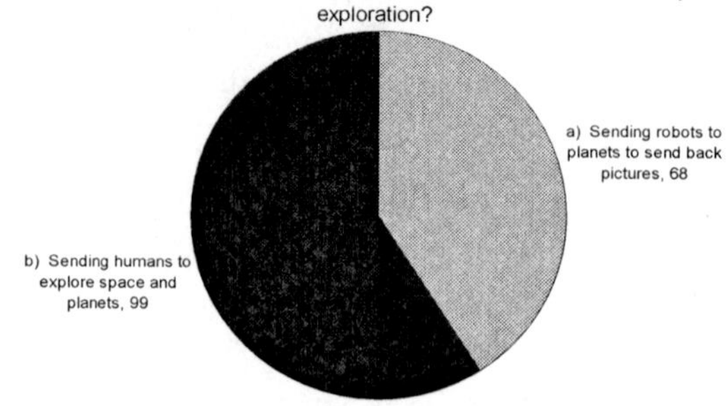

Queisiton 28: Do you plan to work in any field of science in the future?

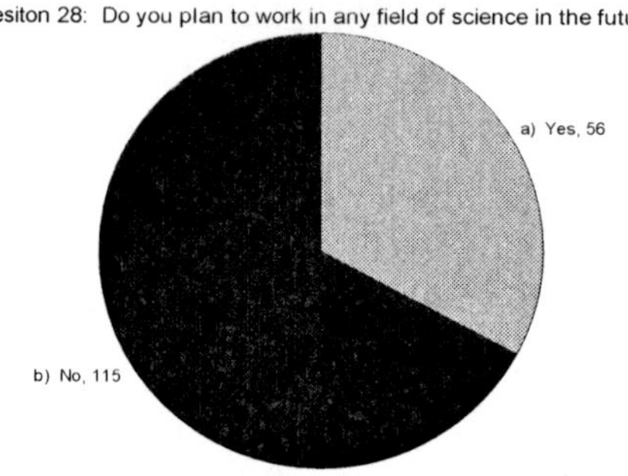

Question 29: Grade Level of Students Surveyed

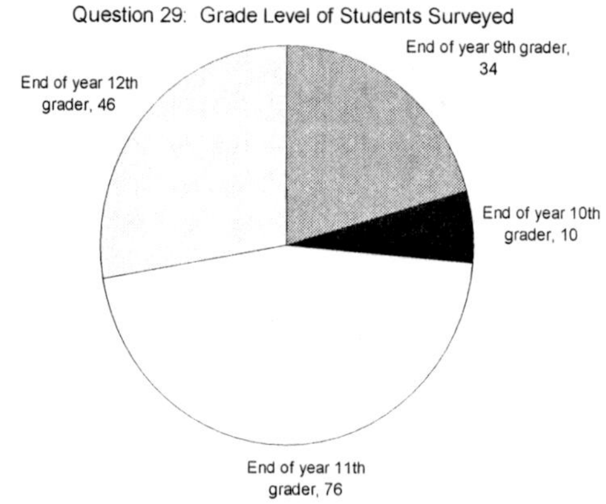

Question 30: What Science classes have you taken or are currently taking in High School?

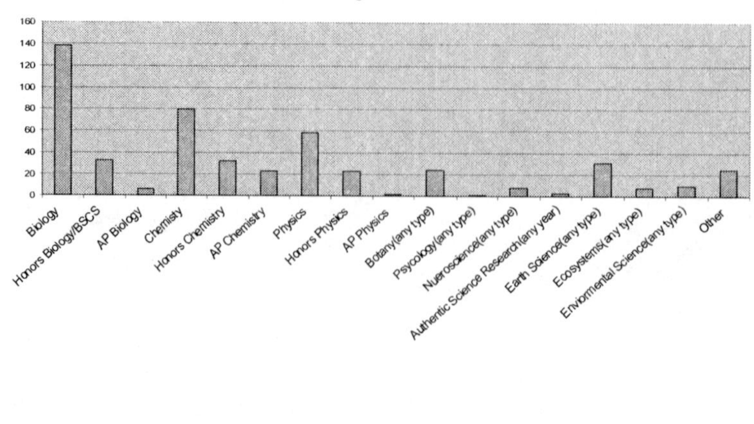

SURVEY RESULTS DISCUSSION

These 30 data sets have each resulted in many different, yet equally important results and conclusions. Different issues were addressed in each question which resulted in many new ideas. Such issues include future space exploration, opinions of NASA, knowledge of space, space elevator support, and so forth. It seems it is necessary to understand what people know and think about space in order to determine how to get people to support the space elevator. This survey has attempted to accomplish this. From the 30 questions, a total of 10 conclusions stand out. These conclusions are:

1. Adults have little to no knowledge of space history, exploration, and so forth.
2. Patriotism greatly influences what people know and believe about space.
3. People believe the space elevator is very expensive and will not be built for a long time.
4. People believe NASA is doing a good job, but do not believe space exploration would help them individually.
5. While most survey respondents might agree to travel on the space elevator, most would also not invest money in the space elevator.
6. People believe space travel should be run by a mix of private and government resources but think humans, not robots, should explore space.
7. People think that orbital debris in space is an issue of concern.
8. People think that the name space elevator is a good one.
9. People like the idea of a sideways elevator on Earth.
10. People think the best way to gain their support and interest for the space elevator is primarily through a good movie, and secondly through any form of advertisement.

These 10 conclusions sum up what the survey determined. It found what people think about a wide range of issues that can help gain support for the space elevator. My recommendations as to how to use these conclusions are:

1. Find ways to get young people interested in space and the space elevator at a young age. And try to get more education on space-related subjects for high school students. Young people need to learn about current developments in space in order to understand NASA's challenges.
2. Make sure that when people hear about the space elevator they understand that it is a real possibility within the next 30 years or less, and is currently projected to cost approximately $10 billion (Dr. Bradley Edwards).
3. Once Carbon Nanotubes are strong enough, if possible, build a sideways elevator (a gondola-like structure over vast areas using carbon Nanotube cables) on Earth to show people that a space elevator is possible.
4. The best way to get young people to support the space elevator is to create a motion picture that can capture the attention of young adults today so that they will learn about, understand, and support the space elevator concept. Advertisements (television, radio, posters, etc.) will help to gain more support.

These are my four recommendations for getting the space elevator concept from where it is now to where it could be with the interest and support of the future generation. Hopefully, these recommendations will be given serious consideration in the future.

There were some survey limitations. This survey was only conducted with a few hundred available high school students at one suburban high income high school. Where the survey is taken and in what kind of situation the test takers live affects the results. Also, because the survey was only given to high school students, it is also necessary to test college level students in order to get even more accurate results and conclusions. This research is the next step toward getting better data to solve the problem of gaining support for the space elevator concept.

CONCLUSION

As described above, a plausible plan of how to build and sustain an Earth space elevator within the next few decades has been thoroughly researched and created. All over the world research continues on all aspects of the space elevator, especially the carbon nanotubes. The key to creating the space elevator is being able to mass produce carbon nanotube composite cable. The SE can be built by the end of this decade if progress continues. There is a lot of research that is trying to improve the concept and solve potential problems. There are now ideas to create future space elevators on the Moon and even Mars.

The second key to building the Earth space elevator, in the near future, is knowledge and support. Right now, as the survey indicates, a lot of people do not know what the space elevator is and once hearing about it, they immediately assume it is not plausible. It is necessary to educate the public about the space elevator through all types of media, especially movies and advertisements. Also, it is necessary to find a way to teach the younger children about the SE because they will be the ones who will be running the world when the first Earth space elevator is built. If the two described keys to the success of the SE are pursued and fulfilled, the space elevator would be constructed in only 20 years, or even less.

ACKNOWLEDGMENT

Prof. George Morgenthaler—Professor at the University of Colorado, Boulder. Professor Morgenthaler introduced me to the International Astronautical Congress and helped edit this chapter.
Dr. Bradley Edwards—Dr. Edwards is one of the main proponents of the space elevator concept and helped me by giving me his chapters and pictures on the space elevator.
Mr. Peter Swan—Mr. Swan has been helped with this project and he has always been very encouraging with my papers and project.
Mr. David Lewis—Mr. Lewis is the teacher of my high school's three-year independent research class, and he has been very helpful in the development of my project.
Mr. and Mrs. Robert Nelson—My parents have always been very supportive, including supporting my efforts to present my project in Vancouver, Canada and Fukuoka, Japan.
Mr. Dan Dzierski—Mr. Dzierski works for the Institute of Scientific Research and has helped me with ideas for papers and updates on the space elevator.

REFERENCES

Carroll, J. (2002). *Space transport development using orbital debris.* Principal Investigator, Tether Applications, Inc. Phase I Study - NASA Institute for Advanced Concepts.

Clarke, A. C. (1978). *The fountains of paradise.* New York: Warner Books.

Edwards, B. (2000). *The space elevator Phase I study.* Eureka Scientific/NASA Institute for Advanced Concepts.

Edwards, B. (2003). *The space elevator Phase II study.* NASA Institute for Advanced Concepts. Retrieved March 3, 2007, from http://www.niac.org

Edwards, B. *The space elevator: An ideal application for the free electron laser.* Eureka Scientific/Institute for Scientific Research.

Edwards, B. *The NIAC space elevator program.* Eureka Scientific/Institute for Scientific Research.

Edwards, B., & Westling, E. A. (2003). *The space elevator: A revolutionary Earth-to-space transportation system*. Houston, TX.

Smitherman, D. V. (2000). *Space elevators—An advanced Earth-space infrastructure for the new millennium.* Huntsville, AL: Marshall Space Flight Center.

Chapter XVII
Commerce in Space:
Infrastructures, Technologies, and Applications

Chris Robinson
University of South Carolina

ABSTRACT

It is hard to track the history and meaning of space art because it holds such widely varied meanings for differing constituencies and, compared with other disciplines, has diverse participation, but little formal history in space development. However, we all seem to be interested in following the exploration and discovery of space, largely through the powerful images that characterize its progress. There are two major constituencies that are worlds apart: the usually consistent formal and popular visual documentation of the development of space, and the intermittent and reluctant interest of the fine or academic arts.

PART 1

The first world of formal and popular visual documentation includes all of the photographs and illustrative documentation of far away places and the vehicles and hardware to get us there. Space agencies have documented every aspect of this development, from the first whole views of our gentle planet and astronauts stepping onto the surface of the moon to the unusual surfaces of Venus and Mars and constantly increasing telescopic information about deep space. These images first act as a means of simply keeping track and recording any anomalies, but because of their sheer number and perspective, contain information that reach at the soul of our curiosity and desire to know. The space shuttle has a collection of onboard cameras that click away in independent observation, recording many possibly interesting or informative elements; most of this film remains undeveloped and unobserved, and that doesn't even start to consider the influence of moving images. The best of these images end up in press releases, on walls of museums, and in entrepreneurial coffee table books about space. They range from beautiful examples of other worlds to "discoveries" of the face of Jesus in a deep space nebula.

There has also been a large collection of illustrative artists, graphic designers, and artistically curious others who find allure, creative motivation, and subject matter in this photographic documentation. Their use of these images as an artistic base commonly transpose the formal documentation of space into recognizable forms of artistic process such as traditional drawing, painting, printmaking, and sculpture.

Space agencies recognize the vital role that the arts play in public understanding, mediation, and the subsequent popular interest, support, and related funding. The National Aeronautics and Space Administration (NASA), as well as the proceeding military organizations, have programs designed to allow artistic participation and promotion. But this is also where we start to see some of the dichotomies that surround space art. While some in these organizations may have an awareness of the significant role that the arts can play, they also realize how frivolous these endeavors may and can be made to appear to a critical eye, especially when they fall so apparently far from the intended practical purpose of the scientific organization. Contemporary space exploration and its related innovations are documented by traditional and historic modes in the arts. The continuing reality is that it is hard for administrators, in ever tightening public budgets, to justify even these activities that appear so unnecessary and are so hard to define, especially pragmatically to those that provide and control the necessary funding.

Like so many other elements of space development, these fledgling programs started in the military services and evolved to the space agencies. The Salmagundi Art Club in New York City has administered some of the programs for United States military and space organizations and the organizations have found ways to established ongoing programs and program support. They represent various forms of characterization and justification and their own descriptions provide a fair and accurate understanding of the underlying scientific development.

Military Programs

The Navy Art Program, founded in 1941 by Griffith Baily Coale, sent artists into the field in an effort to record the drama and complexity of combat through works of art. Later in World War II, Abbot Laboratories of Chicago, with the help of the Associated American Artists (AAA) and the military, commissioned artists to document all aspects of the war. The Navy Combat Program was revived in Korea and Vietnam with the assistance of the Salmagundi Club, later merged with the Naval Historical Center, and artists were also sent to the Persian Gulf and Desert Storm.

While these programs may have been started by creative and ambitious artists employing creative processes to significant cultural occurrences, over the years they seem to have lost that edge or, at least, brought into debate the concepts of meaning and competency. The resulting artwork often depends too much on visual depiction and the inherent beauty and interest in the subject matter, that they often miss fundamental aspects of design and the value of broader meaningful content.

Nonetheless, the Navy Art Collection has over 13,000 works of art covering all aspects of U.S. Naval history, including aviation and space (Naval Historical Center, 2006).

These programs were often buried under the complexity of government jargon and not well publicized or easily accessible, and therefore were open more to those who had some means of connection, or understanding of bureaucracy, access, or persistence vs. those who satisfied any kind of consistent professional measure (The Federal Register, 2004).

Figure 1. Neil Armstrong portrait (Source: Detlev van Ravenswaay)

Figure 2. Mars reconnaissance orbiter for JPL (Source: Dana Berry, SkyWorks)

Figure 3. Sonnenenergiegewinnung im Orbit (Source: Detlev van Ravenswaay)

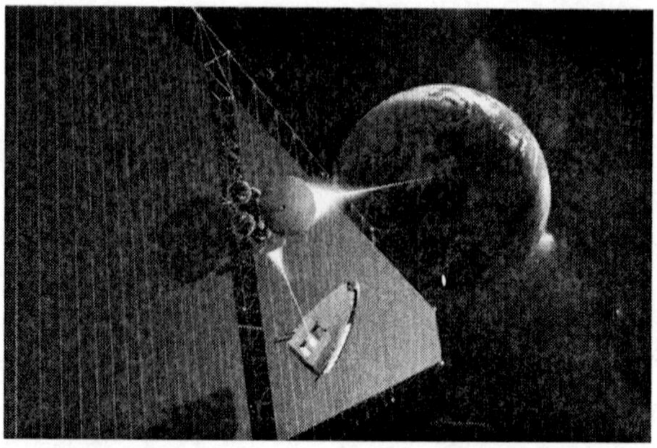

Figure 4. Ptolemeic Model of the Cosmos for the Smithsonian (Source: Dana Berry, SkyWorks)

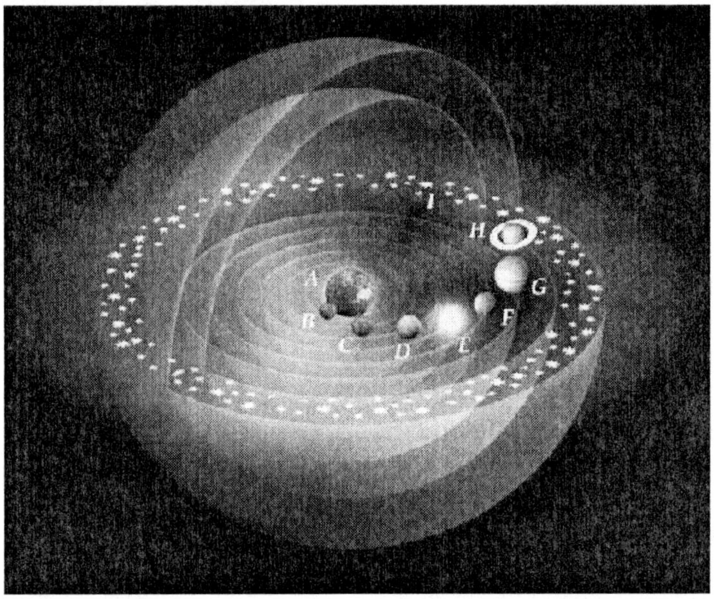

Subsequently, national space agencies followed the design and function of these successful programs to establish initiatives of their own. NASA has continued these traditions of witnessing and interpreting U.S. space activities in a unique melding of art and science. Much of the resulting collection has traveled around the U.S. on Artrain, an unusual, but easily accessible museum on rails (Artistry of Space, 2000).

These programs most often speak about and justify their existence more with an important educational role rather than with reference to the meaning or quality of the art. Meaningful, but hard to defend or describe activities often end up under the guise of education. Professionals in education, versus the arts, then administer the programs and have a significant effect on the appearance, outcome, and evaluation of the resulting work. (Space Art Home, 2006).

More recently, the Smithsonian National Air and Space Museum took an active part in becoming a central collection point and element of scholarship for many versions of this important work (Flight and the Arts, 2006).

The IAAA

These beneficial efforts encouraged a small but dedicated community of illustrators and designers who took on and established the concept of space art. However, most of this work was more directed at the illustrator's art vs. the creative exploration inherent in contemporary visual art. Their discussions were more about the use of materials, accuracy in depiction, and trips to terrestrial substitutes for space-like locations. This type of undertaking has become the popular standard of expectation of other worldly places and the camera like renderings of "Space Art," and their words and descriptions provide an appropriate context.

In 1981, the first formal group of astronomical artists met in conjunction with the Planetary Society, then for the first Space Art Workshop in Hawaii, and subsequently at a second workshop in Death Valley. These locations lent themselves to the landscape of space. The resulting work toured in the United States and Canada, much of it at natural science museums, and the group was established as the International Association of Astronomical Artists (IAAA), incorporating in 1988, and forming working relationships with NASA and space agencies in other countries. The organization continues to hold workshops, has a broad range of participants and skill levels, is broadly recognized as the home of space art, and in words from their Web site, says it continues to "render and paint the silence and grandeur of the heavens" (The History of the IAAA,"2006).

The IAAA's manifesto acknowledges the 19th century tradition of involving artists in geographic exploration, later expanding to, "Armed with science, creativity and imagination, they construct realistic images of visions throughout the Universe, from our Earth to the Stars. Not only realist; surrealist and impressionist styles are equally valuable in this adventurous and innovative field." "From this fertile background of research and imagination comes the body of artwork known generally as the genre of SPACE ART."

"The object of the IAAA, as a non-profit foundation, is to implement and participate in astronomical and space art projects, to promote education about astronomical art and to foster international cooperation in artistic work inspired by the exploration of the Universe." (The IAAA Manifesto, 2006).

PART 2

The other part, at the other end of the spectrum, is a collection of artists who are as or more interested in the development of the visual arts then they may be in space, but who are also interested in and understand what a significant role science and technology and space development plays in our culture. They also understand the equally important role an artist can play in mediating and communicating interesting, but often complex information to an ever scientifically curious public. Here, the artist can act as a mediator of the concepts and complexity of space to a public, while also paying attention to the creative growth and development of the discipline. It seems odd that the public's understanding of space art is generally based on and employs the use of very traditional and conservative methods to illuminate some of the most unusual and novel exploratory activities of our time. These artists, conversely, employ state of the art and experimental artistic means to convey the interest and excitement of space development.

The formal, contemporary art world has never been very interested in endeavors focused on anything other than art itself and generally doesn't value or have too much interest in "space" or any other prefix to what they consider as an already multifaceted discipline described simply as art, other than curiosity and improvement in what makes art socially consequential and good. There is still a significant gap and professional distrust between the two cultures of science and art.

One significant exception was experiments in art and technology (EAT), originally founded in 1966 by engineers Bill Klüver and Fred Waldhauer and artists Robert Rauschenberg and Robert Whitman, the subsequent Art & Technology Program created by Maurice Tuchman at the Los Angeles County Museum of Art (LACMA) in 1967, and the resulting work exhibited at the Pepsi Pavilion at the 1970 Osaka Universal Exhibition at the Japan World Exposition. While these efforts were not about space development in particular, they formalized science and technology as a meaningful part of contemporary art and spawned some of the artists and projects that eventually resulted in innovative space art, and captured the interest of the formal art world (Klüver, 2006).

Even with these objective successes and a collaborative history that can be traced back to Leonardo and Stone Henge, and the subsequent and ever increasing role of digital technology, there remains a subtle, but tangible reluctance and caution about any interaction between the arts and sciences. Most recently and oddly, this is manifest more by the arts than the sciences. The arts have always feared the empiricism of the sciences, as well as the linear logic of technology, though those who take time for a closer look often discover that creativity in both requires a healthy combination of common sense coupled with some uncertainty, risk, and intuitive trust. The sciences have a continued interest and enthusiasm for the arts, albeit sometimes limited knowledge and exposure.

Robert Irwin

California artist, theorist, and thinker Robert Irwin's constant investigations in the characteristics of light, perception, and environment shadowed the general expansion and development of the art world and helped establish a foundation of thinking that continues to guide and infuse this area of art.

Irwin and James Turrell, working with NASA scientist Ed Wortz, focused on their own curiosity and ended up with meaningful space art in many varied forms, from perceptual environments to habitats for long duration space flight. Irwin's reputation and articulate manner gave a much needed credibility to the meaning and artistic value to these investigations.

In 1968, he embarked on a now-famous stretch of research in collaboration with artist James Turrell and a NASA scientist named Ed Wortz. For a period of several months the three of them pursued whatever interested them: They sat in anechoic chambers; they played with light; they discussed ideas. By the time their investigations were done, Irwin had given up on painting altogether. Instead, he began experimenting with nearly invisible acrylic columns, which he installed as refractory presences in the white environment of various galleries. (Lacerte, 2006)

This motivates us to focus on an academic inquiry of what is art and what is space art? Should we include Turrell's Roden Crater or a variety of other artworks designed to provide attention to, be seen, or function from the space above Earth, such as Robert Smithson's Spiral Jetty, Michael Heizer's Double Negative, or Walter De Maria's Lightening Field?

After the social innovations of the1960s, technology remained progressive, but art, like society in general, became cautiously digressive. However, curious and creative artists continued to search for any avenue to explore the new and innovative.

In the 1970s, with ambitious plans for a space shuttle, Werner von Braun's vision of using this space truck to develop other disciplines meaning in space development, and the subsequent NASA Advisory Committee's recommendations that citizen observers include visual and educational observers and print journalists, artists continued to express and exhibit interest in these possibilities.

Sky Art

Many of these artists found affinity in the Sky Art conferences, organized and energetically promoted by Otto Piene, the executive director of the Massachusetts Institute of Technology's (MIT) Center for Advanced Visual Studies (CAVS). MIT was insightful enough to initiate and promote artistic investigation with the creation of CAVS with Gyorgy Kepes as its director in 1967, and MIT's heritage and reputation provided a much-needed additional credibility to the art's role in technology and space. These Sky Art Conferences brought together innovative and sympathetic scientists and engineers, such as Philip Morrison, Paul MacCready, and Harold Edgerton, with CAVS fellows and a small but ambitious group of other international artists interested in the sky and above. Initially, visual participants ranged from flag and kite flyers to those interested in media transmission to aviation and space.

Most of the latter artists were interested in putting earth objects in the rarified environment of space, but the important early work really had more to do with finding access and administrative support. Joe Davis convinced NASA to place one Martian rock on top of another, thereby initiating sculpture on Mars, and Lowry Burgess worked long and hard to pave the way for others when he finally flew the richly charged "boundless cubic lunar aperture," the first official nonscientific or artistic payload, on the space shuttle Discovery.

Burgess' own technical description:

The boundless cubic lunar aperture is a hyper cubic form or set of nested cubes extending itself throughout a nearly 18 billion-year time frame, over 4 billion years into the past and out nearly 14 billion years into the future. At its center is a cubic vacuum chamber enclosed by six holograms of 'nothing'. This central cube actually floats in an outer cube half filled with water and air. The water is distilled water purified on the surface of the southern Dead Sea from waters gathered from the mouths of 18 of the world's greatest and longest rivers (Nile, Mississippi, Amazon, Tigris-Euphrates, Ganges, Murray, Yang Tzu, Danube, etc. and 18 other sources, miraculous water, very old water, glaciers, geysers, wells, springs, etc.? all comprising a wide phenomenology of water. The waters and air are augmented by other atomic elements thus comprising all the elements and trace elements or the periodic table, hence, the 'everything'. Thus the Cube contains the 'everything/nothing' to be released from gravity in outer space. On two outer surfaces of the cube, opposite each other, are 2 holographic poetic texts concerning the concept of 'release', one through light, the other through darkness. In light these texts float above the surface of the cube. This hypercube is the part of the total 'boundless cubic lunar aperture' taken into outer space by NASA in March of 1989. The hypercube was brought back to earth where it becomes the center of a larger framework of forms energies materials and geology. The cube was placed in the center of a section of a 40 million year old petrified sycamore tree which in turn hovers in a permanent magnetic field in a shaft cut within a 400 million year old, large, rock-mass which is the residue of a lake that existed in the area of what would be now Afghanistan 400 million years ago. This unique shelf of stone is beside Sandy Pond (Flint's Pond) close by Walden Pond in Lincoln, Massachusetts, USA.

Both Davis and Burgess were fellows and active participants in Sky Art, and had the strong support of their host institutions of MIT and Carnegie Mellon University. The latter also continues to support these kinds of activities through the Studio for Creative Inquiry.

My own work investigates the role the arts might play in space development. Working on a National Endowment for the Arts environmental light and laser installations for Third Century America at the Kennedy Space Center, I could not help but be impressed with the surrounding space hardware and the role that technology and those first images from space had played in my own, my generation's, and general culture's development. My father had worked for Grumman Aircraft and I spent some of my high school years sitting on a surfboard on Satellite Beach watching various rockets, including many Saturn V's, head skyward.

In 1976, I made arrangements to visit the Marshall Space Center and NASA Headquarters in Washington and learned of plans for the Space Shuttle and the potential inclusion of many disciplines in the project. Scientists at Marshall helped me understand that there were not a few brilliant individuals making the ambitious Saturn V moon program work, but rather a great management process bringing thousands of small but highly refined bits of information together to a master totality; good, sound management.

Figure 5. Boundless cubic lunar aperture (Source: Lowry Burgess)

Figure 6. Boundless cubic lunar aperture (Source: Lowry Burgess)

Figure 7. Launch of Discovery with boundless cubic lunar aperture (Source: Lowry Burgess)

Realizing that the arts had little knowledge of or interest in space development, but that some visual observer might be given an opportunity to fly as a citizen observer, I set out to provide a base of information, so that whoever was selected would be able to make good use of the opportunity. No matter how generous and helpful NASA and related organizations were with their time, I always felt like an outsider being provided a short glimpse, but no real participation in the significant occurrences formulating space development. This led me on a course through military aviation. Having earlier avoided the draft, I now sought recruitment and entry in the Naval Aviation program, and ultimately later to attendance at launches and landings of the shuttle and training on the Weightless Wonder, NASA's now popular zero gravity aircraft. I was interested in both the process of observation and participation in exploring what sensible opportunities might exist for artistic participation. I believed then and continue to believe that artists who are allowed to play an active role in scientific research cannot only perform a significant role in the mediation of complex information and related public understanding, acceptance, and support, but also often offer objective and meaningful insight to the active research and development process.

My initial attempts to access the early launches of the Space Shuttle is an interesting subscript and gives some insight to how tough it was for less conventional or experimental artists to gain credibility. I had an advanced degree in studio art, was teaching art at a major state university, and had the beginnings of a meaningful scholarly and exhibition record with a focus on art and technology, but could not get authorization as an artist to the launch site. Often, it is whom you know as much as what you know, and I was advised to get assigned as a journalist to cover the first launch of the Space Shuttle, Columbia. I had a friend who was an editor of an arts publication, who "assigned" me to cover the launch, and I was provided immediate access although I had no training or capability as a journalist. I later raised this with the appropriate authorities and was finally provided ongoing access and assistance as an artist.

Several others, such as Seattle artist Jim Pridgeon, concurrently wrestled through similar processes, trying to find access for more ambitious, nontraditional, or idea-oriented art. Similar activities were going on in Europe, though there seemed, in the European space community, as with other art activities, a greater acceptance and understanding of the need for sincere artistic or cultural participation. The Sky Art Conference met in Europe, first in conjunction with Ars Electronica[12A] and later in Greece with support from the European Cultural Commission. The more recent arts catalyst in Britain "is an arts organisation that actively makes connections between art and science through commissions and strategic projects" and pursues and supports many space-oriented artistic projects.

Some of the more recent participants in Sky Art, including Frank Pietronigro, have brought an energetic second generation of enthusiasm to the role of space art in space development. Burgess and Pietronigro organized an important Workshop on Space Artist's Residencies and Collaborations held in February 2005 at Carnegie Mellon University West at the NASA Ames Research Center, Moffett Field, California. The resulting report, "The Arts and Space Culture: The Common Ground of Creativity," helped reconnect both sides of space art with exhibitions and a space art track in the most recent 2006 International Space Development Conference. While those of us in the past were overwhelmed with the associated costs and sought the paternal good will and interest of the sponsoring agencies, Pietronigro's entrepreneurial spirit has resulted in productive efforts to raise funds so that the arts can be sincere and active players in zero gravity and space research in the same way as any other experimental entity or discipline (European Space Agency, 2004).

Figure 8. F-16 orientation (Source: Chris Robinson)

Figure 9. Black launch (Source: Chris Robinson)

Figure 10. Cost of freedom (Source: Chris Robinson)

Figure 11. Space Shuttle, Columbia (Source: Chris Robinson)

Figure 12. The End of the Moon (Source: Laurie Anderson)

Laurie Anderson

Performance artist Laurie Anderson served as NASA's first formal artist in residence since the inception of their art program in 1963. This represented a bold new move for NASA, with Anderson being so clearly innovative and different from the illustrative art that normally characterized the program. While Anderson found plenty of substance for the new solo work, "The End of the Moon" and will continue to employ the experiences in her art, her participation was broadly criticized and seen by the organizers as a functional failure (Hull, 2004).

Various artists continue to work at NASA's facilities as illustrators and in other capacities, and usually start out with the highest ambitions, only to end up caught in a collection of other responsibilities, and the need to sustain their employment through adaptation, survival, and an interest in job success.

There is also a new generation of multidisciplinary, Renaissance artists such as Bradley Pitts who has two degrees in aeronautics and astronautics, an undergraduate concentration in visual art from MIT, and a healthy understanding of the power of contemporary art. His ongoing work employs the best of engineering and the visual arts, and he moves comfortably between the two cultures, with work ranging from space suit design to zero gravity performance. It is somewhat easier to find social acceptance of the sciences vs. the arts, but neither is easy, and he is a great example of a rare new breed.

CONCLUSION

It is interesting to think about what really constitutes art, space art, good and meaningful art, real innovation, and general creativity. We are enamored with the creative genius of Paul McCready, and Burt Rutan and Space Ship 1, but while this creativity has the potential to be art, until an artist takes an active role, it simply remains intriguing and insightful descriptive engineering. The question is who and how we take the occurrence and transform its context so that it becomes artistically meaningful to culture.

Much like the development of the Saturn program and landing a man on the moon, what is required is not one single answer or overriding genius, but rather a collection of expertise, which contribute to and help time and history flesh out what constitutes real value in the arts. Like NASA's wide-ranging research investigations and spin-offs, artists now have their hands in many aspects and tangents of space development. The last century of space-related research brought us, and we reveled at, the laser and microchip, and this century it may be the wide reaching influence of the quantum realm and nanotechnology. So, we all collect our experience, make our best efforts, and see what happens.

If you would like to read and know more about the interaction of art and science, and the development of space art, Leonardo, the journal of the International Society for the Arts, Sciences, and Technology, published by the MIT Press, is the leading publication in this area of investigation. Most of the artists, organizations, and topics covered here are easily searchable at the library or on the Internet.

REFERENCES

Airforce Art Program. (2004, August 23). *The Federal Register, 69*(162), 51815. Retrieved March 5, 2007, from http://frwebgate.access.gpo.gov/cgi-bin/getpage.cgi?dbname=2004_register&position=all&page=51815

Artrain, USA, AbsoluteArts.Com, worldwide arts resources. *Artistry of Space: The NASA Art Program*. Retrieved March 5, 2007, from http://www.absolutearts.com/artsnews/2000/04/18/26850.html

Hull, A. (2004, June 30). Moon and stars align for performance artist: Laurie Anderson accepts art commission from NASA. *The Washington Post*, p. A19. Retrieved March 5, 2007, from http://www.washingtonpost.com/wp-dyn/articles/A15916-2004Jun29.html

International Association of Astronomical Artists. *The History of the IAAA*. Retrieved March 5, 2007, from: http://www.iaaa.org/history.html

International Association of Astronomical Artists. *The IAAA Manifesto*. Retrieved March 5, 2007, from http://www.iaaa.org/history.html

Klüver, B. (2000). *Experiments in art and technology (E.A.T.)*. Archive of published documents. Retrieved March 5, 2007, from http://www.fondation-langlois.org/html/e/page.php?NumPage=306

Lacerte, S. 9 evenings and experiments in art and technology: A gap to fill in art history's recent chronicles. *la fondation Daniel Langlois*, (Introduction, p. 1/6). Retrieved March 5, 2007, from http://www.fondation-langlois.org/html/e/page.php?NumPage=1716

Lewis, J. (1993, June). Robert Irwin: A room with a view. *Harper's Bazaar: A room with a view.* Retrieved March 5, 2007, from http://www.airic.com/articles/roomview.html

Naval Historical Center. *The Navy Combat Art Program.* Retrieved March 5, 2007, from http://www.history.navy.mil/branches/org6-4.htm

pietronigro.com. (2004, April 23). *European Space Agency invites U.S. artists to present new plans for art in space.* Retrieved March 5, 2007, from http://www.pietronigro.com/news_2004-04-23.htm

Smithsonian National Air and Space Museum.(2004). *Generous friends: Building an art collection for the National Air and Space Museum.* Flight and the Arts, Gallery 211, New Exhibition, Retrieved March 5, 2007, from: http://www.nasm.si.edu/exhibitions/gal211/gal211.cfm

Space Art Home Page. *Space educator's handbook: The artist looks at space*. Retrieved March 5, 2007, from http://www1.jsc.nasa.gov/er/seh/spaceart.html

Compilation of References

A Med-World AG. (2006). *Strahlenbelastung in der Raumfahrt.* Retrieved March 3, 2007, from http://www.m-ww.de/enzyklopaedie/strahlenmedizin/strahlenbelastungraumfahrt.html. Berlin, Germany.

Abitzsch, S. (1996). Prospects of space tourism. In *Paper presented at the 9th European Aerospace Congress,* Berlin, Germany.

Abitzsch, S. (1998). *Chancen und Entwicklungsmöglichkeiten der Weltraumtouristik*, Doctoral thesis, Technical University Berlin, Germany.

ACE. (2006). Atmospheric chemistry experiment of the Canadian Space Agency. Retrieved February 22, 2007, from http://www.ace.uwaterloo.ca/

Ad hoc Group on Earth Observation (GEO). (2003). *The Ad hoc Group on Earth observation (GEO): Terms of reference.* Retrieved March 5, 2007, from http://earthobservations.org/

Adirim, H., Lo, R., & Paatsch, A. (1999). Next generation high performance low pollution launchers: Problems and solutions. In *Paper presented at the 50th International Astronautical Congress, IAA-99-IAA.3.3.01,* Amsterdam, The Netherlands.

Affens, D. W. (2006, April). NASA Goddard Space Flight Center DASS Ground Station. In *Proceedings of the CS/EWG-2 Meeting*, Montreal, Canada.

Agreement Governing the Activities of States on the Moon and Other Celestial Bodies (The Moon Treaty). Retrieved March 2, 2007, from http://www.unoosa.org/oosa/SpaceLaw/moon.htm[1]

Airforce Art Program. (2004, August 23). *The Federal Register, 69*(162), 51815. Retrieved March 5, 2007, from http://frwebgate.access.gpo.gov/cgi-bin/getpage.cgi?dbname=2004_register&position=all&page=51815

Alcazar Diaz, J., & Velasco, T. (2004, October). *The GALILEO programme: Risks and challenges.* Paper presented at the 55th International Astronautical Congress at Vancouver.

Aldridge, E. C., Jr., Fiorina, C. S., Jackson, M. P., Leshin, L. A., Lyles, L. L., Spudis, P. D. et al. (2004). Report of the president's commission on implementation of United States exploration policy. Retrieved March 3, 2007, from http://www.nasa.gov/mission_pages/exploration/main/index.html

Aldridge, P. (2004). *A journey to inspire, innovate, and discover.* Report of the President's Commission on Implementation of United States Space Exploration Policy. Washington, DC: U.S. Government Printing Office.

Alic, D., Brion, J. J., Monte, G. E., & Piris Botalla, L. (2005, May). *Nanopehuensat. Telemetría Digital de Datos y Transmisión por Voz*. Paper presented at the Congreso Argentino de Tecnología Espacial, Córdoba, Argentina.

Allan, T. (1998). Watersheds and problemsheds: Explaining the absence of armed conflict over water in the Middle East. *MERIA Journal*, *2*(1), 49-51.

American Institute for Astronautics and Aeronautics (AIAA). (2001, March). The working group on the contribution of space systems to the development and implementation of multilateral environmental agreement (Working Group Rep.). In *Proceedings of the 6th International Space Cooperation Workshop*, Seville, Spain.

Arianespace. (2006, April 25). *Arianespace reports 2005 results: A very successful year* (Press release). Retrieved March 2, 2007, from http://www.arianespace.com/site/news/releases/06_04_25_release_index.html

Arias, A., Pelayes, N., Cajarabilla, M., Márquez, F., & Quiroga, J. J. (2005, May). *Nanopehuensat: Integración de la Computadora de Abordo.* Paper presented at the Congreso Argentino de Tecnología Espacial, Córdoba, Argentina.

Arias, A., Pelayes, N., Orejas, M., Giacomelli, M., Mare, J., Odello, L., & Quiroga, J. J. (2003, May). *Administración de Energía y Mediciones de una Plataforma de Ensayo Espacial. Nanopehuensat.* Paper presented at the Congreso Argentino de Tecnología Espacial, Neuquén, Argentina.

Arnould, J. (2001). *La seconde chance d'Icare. Pour une éthique de l'espace*. Paris: Cerf.

Arroyo D., Brion J. J., Eggers, S., Marcellino, A., Mollo, D., & Monte, G. (2003, May). *Nanopehuensat Computadora de Abordo para Transmisión de Datos.* Paper presented at the Congreso Argentino de Tecnología Espacial, Neuquén, Argentina.

Artrain, USA, AbsoluteArts.Com, worldwide arts resources. *Artistry of Space: The NASA Art Program.* Retrieved March 5, 2007, from http://www.absolutearts.com/artsnews/2000/04/18/26850.html

ATSC. (2000). ATSC Data Broadcast Standard (A/90). Washington, D.C.: Advanced Television Systems Committee.

ATSC. (2005). ATSC Digital Television Standard (A/53). Washington, D.C.: Advanced Television Systems Committee.

Australian Government (2004). *Longest rivers: Geoscience Australia.* Retrieved February 15, 2007, from http://www.ga.gov.au/education/facts/landforms/longrive.htm

Avagnina, D., Dovis, F., Ghiglione, A., & Mulassano, P. (2002). Wireless network based on high-altitude platforms for the provision of integrated navigation/communication services. *IEEE Communication Magazine*, *40*(2), 119-125.

Bamberg, G., & Coenenberg, A. G. (1996). *Betriebwirtschaftliche Entscheidungslehre*, ISBN 3-8006-2076-6, pp. 105-113. Munich, Germany: Vahlen.

Bekey, I. (1998). Economically viable public space travel. In *Paper presented at the 49th International Astronautical Congress*, Melbourne, Australia.

Bell, R.(1998). *Les Péchés capitaux de la haute technologie. Superphénix, Eurotunnel, Ariane 5...*, Paris: Seuil.

Berry, J., Conger, R., & Kulpa, J. (2001, April 17-20). The sprite mini-lift vehicle: Affordable launcher for small spacecraft missions. In *Proceedings of the 19th AIAA International Communications Satellite Systems Conference*, Toulouse, France.

Bildt, J., Peyrelevade, J., & Späth, L. (2000). *Towards a space agency for the European Union. Report for the Director-General of the European Space Agency.* Retrieved March 2, 2007, from http://esamultimedia.esa.int/docs/annex2_wisemen.pdf

Bilstein, R. E. (2003a). *Testing aircraft, exploring space: An illustrated history of NACA and NASA.* Baltimore, MD: Johns Hopkins University Press.

Bilstein, R. E. (2003b). *Stages to Saturn: A technological history of the Apollo/Saturn launch vehicles.* Melbourne, FL: University Press of Florida.

Bogdanov, V., Ageev, V., Belousov, A., Buyanov, A., & Terekhin, V. (1984, April). Ground segment of Cospas System. In *Proceedings of the International Sympo-*

sium on Satellite Aided Search and Rescue, Toulouse, France.

Border, J., Kojo, M., Griner, J., Montenegro, G., & Shelby, Z. (2001). Performance Enhancing Proxies Intended to Mitigate Link-Related Degradations. RFC 3135.

Bush, G. (2004). *Renewed spirit of discovery: The president's vision for U.S. space exploration.*

Caffrey, R. H., Coreil, P., & Demcheck, D. (2002). *Mississippi River water quality: Implications for coastal restoration.* Interpretive Topic Series on Coastal Wetland Restoration in Louisiana (National Sea Grant Library No. LSU-G-02-002).

Carroll, J. (2002). *Space transport development using orbital debris.* Principal Investigator, Tether Applications, Inc. Phase I Study - NASA Institute for Advanced Concepts.

Casner, S., & Jacobson, V. (1999). Compressing IP/UDP/RTP Headers for Low-Speed Serial Links. RFC 2508.

Cayla, G., Cohen, S., & Guigon, D. (2005). *WiMax: An efficient tool to bridge the digital divide.* Retrieved February 22, 2007, from www.wimaxforum.org

Chakroborty, S., Wertz, J. R., & Conger, R. (2004, October 4-8). Responsive access to space. The scorpius low-cost launch. In *Proceedings of the 55th International Astronautics Federation (IAF) Congress*, Vancouver, Canada.

Chandra, A. H., & Ghosh, S. K. (2006). *Image interpretation, remote sensing and geographical information system* (p. 77). New Delhi: Narosa Publishing House.

Chang, K.-T. (2003). *Introduction to geographic information systems* (p. 7). New Delhi: Tata McGraw-Hill Publishing.

Chatterji, A. K. (2002). *Ecosystem approach in environmental management: Introduction to environmental biotechnology* (p. 9). New Delhi: Prentice Hall of India.

Chinese Defence Today. (2004). *BEIDOU-1 satellite navigation system.* Retrieved February 22, 2007, from http://www.sinodefence.com/strategic/spacecraft/beidou1.asp

Chodorek, A., & Chodorek, R. R. (2000). Characterization of the MPEG 2 video traffic generated by DVD applications. In *Proceedings of the 1st IEEE European Conference on Universal Multiservice Networks* (pp. 62-70). Piscataway, NJ: IEEE Press.

Chodorek, A., & Chodorek, R. R. (2005). PGM-based reliable multicast data dissemination in satellite IP networks. In *Proceedings of the International Workshop on Satellite and Space Communications.* IEEE Press.

Chodorek, A., & Papir, Z. (2000). Selfsimilarity of an MPEG 2 video stream generated by DVD applications. In *Protocols for Multimedia Systems (PROMS'2000)* (pp. 389-399). Cracow, Poland: Faculty of Electrical Engineering, Automatics, Computer Science and Electronics Press.

Churchill, S. (1999). Introduction to human space life sciences. In A. Houston & M. Rycroft (Eds.), *Keys to space* (pp. 18-13). USA: McGraw-Hill.

Cianca, E., Prasad, R., De Sanctis, M., De Luise, A., Antonini, M., Teotino, D., & Ruggieri, M. (2005). Integrated satellite-HAP systems. *IEEE Communications Magazine*, *43*(12), 33-39.

Clarke, A. C. (1945). Extra terrestrial relays. Can rocket stations give world-wide radio coverage? *Wireless World*, 305-308.

Clarke, A. C. (1978). *The fountains of paradise.* New York: Warner Books.

Clement, J., Holbrook, P., & Staman, M. (Eds.). (1996, June 25-28). *Extending Internet access to rural areas and small communities in the upper Midwestern United States.* Montreal Convention Center, Montreal, Canada.

Coalition for Space Exploration. (2004). *CSA joins Coalition for Space Exploration.* Retrieved March 3, 2006, from http://www.californiaspaceauthority.org

Coase, R. H. (1959). The Federal Communications Commission. *Journal of Law and Economics,* 1-40.

Collins, P. (2002. June 4-7). Preparing for passenger space travel, the future of human spaceflight. In M. Rycroft (Ed.), *Beyond the ISS: The future of human spaceflight.*

Proceedings of an International Symposium (pp.165-172), Strasbourg. Kluwer Academic Publishers.

Collins, P., & Yonemoto, K. (1998). Legal and regulatory issues for passenger space travel. In *Paper presented at the 49th IAF Congress,* Melbourne, Autralia. Retrieved March 3, 2007, from http://www.spacefuture.com/archive/legal_and_regulatory_issues_for_passenger_space_travel.shtml

Collins, Stockmans, & Maita. (2004). *Demand for space tourism in America and Japan and its implications for future space activities.* Retrieved March 3, 2007, from http://www. spacefuture.com/archive/demand_for_space_tourism_in_america_and_japan.shtml. Tokyo, Japan.

Committee on an Assessment of Balance in NASA's Science Programs. (2006). *An assessment of balance in NASA's science programs.* National Research Council. National Academy Press.

Committee on Earth Observation Satellite (CEOS). (2002). *CEOS: Overview.* Retrieved March 5, 2007, from http://www.ceos.org/

Convention for the Protection of the Ozone Layer (Vienna Convention), March 22, 1985, 1513 U.N.T.S 293

Convention on Biological Diversity. (UNCBD). June 5, 1992, 1760 U.N.T.S. 79

Convention on Wetlands of International Importance Especially as Waterfowl Habit (Ramser Convention), February 2, 1971, 11.I.L.M.963

Cospas-Sarsat (2006a, February). Information bulletin (No. 18). Retrieved February 22, 2007, from www.cospas-sarsat.org

Cospas-Sarsat (2006b). Retrieved February 22, 2007, from www.cospas-sarsat.org

Cospas-Sarsat (2006c, June). Status of the European MEOLUT prototype. In *Proceedings of the GJU/ESA at CS/JC-20 Meeting,* Montreal, Canada.

Cospas-Sarsat (2006d, June). SAR/GLONASS system status. In *Proceedings of the CS/JC-20 Meeting,* Montreal, Canada.

Cospas-Sarsat Document (2002). *International Cospas-Sarsat Programme Agreement.* July 1, 1988, signed in Paris, France.

Cospas-Sarsat Document System Data (No. 31). (2005, December) Retrieved February 22, 2007, from www.cospas-sarsat.org

Cospas-Sarsat Introduction to the Cospas-Sarsat System, C/S G.003, Issue 5 – Revision 1, October 1999 www.cospas-sarsat.org

Cospas-Sarsat Project Report (1985, September). Document C/S R.001. Retrieved February 22, 2007, from www.cospas-sarsat.org

Creola, P. (2001). Some comments on the ESA/EU space strategy. *Space Policy, 17,* 87-90.

CSIC—Coordination Scientific Information Center, Russian Federation, Ministry of Defense. (2002). *GLONASS Interface and Control Document Version 5.0.* Retrieved February 22, 2007, from http://www.glonass-center.ru/ICD02_e.pdf

Dale, W. N. (2001). *Middle East water problems.* Retrieved February 15, 2007, from http://www.unc.edu/depts/diplomat/archives_roll/2001_07-09/dale_water/dale_water.html

Davis, C., Hoban, S., & Penhoet, B. (1999). Moving pictures: How satellite, the Internet, and international environmental law can help promote sustainable development. *Statson Law Review, 28,*1153.

De León, P. (1999, July). *Non-governmental and educational space activities in Argentina.* Paper presented at the Unispace III, Space Generation Forum, Vienna, Austria.

De León, P., & Lassig, J. (1999, September). *The impact in the space activities on education and public outreach (The Argentine experience).* Paper presented at the International Simposium on Commercial and Non-governmental Space Activities, Rio De Janeiro, Brazil.

De Zan, H. J., & Simone, D. D. (2005, May). *Estudio y Análisis de las Influencias del Ambiente Espacial en las Comunicaciones Satelitales.* Paper presented at the

Congreso Argentino de Tecnología Espacial, Córdoba, Argentina.

Dean, E. B. et al. (1986). Cost risk analysis based on perception of the engineering process. In *Paper presented at the Annual Conference of the International Society of Parametric Analysts*, Kansas City, USA.

Deering, S., & Hinden, R. (1998). Internet Protocol, Version 6 (IPv6) Specification. RFC 2460.

Degermark, M., Nordgren, B., & Pink, S. (1999). IP Header Compression. RFC 2507.

Diaz, A. (2004). Renewed commitment to excellence: An assessment of the NASA agency-wide applicability of the Columbia accident investigation board report.

DIVERSITAS (2004, May 20). *The global water system project: Scientific framework and implementation activities*. Retrieved February 15, 2007, from http://www.gwsp.org/downloads/GWSP.gws-framework-20-05-04.pdf

Doolittle, J., & Glines, C. V. (1991). *I could never be so lucky again: An autobiography*. New York: Bantam Publishing.

Dorner, D. (1996). *The logic of failure* (R. Kimber & R. Kimber, Trans.). New York: Metropolitan Books.

DSBTF—Defence Science Board Task Force. (2005, October). *The future of the global positioning system*. Report of the Office of the Under Secretary of Defense for Acquisition, Technology and Logistics. Washington D.C.: Department of Defense.

Duncuing, J., & Pierotti, P. (2001). PACTES: A project for improving the operational management of the flooding risk, by integration of Earth observation and advanced modeling techniques. In *Proceedings of the 52nd International Astronautical Federation (IAF) Congress*, Toulouse, France.

Dupas, A. et al. (2001). A Franco-German view of Europe's ambition in space for the 21st century. *Space Policy, 17*, 103-110.

Duros, E., Dabbous, W., Izumiyama, H., Fujii, N., & Zhang, Y. (2001). A Link-Layer Tunneling Mechanism for Unidirectional Links. RFC 3077.

Dyson, F. (1999). *The Sun, the Genome, the Internet*. New York: Oxford University Press.

EADS Space. (2005). *Astro +: Using space to make Europe a safer place*. Retrieved March 2, 2007, from http://www.space.eads.net/familiess/a-safer-world/futuredefnce/astro

EADS-Space. (2005). *Orbiting Internet satellites to bridge the digital divide*. Retrieved February 22, 2007, from http://www.space.eads.net/press-center/editorial-features/428-en

Edwards, B. (2000). *The space elevator Phase I study*. Eureka Scientific/NASA Institute for Advanced Concepts.

Edwards, B. (2003). *The space elevator Phase II study*. NASA Institute for Advanced Concepts. Retrieved March 3, 2007, from http://www.niac.org

Edwards, B. *The NIAC space elevator program*. Eureka Scientific/Institute for Scientific Research.

Edwards, B. *The space elevator: An ideal application for the free electron laser*. Eureka Scientific/Institute for Scientific Research.

Edwards, B., & Westling, E. A. (2003). *The space elevator: A revolutionary Earth-to-space transportation system*. Houston, TX.

Elbert, B.R. (2001). *The satellite communication ground segment and Earth station handbook*. Boston: Artech House.

Ellis, G. F. R. (Ed.). (2002). *The far future universe*. Philadelphia, PA: Templeton Foundation Press.

EM-DAT. (2006). The OFDA/CRED international disaster database—Université Catholique de Louvain—Brussels—Belgium. Retrieved February 22, 2007, from www.em-dat.net

EOH. (2005). Earth observation handbook. The committee on Earth observation satellites of the European Space Agency (ESA). Retrieved February 22, 2007, from http://www.eohandbook.com

ESA. (2001). *General policies and procedures, process, pricing and conditions for sale*.

ESA. (2005). *How much does it cost?* Retrieved March 3, 2007, from http://www.esa.int/esaHS/ESAQHA0V-MOC_iss_0.html

ETSI. (1997). *Digital video broadcasting (DVB): Framing structure, channel coding and modulation for 11/12 GHz satellite services* (European Standard Rep. No. EN 300 421 V1.1.2). Sophia Antipolis Cedex, France: European Telecommunications Standards Institute.

ETSI. (2004). *Digital video broadcasting (DVB): DVB specification for data broadcasting* (European Standard Rep. No. EN 301 192 V1.4.1). Sophia Antipolis Cedex, France: European Telecommunications Standards Institute.

ETSI. (2005a). *Digital video broadcasting (DVB): User guidelines for the second generation system for broadcasting, interactive services, news gathering and other broadband satellite applications (DVB-S2)* (Tech. Rep. No. TR 102 376 V1.1.1). Sophia Antipolis Cedex, France: European Telecommunications Standards Institute.

ETSI. (2005b). *Digital video broadcasting (DVB): Interaction channel for satellite distribution systems* (European Standard Rep. No. EN 301 790 V1.4.1). Sophia Antipolis Cedex, France: European Telecommunications Standards Institute.

Eurocontrol. (2000). *Guidelines for the economic appraisal of EATMP projects: The effective use of cost-benefit studies.* PLC.ETL.STO7.1000-GUI-01-00.

European Commission. (1999, February 10). *Communication on Galileo.* COM (1999) 54 final. Not published in the Official Journal.

European Commission. (2002). *eEurope 2005 Action Plan: An Information Society for all (Digital Divide).* Communication from the Commission to the Council, the European Parliament, the European Economic and Social Committee and the Committee of the Regions. COM (2002) 263 final.

European Commission. (2002). *Strategic Aerospace Review for the 21st century (STAR 21)—Creating a coherent market and policy framework for a vital European industry.* Report by the European advisory group on aerospace.

European Commission. (2003). *A coherent framework for aerospace—A response to the STAR 21 report.* Communication from the Commission to the Council, the European Parliament, the European Economic and Social Committee and the Committee of the Regions. COM (2003) 600 final.

European Commission. (2003). *European Defence—Industrial and Market Issues—Towards an EU defence Equipment Policy.* Communication from the Commission to the Council, the European Parliament, the European Economic and Social Committee and the Committee of the Regions. COM (2003) 113, final.

European Commission. (2003). *Guidelines in criteria and modalities of implementation of Structural Funds in support of electronic communication (broadband access).* Commission Staff working paper. SEC (2003) 895 EU Com July 2003.

European Commission. (2003). *Investing in research: An action plan for Europe.* Communication from the Commission to the Council, the European Parliament, the European Economic and Social Committee and the Committee of the Regions. COM (2003) 226 final.

European Commission. (2003). *White Paper on Space: A new European frontier for an expanding Union—An action plan for implementing the European space policy.* COM (2003) 673, released 11 November 2003.

European Commission. (2004). Commission decision on the implementation of the preparatory action on the enhancement of the European industrial potential in the field of security research 2004-2006. C (2004) 249 final.

European Commission. (2004). *Financial Perspectives 2007-2013.* Communication from the Commission to the Council and the European Parliament. COM (2004) 487 final.

European Commission. (2004). Green Paper on Defence Procurement. COM (2004) 608, final.

European Commission. (2005). *Report of the Panel of Experts on Space and Security*. Retrieved March 2, 2007, from: http://ec.europa.eu/comm/space/news/article_2262.pdf

European Commission. *Financial programming and budget: Financing the budget*. Retrieved March 2, 2007 from http://ec.europa.eu/budget/budget_detail/financing_en.htm

European Commission. *Sixth Framework Programme: 2002-2006*. Retrieved March 2, 2007, from http://europa.eu.int/comm/research/fp6/index_en.cfm?p=0_instruments

European Space Agency (ESA). (2004, December 20). *ESA presents space solutions at post-Kyoto climate summit*. Retrieved March 5, 2007, from http://www.esa.int/esaEO/

European Space Agency (ESA). (2005a, November 11). *Satellite support Kyoto Protocol through forest mapping service*. Retrieved March 5, 2007, from http://www.esa.int/esaEO/

European Space Agency (ESA). (2005b). *How can environmental conventions benefit from space?* [CD-ROM].

European Space Agency (ESA). (n.d.). *Treaty enforcement services using Earth observation (TESEO)*. Retrieved March 5, 2007, from http://styx.esrin.esa.it:5000/teseo/about.html

European Space Agency. (2006, February 24). *ESA confirms CryoSat recovery mission*. Retrieved March 2, 2007, from http://www.esa.int/SPECIALS/Cryosat/SEM7WFMVGJE_0.html

European Space Agency. Convention for the Establishment of a European Space Agency, Paris, done 30 May 1975, entered into force 30 October 1980; 14 ILM 864 (1975).

European Space Agency. *The European Space Exploration Programme Aurora*. Retrieved March 2, 2007, from http://www.esa.int/SPECIALS/Aurora/ESA9LZPV16D_0.html

European Union (2001, January 22). Council decision setting up the political and security committee 2001/78/CFSP. *Official Journal of the European Communities L 27/1*, 30 January 2001.

European Union Institute for Security Studies. (May 2005). *Defence Procurement in the European Union—The current debate* (Report of an EUISS Task Force. ISBN No. 92-9198-073-0).

European Union. (1992, July 29). Treaty on European Union. *Official Journal of the European Communities, C 191*.

European Union. (1997, November 10). Treaty of Amsterdam amending the Treaty on European Union: The treaties establishing the European communities and certain related acts. *Official Journal of the European Communities, C 340*.

European Union. (2000, November 16). Council resolution on a European space strategy. *Official Journal of the European Communities, C 37, 23/12/2000*.

European Union. (2001, March 10). Treaty of Nice amending the Treaty on European Union, the treaties establishing the european communities and certain related acts. *Official Journal of the European Communities, C 80*.

European Union. (2002). A regulatory framework for radio spectrum policy in the European Community (towards a harmonised market for space services in the EU. Decision of the European Parliament and of the Council, 676/2002/EC. *Official Journal of the European Communities L 108, 24/04/2002*.

European Union. (2002). The treaty establishing the European community. *Official Journal of the European Communities C 325*, 24 December 2002.

European Union. (2003). Council decision on the signing of the Framework Agreement between the European community and the European Space Agency. 12858/03 RECH 152 OC 589.

European Union. (2004). *Headline Goal 2010*. Retrieved March 2, 2007, from http://ue.eu.int/uedocs/cmsUpload/2010%20Headline%20Goal.pdf

European Union. (2004). Treaty establishing a constitution for Europe. *Official Journal of the European Communities C 310,* 16 December 2004.

European Union. (2005, September 7). *The Galileo family is further expanding: EU and India seal their agreement.* Press release IP/05/1105.

European Union. (January 22, 2001). Council decision on the establishment of the military staff of the European Union, 2001/80/CFSP. *Official Journal of the European Communities L 27/7,* 30 January 2001.

European Union. (January 22, 2001). Council decision setting up the military committee of the European Union 2001/79/CFSP. *Official Journal of the European Communities L 27/4,* 30 January 2001.

European Union. (September 24, 2004). Competitiveness (Internal market, industry and research). Press release of the 2605th Council meeting. Brussels. Press 269. 12487/04.

Eurospace. (2004). *European space directory* (19th ed.). Eurospace, Paris.

Fairhurst, G., & Collini-Nocker, B. (2005). Unidirectional Lightweight Encapsulation (ULE) for Transmission of IP Datagrams over an MPEG-2 Transport Stream (TS). RFC 4326.

Fairhurst, G., & Montpetit, M.-J. (2006). Address Resolution for IP Datagrams over MPEG-2 Networks. Internet-Draft, draft-ietf-ipdvb-ar-03.

Fernández, R., Nolly, O. A., Quiroga, J. J., Sierra, E. A., & Monte, G. E. (2005, May). *¿Tolerancia O Evitación? ... Esa Es La Cuestión.* Paper presented at the Congreso Argentino de Tecnología Espacial, Córdoba, Argentina.

Fernández, R., Nolly, O., Sierra, E., Quiroga, J. J., & Monte, G. E. (2003, May). *Fuentes de Energía para Misiones Espaciales.* Paper presented at the Congreso Argentino De Tecnología Espacial, Neuquén, Argentina.

Flatow, F. S., Gal, C., & Hayes, E. J. (1984, April). 406 MHz ELTs/EPIRBs. In *Proceedings of the International Symposium on Satellite Aided Search and Rescue*, Toulouse, France.

Floyd, S. (2004). Limited Slow-Start for TCP with Large Congestion Windows. RFC 3742.

Floyd, S., Allman, M., Jain, A., & Sarolahti, P. (2006). Quick-Start for TCP and IP. Internet-Draft, draft-ietf-tsvwg-quickstart-02.

Forward, R. L. (1988). *Future magic.* New York: Avon Books.

Gabrynowicz, J.I. (1999). Expanding global remote sensing services: Three fundamental considerations. In *Proceedings of the Workshop on Space Law in the Twenty-first Century, UNISPACE III* (pp. 100–104). Vienna, Austria: United Nations.

Galileo Joint Undertaking. (2003). Business in satellite navigation: An overview of market developments and emerging applications. Retrieved March 2, 2007, from http://www.sbf.admin.ch/htm/services/publikationen/international/raumfahrt/business_in_satnav_en_march2003.pdf

Garono, J. R., McFall, K., Burke, J., & Sutherland, B. (2003) *Critical habitat on the lower Columbia River.* Geospatial Solutions. Retrieved February 15, 2007, from http://www.geospatial-online.com/geospatialsolutions/article/articleDetail.jsp?id=76575

Gaubatz, W. (1998). Reusable space transportation – The key infrastructure elements in opening the space frontier to the public. In *Paper presented at the 21st ISTS,* Sonic City, Japan. Retrieved March 3, 2007, from http://www.spacefuture.com/archive/reusable_space_transportation_the_key_infrastructure_element.shtml

Gaubert, A. (2006). Is there really any duplication in Europe's space activities? *Space Policy, 22,* 1-2.

Gayrard, J. (2003a, April 15-19). Satellite based systems for communication recovery in emergency situations. In *Proceedings of the 21st AIAA International Communications Satellite Systems Conference and Exhibit*, Yokohama, Japan.

Gayrard, J. (2003b, September 29-3). Roles of communication broadband satellites in public protection and disaster relief. In *Proceedings of the 54th IAC International Astronautical Congress*, Bremen.

Gayrard, J., & Blanc, F. (2004, May 9-12). Ka-band satellite system for public protection and disaster relief operations. In *Proceedings of the 22nd AIAA International Communications Satellite Systems Conference and Exhibit*, Monterey, CA.

Gehrman, H. W., Jr. (2003). Columbia accident investigation board report. Washington, D.C.: Government Printing Office.

Geoffrey, I.C. (2001). Researching the space tourism market. In *Paper presented at the Annual Conference of the Travel and Tourism Research Association.* Retrieved March 3, 2007, from http://www.spacefuture.com/archive/researching_the_space_tourism_market.shtml.

Gilmore, D. G. (1994). *Satellite thermal control handbook.* Washington, D.C.: The Aerospace Corporation Press.

Giordano, M., & Zhu, Z. (2004). *Water management in the Yellow River Basin: Background, current critical issues and future research needs.* Retrieved February 15, 2007, from http://www.iwmi.cgiar.org/Assessment/files/pdf/publications/ResearchReports/CARR3.pdf

GJU—GALILEO Joint Undertaking. (2005). *Business in satellite navigation: An overview of market developments and emerging applications* (2nd ed.). Brussels: GALILEO Joint Undertaking and the European Commission.

Global Climate Observing System (GCOS). (2003, October). *The Implementation Plan for the global observing system for the climate in support of the UN-FCCC.* Retrieved March 5, 2007, from http://www.wmo.int/web/gcos/gcoshome.html

Goehlich, R. A. (2003). *A representative program model for developing space tourism* (ISBN 3-936846-29-4). Berlin, Germany: Wissenschaftlicher Verlag Berlin.

Goehlich, R. A. (2005). A ticket pricing strategy for an oligopolistic space tourism market. *Space Policy Journal, 21*(4), 293-306. Amsterdam, The Netherlands: Elsevier Science.

Goehlich, R. A. (2006). *Spaceships—A reference guide to international reusable launch vehicle concepts from 1944 to the present* (ISBN 1-894959-50-7). Burlington, Canada: Apogee Books, CG Publishing.

Gonzalo, J., Martínez, S., & Martín, G. (2004, April 19-22). REMFIRESAT: Real time emergency management for forest fire services via satellite. In *Proceedings of the 2nd Symposium on Fire Economics, Planning and Policy*, Cordoba, Spain.

Goudy, P. H., King, J., & Kissel, F. (1984, April). Sarsat 406 MHz System technical evaluation. In *Proceedings of the International Symposium on Satellite Aided Search and Rescue*, Toulouse, France.

Government Accounting Office. (2002). *Major management challenges and program risks: National Aeronautics and Space Administration* (No. GAO-03-114). Washington, D.C.: General Accounting Office.

Government Accounting Office. (2003). *Space shuttle—Actions needed to better position NASA to sustain its workforce through retiremen*t (No. GAO-05-230) Washington, D.C.: General Accounting Office.

Grace, D., Daly, N., Tozer, T., Burr, A., & Pearce, D. (2001). Providing multimedia communications services from high altitude platforms. *International Journal on Satellite Communication, 19*, 559-580.

Group on Earth Observation (GEO). (2004, April). *From observation to action-achieving comprehensive, coordinated, and sustained Earth observations for the benefit of humankind, framework for a 10-year implementation plan (Framework Document),* as adopted by Earth Observation Summit, Tokyo, Japan.

Group on Earth Observation (GEO). (2005a). *Earth Observation Summit: Science and technology for sustainable development G8 action plan.* Retrieved March 5, 2007, from http://earthobservations.org/summit/g8_action.html

Group on Earth Observation (GEO). (2005b, February). *The global Earth observation system of systems (GEOSS) 10-year implementation plan (10-year implementation plan),* as endorsed by Earth Observation Summit III, Brussels, Belgium.

Group on Earth Observation (GEO). (2005c, February). *GEOSS 10-year implementation plan reference document* (Reference Document). Retrieved March 5, 2007, from http://www.earthobservations.org/doc/10-Year Plan Reference Document (GEO R 1000) .html

Group on Earth Observation (GEO). (2006a). *About GEO.* Retrieved March 5, 2007, from http://www.earthobservations.org/about/about_GEO.html

Group on Earth Observation (GEO). (2006b). *About GEO: Committees & working groups.* Retrieved March 5, 2007, from http://www.earthobservations.org/about/entities/about_cmtes_wgs.html

Group on Earth Observation (GEO). (2006c). *Group on Earth observation work plan for 2006 (Version-4).* Retrieved March 5, 2007, from http://www.earthobservations.org/docs/GEO-II/GEO_0204-4_060330.pdf

Guha, P. K. (2003). *Interpretation of remote sensing data: Remote sensing for the beginner* (p. 70). New Delhi: Affiliated East—West Press.

Hartog, J., Carbonell, A., & Jonker, N. (2000). *On a simple survey measure of individual risk aversion* (pp. 1-6) (CESifo Working Paper No. 363). University of Amsterdam, The Netherlands.

Holton, J.R. (Ed.). (1992). *An introduction to dynamic meteorology* (3rd ed.) International Geophysics Series (vol. 48). San Diego, CA: Academic Press.

Hori, H. (1990). Macro-Engineering: A view from Japan. *Technology in Society, 12*(1), 45-63.

Horwitch, M. (1990). From unitary to distributed objectives. *Technology in Society, 12*(2), 173-195.

Houghton, J.(2004). *Why should we be concerned? Global warming, the complete briefing* (p. 200). Cambridge, UK: Cambridge University Press.

Hudgins, E. L. (2006). *Space policy and space tourism.* Retrieved March 3, 2007, from http://www.cato.org/testimony/ct-eh062601.html. Washington, DC: CATO Institute.

Hull, A. (2004, June 30). Moon and stars align for performance artist: Laurie Anderson accepts art commission from NASA. *The Washington Post*, p. A19. Retrieved March 5, 2007, from http://www.washingtonpost.com/wp-dyn/articles/A15916-2004Jun29.html

Hunt, C. D., & van Pelt, M. O. (2004). Comparing NASA and ESA cost estimating methods for human missions to Mars. In *Paper presented at the ISPA*, Frascati, Italy.

IBRD. (2006). International 406 MHz Beacon Registration Database—IBRD Web site. Retrieved February 22, 2007, from www.406registration.com

Inspector General's Office. (2001). *Faster, better, cheaper: Policy, strategic planning, and human resource alignment* (Audit Rep. No. IG-01-009).

Integrated Global Observing Strategy (IGOS). (2002). *IGOS brochure.* Retrieved March 5, 2007, from http://ioc.unesco.org/

International Association of Astronomical Artists. *The History of the IAAA.*Retrieved March 5, 2007, from: http://www.iaaa.org/history.html

International Association of Astronomical Artists. *The IAAA Manifesto.* Retrieved March 5, 2007, from http://www.iaaa.org/history.html

Internet-World-Stats (Artist). (2006). *Internet users and population statistics for Africa.*

ISU (2000). *Space tourism: From dream to reality.* Chile: ISU SSP.

ISU (2001). *Space tourism: A literature review.* Strasbourg, France: ISU MSS.

ITSO. (2002). *Global broadband satellite infrastructure initiative.* Retrieved February 22, 2007, from http://www.itso.int/php_docs/BroadbandInitiative-FINAL-english-17Dec02.htm

ITU. (2002). *Handbook on satellite communications.* New York: Wiley.

ITU. (2005). *WSIS outcome document.* Retrieved February 22, 2007, from http://www.itu.int/wsis/outcome/booklet.pdf

Jacobson, V., Braden, R., & Borman, D. (1992). TCP Extensions for High Performance. RFC 1323.

Janos, L., & Rich, B. R. (1996), *Skunk works: A personal memoir of my years at Lockheed.* Back Bay Books.

Japan Aerospace Exploration Agency (JAXA). (2005, November). Greenhouse gases monitoring from the GOSAT satellite. In *Proceedings of the Microwave Workshops and Exhibition*, Yokohama, Japan.

Jason, S., Pulinets, S., Silva Curiel, A., & Sweeting, M. (2002, August). Earthquake forecast science research with small satellite. In *Proceedings of the USU 13th Annual Small Satellites Conference* (SSC02-IX-5), Utah State University, Logan, UT.

Jensen, J. R. (2004). *Remote sensing the urban landscape: Remote sensing of the environment* (p. 460). New Delhi: Pearson Education (Singapore).

Johnson, C., & Smith, M. (1990). *Kelly: More than my share of it all.* Smithsonian Institute Press.

Johnson, R. D., & Hobrow, C. (Eds.). (1977). *Space settlement: A design study.* Washington D.C.: NASA.

Joseph, G. (2005). *Applications of remote sensing for Earth resources management: Fundamentals of remote sensing* (p. 389). Hyderabad:University Press.

Kalpakis, K., Markowitz, K., & Jamar, S. (n.d.). *Environmental legal information systems (ERIS): A view from space: Digital Earth applications in environmental law andresource management.* Retrieved March 5, 2007, from http://earthpace.com/conference/confsumindex.html

Karapantazis, S., & Pavlidou, F. (2005). Broadband communications via high-altitude platforms: A survey. *IEEE Communication Surveys & Tutorials*, *7*(1), 1-35.

Kaushik, A., & Kaushik, C. P. (2006). *Environmental studies—A multidisciplinary subject: Perspectives in environmental studies* (p. 3). New Delhi: New Age International Publishers.

Keightley, R. J. (1987, May). Simulation of the Canadian GOES Signal Processor. *IEEE Transactions on Aerospace and Electronic Systems, AES-23*(3), 361-370.

Kelly Space and Technology. (2001). *Space transportation market demand 2010-2030.* Retrieved March 3, 2007, from http://www.kellyspace.com. San Bernardino, USA.

Kelly, T. (2005). *Twenty years of measuring the missing link.* International Telecomunication Union. Retrieved February 22, 2007, from http://www.itu.int/osg/spu/sfo/missinglink/index.html

Kerr, R. A. (2004). NASA reigning in PI-led planetary missions. *Science, 305*(5683), 467.

Kim, H. S. (2003). *Sustainable development and the South-to-North Water Transfer Project in China.* Retrieved February 15, 2007, from http://fred.ccsu.edu:8000/archive/00000054/01/etd-2003-12.pdf

Kimball, L.A. (1996). Treaty implementation: Scientific and technical advice enters a new stage. *Studies in transnational legal policy, 28.*

King, J. V. (2006b, June). Status of Canada's prototype MEOLUT development. In *Proceedings of the CS/JC-20 Meeting*, Montreal, Canada.

King, J. V., Hayes, E. J., & Jutras, G. J. (1983, September). Results of Cospas-Sarsat system performance testing (Paper No. 83074). In *Proceedings of the 1983 International Electrical, Electronics Conference and Exhibition- IEEC&E '83* (pp. 180-183). IEEE Catalogue No. 83CH1955-4, Toronto, Canada

King, J.V. (2006a, April). Coverage of MEOSAR compared to LEOSAR: Preliminary study. In *Proceedings of the CS/EWG-2 Meeting*, Montreal, Canada.

Klüver, B. (2000). *Experiments in art and technology (E.A.T.).* Archive of published documents. Retrieved March 5, 2007, from http://www.fondation-langlois.org/html/e/page.php?NumPage=306

Kolovos, A. (2002). Why Europe needs space as part of its security and defence policy. *Space Policy, 18,* 257-261.

Kraak, M.-J., & Orheling, F. (2004). *GIS applications: Which map to use? In Cartography: Visualization of geospatial data* (p. 51). Delhi: Pearson Education (Singapore).

Kranz, G. (2000). *Failure is not an option: Mission control from Mercury to Apollo 13 and beyond.* New York: Simon and Shuster.

Krige, J., & Russo, A. (2000a). The story of ESRO and ELDO 1958-1973. *A history of the European Space Agency 1958-1987* (vol. I). Retrieved March 2, 2007, from http://www.esa.int/esapub/sp/sp1235/sp1235v1web.pdf

Krige, J., & Russo, A. (2000b).The story of ESA 1973-1987. *A history of the European Space Agency 1958-1987* (vol. II). Retrieved March 2, 2007, from http://www.esa.int/esapub/sp/sp1235/sp1235v1web.pdf

Kuriyama, I. (2002). *The contribution of Earth observation in the implementation and compliance of multilateral environmental agreements.* Unpublished master's thesis, International Space University, Strasbourg, France.

Kuriyama, I. (2005). Supporting multilateral environmental agreement with satellite Earth observation. *Space Policy, 21*(2), 151-160.

Kyoto Protocol to the United Nations Framework Convention on Climate Change (Kyoto Protocol), FCCC/CP/1997/L.7/ Add.1, December 11, 1997

Lacerte, S. 9 evenings and experiments in art and technology: A gap to fill in art history's recent chronicles. *la fondation Daniel Langlois*, (Introduction, p. 1/6). Retrieved March 5, 2007, from http://www.fondation-langlois.org/html/e/page.php?NumPage=1716

Lafferranderie, G. (1993). Espace juridique et juridiction de l'espace. In A. Esterle (Ed.), *L'Homme dans l'espace*. Paris: P.U.F.

Lainas, T. (2005, September 18). Mines on Mars. *To Vima, 107.*

LandSat. (2006). Landsat Project—U.S. Department of the Interior. Retrieved February 22, 2007, from http://landsat.usgs.gov

Lassig, J., Keil, G., Fernández, R., & Quiroga, J. J. (1999, October). *Space education at the National University of Comahue: A means for the integration with multiple proposes.* Paper presented at the 50th International Astronautical Congress, Amsterdam, The Netherlands.

Lassig, J., Quiroga, J. J., Keil, G., Fernández, R., & Jurasics, A. (2003). *Pehuensat. elementos de Diseño de pequeños satélites.* Neuquén, Argentina: Artes Gráficas Limay.

Lesgards, R. (1998). *Conquête spatiale et démocratie.* Paris: Presses de Science Po.

Levesque, D., Hodgkins, K., & Drover, P. (1984, April). The development of an institutional structure for the future global satellite system for search and rescue. In *Proceedings of the International Symposium on Satellite Aided Search and Rescue*, Toulouse, France.

Levin, H. J. (1971). *The invisible resource: Use and regulation of the radio spectrum.* Baltimore, MD: John Hopkins Press for Resources for the Future.

Lewis, J. (1993, June). Robert Irwin: A room with a view. *Harper's Bazaar: A room with a view.* Retrieved March 5, 2007, from http://www.airic.com/articles/roomview.html

Lewis, J. S. (1997). *Mining the sky: The untold riches from the asteroids, comets and planets.* New York: Basic Books.

Lewis, R., Kennedy, M., Ghashghai, E., & Bitko, G. (2005). *Building a multinational global navigation satellite system: An initial look.* RAND Corporation.

Li, J. (2002). *Application of remote sensing to water resources management in arid regions of China.* Retrieved February 15, 2007, from http://www.gisdevelopment.net/aars/acrs/2002/pos3/283.pdf

Lindstrom, G., & Gasparini, G. (2003). *The Galileo satellite navigation system and its security implications* (Occasional Paper No. 44). Paris: European Union Institute for Security Studies.

Livingston, D. (2002). $pace: The final financial frontier? In R. Zubrin & F. Crossman (Eds.), *On to Mars:*

Colonizing a New World (pp.59-63). Apogee Books Space Series.

Livingston, D. M. (2000). *Space tourism and RLVs: You can't have one without the other!* Tiburon, USA: Livingston Business Solutions.

Lo, R., & Paatsch, A. (1998). *Ermittlung der durch den Weltraumtourismus verursachten Emissionen*. Berlin, Germany: IRTS.

Lo, R., & Paatsch, A. (1999). *Raumfahrtinduzierte Belastungen der Atmosphäre unter besonderer Berücksichtigung des Weltraumtourismus: Berechnung des quasistationären Anfangszustandes der Abgaswolke entlang der Aufstiegsbahn von Raketen*. Berlin, Germany: IRTS.

Lorance, R. B., & Wendling, R. V. (1999). Basic techniques for analyzing and presentation of cost risk analysis. In *Paper presented at the Association for the Advancement of Cost Engineering International Annual Conference*, Denver, USA.

Macauley, K. M. (2004, June). *Is the vision of the Earth observation summit realizable?* A case study prepared for the American Meteorological Society 2004 summer policy colloquium.

Márquez, F., Cajarabilla, M., Quiroga, J. J., Arias, A., & Pelayes, N. (2005, May). *Nanopehuensat-1: Encendido Automático de Satélite por Detección de Aceleración, con Eliminación de Ruido*. Paper presented at the Congreso Argentino de Tecnología Espacial, Córdoba, Argentina.

Martin, M. W., & Schinzinger, R. (1996). *Ethics in engineering*. New York: McGraw-Hill.

Matar, M. A. (2005, June 9-11). Space-based facilities for environmental disaster monitoring, assessment and management. In *Proceedings of the 2nd International Conference on Recent Advances in Space Technologies (RAST)*, Istanbul, Turkey.

McClain, M. E. (Ed.). (2002). *The ecohydrology of South American rivers and wetlands* (IAHS Special Publication No. 6). Wallingford, Oxfordshire, UK: International Association of Hydrological Sciences.

McCullough, S. (2002). Financing a manned mission to Mars? In R. Zubrin & F. Grossman (Eds.), *On to Mars, Colonizing a New World*. Apogee Books Space Series.

McCurdy, H. E. (2003). *Faster, better, cheaper: Low-cost innovation in the U.S. space program*. Baltimore, MD: The Johns Hopkins University Press (See also the other books in The Johns Hopkins University Press Series on the History of NASA).

McGunigal, T. E., McKinnon, D. J., Brachet, G., & Zurabov, Y. (1984, April). Cospas/Sarsat- An overview. In *Proceedings of the International Symposium on Satellite Aided Search and Rescue*, Toulouse, France.

Meckling, W. H. (1968). Management of the frequency spectrum. *Washington University Law Quarterly*, 26-34.

Mehta, A. (2006, May 12). Cost benefit analysis for SARSAT program. In *Proceedings of the Beacon Manufacturers Workshop*, Newport Beach, CA, USA.

Menzel, O., Wagner, D., & Miloucheva, I. (2006). Support of mobile receivers with unidirectional links in MIPv6. Internet-Draft, draft-menzel-udlr-mipv6-00.

Mingolo, F. G., Moya, C. D., & Simone, D. D. (2005, May). *Nanopehuensat-1. Análisis y Diseño de Antenas para Satélites Parásitos*. Paper presented at the Congreso Argentino de Tecnología Espacial, Córdoba, Argentina.

Miura, R., & Suzuki, M. (2003). Preliminary flight test program on telecom and broadcasting using high altitude platform stations. *International Journal on Wireless Personal Communications*, *24*(2), 341-61.

Montpetit, M.-J., Fairhurst, G., Clausen, H., Collini-Nocker, B., & Linder, H. (2005). A Framework for Transmission of IP Datagrams over MPEG-2 Networks. RFC 4259.

Montreal Protocol on Substances that Deplete the Ozone Layer (Montreal Protocol), September 16, 1987, 1522 U.N.T.S. 3

Moore, P. G. (1983). *The business of risk* (ISBN 0-521-28497-X) (pp. 97-115). Cambridge, United Kingdom: Cambridge University Press.

Moroney, S. (2002, April). *Satellite and VSAT: Innovative uses for rural telephony and Internet development.* Paper presented at the Expert Workshop on Improving IP Connectivity in LDCs, Geneva.

Morris, J., & Morring, F. (2006, June). Forewarned: Griffin cautions that industry inefficiency could damage U.S. space program. *Aviation Week and Space Technology.*

Murray, A. (2004, March 8). *The Lisbon Scorecard IV: The status of economic reform in the enlarging EU* (Press release). Centre for European Reform (CER).

Murray-Darling Basin Commission (2004). Retrieved February 15, 2007, from http://www.mdbc.gov.au

NASA. (2001). *General policies and procedures, process, pricing and conditions for sale.*

NASA. (2001). *The lure of hematite.* Retrieved March 2, 2007, from http://science.nasa.gov/headlines/y2001/ast28mar_1.htm

NASA. (2003). *Taking business to the stars: Nasa's commercial space centers.* Retrieved March 3, 2007, from http://spd.nasa.gov/csc.html

NASA. (2004). *The vision of space exploration.* Retrieved March 3, 2007, from www.nasa.gov

NASA. (2005). *Prospecting for lunar water.* Retrieved March 2, 2007, from http://science.nasa.gov/headlines/y2005/28apr_lro.htm?list113731

National Aeronautics and Space Administration. *A brief history of NASA: Launching NASA.* Retrieved March 2, 2007, from http://history.nasa.gov/factsheet.htm

National Research Council. (2005). *Principal-investigator-led mission in the space sciences.* Washington D.C.: National Academies Press.

Naval Historical Center. *The Navy Combat Art Program.* Retrieved March 5, 2007, from http://www.history.navy.mil/branches/org6-4.htm

Nelson, K. (2001, February). NASA Ames Center looks at problem of drilling on Mars. *Petroleum News.* Retrieved March 2, 2007, from http://www.petroleumnews.com/pnarchpop/010228-49.html

O'Neill, G. K. et. al. (1979). *Space resources and space settlements.* Washington, D.C.: NASA.

OECD. (2005). Financing ICTs for development efforts of DAC members. Organization for Economic Cooperation and Development, Development Assistance Committee.

Office of the Inspector General. (2001, March 31). *Faster, better, cheaper: Policy, strategic planning, and human resource alignment* (No. IG-01-009).

Oh, E. S. (2001, July 5-7). The international efforts in emergency telecommunications: An overview and analysis. In *Proceedings of the 6th Asia-Pacific Regional Conference of International Telecommunications Society*, Hong Kong, China.

Oliver, J. E., & Hidore, J. J. (2002). *The warming of planet Earth: Climatology* (p. 300). Delhi: Pearson Education (Singapore).

Oliver, M. (2003). River Murray: The big picture. River Murray Urban Users Local Action Planning Committee, 2003. Retrieved February 15, 2007, from http://www.murrayusers.sa.gov.au

ONU. (1998, May). *Misiones con Pequeños Satélites.* Documento preparado por la Oficina de Asuntos del Espacio Ultraterrestre de la Secretaría, para la Tercera Conferencia de las Naciones Unidas Sobre la Exploración y Utilización del Espacio Ultraterrestre con Fines Pacíficos (UNISPACE III – en junio 99). A/CONF.184/BP/9 del 26 de mayo de 1998.

Organisation for Economic Co-operation and Development. (2006). *Global forum on space economics: Descriptive overview.* Retrieved March 2, 2007, from http://www.oecd.org/dataoecd/4/17/36468669.pdf

Pace, P., Aloi, G., De Rango, F., Natalizio, E., Molinaro, A., & Marano, S. (2004, May 17-19). An integrated satellite-HAP-terrestrial system architecture: Resources allocation and traffic management issues. In *Proceedings of the 59th IEEE Vehicular Technology Conference—VTC Spring*, Milan, Italy.

Panda, B. C.(2005). *Principles of geographic information system (GIS): Remote sensing principles and applications* (p. 208). New Delhi: Viva Books.

Partnership. (2002). *Securing the Linchpin ICT for teaching, learning and research*. The Partnership for Higher Education in Africa. Retrieved February 22, 2007, from http://www.foundation-partnership.org/linchpin/index.php

Pasco, X. (2003). A new role for a new millennium? The changing nature of space activities. *Space Policy, 19*, 15-22.

Penders, M., & Ronald, D. (n.d.). *Environmental Protection International: The hanging chads of international environmental law: Towards the better use of technology and information management to detect violations of international agreements, combat transnational crime, and reform environmental law and trade, Publication #1*. Retrieved March 6, 2007, from http://www.epinetwork.com/publication_1.html

Petrovski, I. (2003, June). QZSS—Japan's new integrated communication and positioning service for mobile users. *GPS World, 14*(6), 24-29.

pietronigro.com. (2004, April 23). *European Space Agency invites U.S. artists to present new plans for art in space*. Retrieved March 5, 2007, from http://www.pietronigro.com/news_2004-04-23.htm

Plattard, S. (2005). Going further: Propositions to build up a European Space Policy. *Eurofuture, European Space Policy Institute*, 16-19.

Postel, J. (Ed.). (1981). Internet Protocol. RFC 791.

Pournelle, J. (Ed.). (1979). *The endless frontier* (p. 14). New York: Ace Books.

Principles relating to remote sensing of the Earth from outer space (Remote Sensing [RS] Principles), G.A. Res. 41/65, 42 UN GAOR Annex (95th plenary meeting) at 2 UN Doc. A/RES/41/65 (1987).

Quiroga, J. J., Fernández R., Estévez S., Mare J., Odello L., & Sierra, E. (2000b, October). *Ensayo de Baterías y Paneles Solares Estándar para Ser Utilizados en Nano-satélites*. Paper presented at the Congreso Argentino De Tecnología Espacial, Buenos Aires, Argentina.

Quiroga, J. J., Fernández, R., Keil, G., Jurasic, A., Sierra, E. A., De León, P., & Álvarez, A. (2000a, October). *Small satellites Pehuensat Program*. Paper presented at the 51st Internacional Astronautical Congress, Rio De Janeiro, Brazil.

Quiroga, J.J., Lassig J., Keil, G., Monte, G.E., Fernández, R., Simone, D. et al. (2004, October). *Pehuensat-1: Development Status*. Paper presented at the 55th International Astronautical Federation Congress, Vancouver, Canada.

Raja Rao, K. N. (2005). *An overview of space and satellite: Fundamentals of satellite communication* (p. 1). New Delhi: Prentice Hall of India.

Rammos, E., Eldridge, N., Wu, S., Brownsword, P., Knight, T., Winder, M. et al. (2000, April 10-12). REMSAT: A demonstration of the use of satellite techniques to help in the battle against forest fires. In *Proceedings of the International Symposium GEOMARK 2000*, France.

Rao, U. R. (1996). *Space and disaster management: Space technology for sustainable development* (p. 387). New Delhi: Tata McGraw-Hill Publishing.

Reding, V. (2005). *Challenges for TV—convergence—IPTV*. Paper presented at the 8th EIM Conference, Vilnius.

Renner, R. C., & Kozminchuk, B. W. (1984, April). 121.5/243 MHz ELT identification. In *Proceedings of the International Symposium on Satellite Aided Search and Rescue*, Toulouse, France.

RESPOND. (2004). RESPOND alliance of European and international organizations. Retrieved February 22, 2007, from http://respond-int.org/Respond/

Revnivykh, S. (2006). GLONASS: Policy, status and modernization plan. In *Proceedings of the Munich Satellite Navigation Summit*, Munich.

Rio Declaration on Environment and Development (Rio Declaration).(1992, June 13). U.N. Doc/A/CONF.151/26.

Roberts, L. (1996). Space business incentives: It's time to act. *Ad Astra, 8*(2).

Rogalski, V., & Krupen, V. (1984, April). The Cospas System tests technical results. In *Proceedings of the International Symposium on Satellite Aided Search and Rescue*, Toulouse, France.

Rosaviakosmos. (2001). *ISS commercial program rosaviakosmos policies and procedures.*

Rosenqvist, A. (Ed.). (2003, July). Support to multinational environmental conventions and terrestrial carbon cycle science by ALOS and ADEOS-II—the Kyoto & Carbon Initiative. In *Proceedings of the International Geoscience and Remote Sensing Symposium (IGARSS'03),* Toulouse, France.

Rosenqvist, A., Imhoff, M., Milne, A., & Dobson, C. (Ed.). (1999, October). *Remote sensing and Kyoto Protocol : A review of available and future technology for monitoring treaty compliance* (Workshop report). Michigan.

Salomon, J. J. (2000). *Survivre à la science. Une certaine idée du futur.* Paris: Albin Michel.

Sands, P. (1995). *Principles of international environmental law I: Frameworks, standards and implementation.* Manchester, United Kingdom: Manchester University Press.

Schmidt, S., & Zubrin, R. (Eds.). (1996). *Islands in the sky: Bold new ideas for colonizing space.* Wiley Popular Science.

Schulzrinne, H., Casner, S., Frederick, R., & Jacobson, V. (2003). RTP: A Transport Protocol for Real-Time Applications, RFC 3550.

Scientific Committee on Antarctic Research. *The Antarctic Treaty System: An introduction.* Retrieved March 2, 2007, from http://www.scar.org/treaty

Scot, G. P. (2000, April 3-6). The contribution of VEGETATION/ SPOT 4 products to remote sensing applications for food security, early warning and environmental monitoring in the IGAD sub-region. In *Proceedings of the Vegetation 2000 Conference*, Lake Maggiore, Italy.

Sedell, J. R., Bennett, K., Steedman, R., Foster, N., Ortuno, V., Campbell, S., & Achouri M. (2002, October). Integrated watershed management issues in North America. In *Proceedings of the 21st Session of the North American Forestry Commission, Food and Agriculture Organization of the United Nations*, Hawaii, USA.

Shacham, A., Monsour, R., Pereira, R., & Thomas, M. (1998). IP Payload Compression Protocol (IPComp). RFC 2393.

Shiklomanov, I. A. (2000). *Appraisal and assessment of world water resources. Water International, 274*(1), 11-32.

Smith, L. J., & Hörl, K.-U. (2004). Institutional challenges for space activities in Europe. In *Proceedings of the 47th International Institute of Space Law Colloquium*, Vancouver. IAC-04-IAA.5.13.1.04.

Smith, M. (2005). *U.S. space programs: Civilian, military and commercial* (CRS Issue Brief for Congress, Code IB92011). The Library of Congress: Congressional Research Service (CRS).

Smitherman, D. V. (2000). *Space elevators—An advanced Earth-space infrastructure for the new millennium.* Huntsville, AL: Marshall Space Flight Center.

Smithsonian National Air and Space Museum.(2004). *Generous friends: Building an art collection for the National Air and Space Museum.* Flight and the Arts, Gallery 211, New Exhibition, Retrieved March 5, 2007, from: http://www.nasm.si.edu/exhibitions/gal211/gal211.cfm

Space Art Home Page. *Space educator's handbook: The artist looks at space.* Retrieved March 5, 2007, from http://www1.jsc.nasa.gov/er/seh/spaceart.html

Space Daily. (2003). *Government space budgets to continue growth.* Retrieved March 3, 2007, from http://www.spacedaily.com/news/satellite-biz-03zzzl.html

Space Studies Board. (2004). *Steps to facilitate principal-investigator-led Earth science missions.* Washington, D.C.: National Academies Press.

Space Studies Board. (2005). *Earth science and applications from space: Urgent needs and opportunities to serve the nation.* Washington, D.C.: National Academies Press.

Space. (2005). Mir milestones. Retrieved March 3, 2007, from http://www.space.com/missionlaunches/missions/mir/_landmarks/_010307.html

Spear, T. (2000). *NASA faster, better, cheaper task final report.* Washington, D.C.: Government Printing Office.

Steiner, R., Tirivaya, N., Jensen, M., & Gakio, K. (2004). *Connectivity in African tertiary institutions.* Retrieved February 22, 2007, from http://www.ifc.org/ifcext/bsn.nsf/AttachmentsByTitle/ATICS_2004_Report/$FILE/ATICS+2004+Report.pdf

Stevinson, H. T., & Baker, D. A. (1971). Locating people in high latitudes (Laboratory Tech. Rep. No. LTR-FR-32 and–32A). National Aeronautical Establishment. Ottawa, Canada: National Research Council of Canada.

STREAM Team (2004). *Space technologies for the research of effective water management* (ISU Team Project Report). Adelaide, Australia: International Space University.

Sykes, A. (1990). Macro projects. *Technology in Society, 12*(2), 157-172.

Tascione, T. (1988). *Introduction to the space environment.* Malabar, USA: Orbit Book Company.

Thakur, I. S. (2006). *Introduction: Environmental biotechnology* (p.9). New Delhi: I.K. International.

The Socioeconomic Data and Application Center, Center for International Earth Science Information Network (CIESIN), Columbia University. (2000, December). *Remote sensing and environmental treaties: Building more effective linkages* (Report of a Workshop). Washington, DC, USA: Author.

Thornton, E.A. (1996). *Thermal structures for Aerospace applications.* Washington, D.C.: AIAA (American Institute of Aeronautics & Astronautics).

Tkatchova, S. (2005). *From space exploration to commercialization.* Unpublished doctoral dissertation, TU Delft, Delft.

Tkatchova, S., & van Pelt, M. (2004). ISS commercialization paths the way for tomorrow's future commercial missions. In *Final Papers 55th IAC 2004* (p. 567), Vancouver, Canada.

Toumi, A. (2004). *A global broadband satellite system to serve development.* International Telecomunication Union. Retrieved February 22, 2007, 2006, from http://www.itu.int/newsarchive/wrc03/documents/broadband.html

Tower, C. (1983, April). Rescue from outer space. *Reader's Digest.* 49-54.

Townsend, A. M., & Moss, M. L. (2005). *Telecommunications infrastructure in disasters: Preparing cities for crisis communications* (Tech. Rep.) Robert F. Wagner Graduate School of Public Service, New York University.

Tozer, T., & Grace, D. (2001). High-altitude platforms for wireless communications. *Electronics and Communication Engineering Journal, 13*(3), 127-137.

Treaty on Principles Governing the Activities of States in the Exploration and Use of Outer Space, including the Moon and Other Celestial Bodies (Outer Space Treaty), January, 27, 1967, 610 U.N.T.S. 205

Tsujino, T. (2005, July). Effectiveness of the Quasi-Zenith satellite system. *Ubiquitous Positioning, Science & Technology Trends, Quarterly Review* (16). Japan: National Institute of Science and Technology Policy.

U.S. Congress. (2006). *House passes legislation setting space tourism guidelines.* Retrieved March 3, 2007, from http://www.space.com/news/tourism_congress_040305.html. New York, USA.

U.S.-European Collaboration in Space Science. (1998). European Science Foundation (National Research Council Joint Report). Washington D.C.: National Academy Press. ISBN 0-309-05984-4.

UNESCAP—Report. (2002). *Towards a policy framework for integrating space technology applications for sustainable development on the information superhighway.*

United Kingdom Office of Science and Innovation. *FP7 overview.* Retrieved March 2, 2007, from http://fp6uk.ost.gov.uk/fp7/page.aspx?sp=2572

United Nations Committee on the Peaceful Uses of Outer Space. Retrieved March 2, 2007, from http://www.unoosa.org/oosa/COPUOS/copuos.html

United Nations Convention to Combat Desertification in Countries Experiencing Serious Drought and /or Desertification, Particularly in Africa (UNCCD), October 14, 1994, 1954 U.N.T.S.3

United Nations Environment Programme, Division of Early Warning and Assessment, Global Resource Information Database (UNEP-GRID)—North America. (n.d.) *Satellite systems for land cover monitoring: Navigating through the maze* [Brochure].

United Nations Environment Programme, Division of Environmental Policy Implementation (UNEP-DEPI). (2002). *Compliance and enforcement.* Retrieved March 6, 2007, from http://www.unep.org/DEPI/Compliance-and-Enforcement/

United Nations Environmental Programme (UNEP). (2001a). *The Guidelines on Compliance with and Enforcement of Multilateral Environmental Agreements*, adopted by the 7th special sessions of the Governing Council of United Nations Environmental Programme/Global Ministerial environment Forum in its decision SS. VII/4 of February 15, 2002 in UNEP(DEPI)/MEAs/WG.1/3, Annex II.

United Nations Environmental Programme (UNEP). (2001b). Partow, H. *The Mesopotamian Marshlands: Demise of an Ecosystem.* Nairobi, Kenya: Division of Early Warning and Assessment, United Nations Environment Programme.

United Nations Framework Convention on Climate Change (UNFCCC), May 9, 1992, 1771 U.N.T.S. 107

United Nations Framework Convention on Climate Change (UNFCCC), the Conference of the Parties (COP). (2003). *Global observing systems for climate.* Decision 11/CP.9 in FCCC/CP/2003/6/Add.1

United Nations Framework Convention on Climate Change (UNFCCC), the Subsidiary Bodies for Scientific and Technological Advice (SBSTA). (2005a). *Research needs relating to the convention.* UNFCCC/SBSTA/2005/4. P.14, para 71-83.

United Nations Framework Convention on Climate Change (UNFCCC), the Subsidiary Bodies for Scientific and Technological Advice (SBSTA). (2005b). *Research and systematic observation.* Draft Conclusions proposed by the Chair. UNFCCC/SBSTA/2005/L.17

United Nations Office of Outer Space Affairs (UNOOSA). (1998, May). *Management of Earth resources* (Background paper 3, A/CONF.184/BP/3).

United Nations University. (1999, July). *Interlinkages: Synergies and coordination between multilateral environmental agreements* [Report of a conference]. Tokyo, Japan: Author

United Nations. (1999). *Report of the Third United Nations Conference on the Exploration and Peaceful Use of Outer Space (UNISPACE III)* (A/CONF.184/6), Vienna, Austria: Author

US Coast Guard Navigation Center. (2001). *GPS standard positioning service (SPS) performance standard final.* Retrieved February 22, 2007, from www.navcen.uscg.gov/gps/geninfo/2001SPSPerformanceStandardFINAL.pdf

US DoT (2002, January 29). Treatment of value of life and injuries. U.S. Department of Transportation's Inspector General Memorandum.

US Environmental Protection Agency (2001, November). American heritage rivers progress report.

VEGETATION. (2006). VEGETATION programme. Retrieved February 22, 2007, from http://vegetation.cnes.fr

Verheugen, G. (2005, February 17). Speech on the importance of space. In *Proceedings of the International Space Conference*, Brussels. Retrieved March 2, 2007, from http://europa.eu.int/rapid/pressReleasesAction.do?reference=SPEECH/05/99&format=HTML

Vishnevsky, V. M., Tereshchenko, B. N., & Lyakhov, A. I. (2003, October 19-22). Wireless communication networks using HAPs on the base of tethered balloons. In *Proceedings of the 6th International Symposium on Wireless Personal Multimedia Communications WPMC*, Yokosuka, Japan.

Von der Dunk, F. (2003). Towards one captain on the European spaceship: Why the EU should join ESA. *Space Policy, 19,* 83-86.

Von Kries, W. (2003). Which future for Europe's space agencies? *Space Policy, 19,* 157-161.

VTT, GAMMA Remote Sensing Research and Consulting AG, European Forest Institute, & Stra Enso Forest Consulting. (2003, May). *Treaty enforcement services using Earth observation (TESEO) theme carbon—KYOTO Protocol to the UNIFCCC* (Final Report, ESA/ESRIN Contract No. 15557/01/I-LG). Espoo, Finland.

Waldrop, M. M. (1983). The commercialization of space. *Science, 1354.*

Weiss, C., Jr. (1988). Sustainable macro projects. *Technology in Society, 10*(1), 375-384.

Werstiuk, H. L. (1983, September). Cospas-Sarsat system overview (Paper No. 83071). In *Proceedings of the 1983 International Electrical, Electronics Conference and Exhibition—IEEC&E '83* (pp.168-171). IEEE Catalogue No. 83CH1955-4, Toronto, Canada.

Werstiuk, H. L., Ludwig, D., Trudell, B. J., & Selivanov, A. S. (1984, April). The Cospas-Sarsat system. In *Proceedings of the International Symposium on Satellite Aided Search and Rescue*, Toulouse, France.

Wiesmann, A., Wegmuller, U., Haeberlin, Y., Retiere, A., Senegas, O., Strozzi, T., & Werner, C. (2004, September 20-24). SAR based products for the implementation of humanitarian aid and development assistance projects within the UNOSAT project. In *Proceedings of the IEEE International Geoscience and Remote Sensing Symposium IGARSS*, Anchorage, AK.

Winter, A. E. (1978). Assessment of potential ground stations. Communications Research Centre Memorandum, Ottawa, Canada.

Wollersheim, M. (1999). Considerations towards the legal framework of space tourism. In *Paper presented at the 2nd International Symposium on Space Tourism*, Bremen, Germany. Retrieved March 3, 2007, from http://www.spacefuture.com/archive/considerations_towards_the_legal_ framework_of_space_tourism.shtml

Wood, L. (2005). Testing a router onboard a satellite in space. In *Proceedings of the End-to-End Challenges of Broadband via Satellite* (pp. 83-85), IEE Savoy Place, London, England.

World Trade Organisation. (1994, April 15). Agreement on Government Procurement. Official Journal of the European Communities 1994 L336/273.

X Prize Foundation. (2006). *Ansari X Prize.* Retrieved March 3, 2007, from http://www.xprizefoundation.com. St. Louis, USA.

Young, G. J., Dooge, J. C. I., & Rodda, J.C. (1994). *Global water resource issues.* UK: Cambridge University Press.

Young, O. R. (1986). International regimes: Toward a new theory of institutions. *World Politics, 39*(1), 105.

Young, O. R. (1989). *International cooperation: Building regimes for natural resources and the environment* (p. 32). New York: Cornell University Press.

Zhang, Y. (Ed.). (2003). *Internetworking and computing over satellite networks.* Boston: Kluwer Academic.

Zogby International. (2002). *Futron/Zogby public space travel poll.* Retrieved March 3, 2007, from http://www.zogby.com. New York, USA.

Zubrin, R. M. (1997). *The case for Mars.* USA: Free Press.

Zubrin, R. M., Baker, D. A., & Gwynne, O. (1991). *Mars direct: A Simple, Robust, and cost effective architecture for the Space Exploration Initiative* (AIAA-91-0328).

Сафронов, М. С., & Якименко, И.Р. (2006). НекаторыIе перспективыI космической деятельности на пороге освоения ресурсов ЛуныI и дальнего космоса. Федеральное Космическое Агенство. Retrieved March 3, 2007, from http://www.roscosmos.ru/luna.html

About the Contributors

Phillip Olla is a professor at the School of Business at Madonna University in Michigan. His research interests include knowledge management, space science, mobile telecommunication, and health informatics. Over the last decade, in addition to university level teaching, Dr. Olla has worked as an independent information technology consultant. His experience is primarily in the Internet and telecommunication and space industry. Dr. Olla has worked on a wide variety of pioneering projects in conjunction with mobile network operators and mobile service providers, including British Telecom, Hutcinson 3G, T-Mobile, and IBM Global Services. He received his PhD from the Department of Information Systems and Computing at Brunel University, UK. He is a member of the editorial board for the *Industrial Management & Data Systems Journal*, and is currently the book review and software review editor for the *International Journal of Health Care Information Systems*, and is also a member of the editorial advisory & review board for the *Journal of Knowledge Management Practice*.

* * *

Gianluca Aloi received an MS in computer engineering from the University of Calabria, Italy (1999), and a PhD from the University of Calabria, Italy (2002). Currently, he is an assistant professor at the University of Calabria where, since 1999, he has worked with the telecommunications research group, and he is involved in several projects concerning wireless communications. His research interests include enhanced wireless and satellite systems, mobility, traffic, and resource management, QoS support in heterogeneous communications networks and interworking of wireless and wired networks.

Jacques Arnould, born in 1961, is Dominican and an agricultural engineer with a PhD in history of sciences and a PhD in theology. He is taking an active interest in the interrelation between sciences, cultures, and religions, with a particular interest for two set of themes: the first related to the life sciences and his evolution; the second related to space conquest. To the first he devoted several works and publications on the historical and theological aspect. To the second, he is the French Space Agency (CNES) expert in charge of ethical, social, and cultural aspect of space activities.

Luigi Boccia was born in Lungro (CS), Italy, in 1975. He received his degree in information technology engineering from the University of Calabria, Italy, and a PhD in electronic engineering from the University "Mediterranea" of Reggio Calabria, Italy (2000 and 2003, respectively). In 2000, he has been with the European Stage Agency at the Attitude Determination and Orbit Control section. Currently, he holds the position of assistant professor at the Dipartimento di Elettronica, Informatica e Sistemistica of the University of Calabria. His research interests include microstrip antennas and millimeter wave circuits.

Angie Bukley has nearly 25 years of professional experience in space and defense systems. She holds a PhD in electrical engineering, specializing in control theory, from the University of Alabama in Huntsville. Dr. Bukley is the associate dean for research and graduate studies in the Russ College of Engineering and Technology at Ohio University. From October 1998 through October 2003, she was the director of laser applications for The Aerospace Corporation in Albuquerque, New Mexico. Before that, Dr. Bukley spent 7 years at the NASA Marshall Space Flight Center in Huntsville, Alabama, where she directed the Large Space Structures Controls Laboratory and worked on remote sensing applications for environmental and disaster management.

Agnieszka Chodorek received an MS in electrical engineering from the Kielce University of Technology in Kielce, Poland (1991), and a PhD in telecommunications from the AGH University of Science and Technology in Krakow, Poland (2001). She is an assistant professor in the Department of Telecommunications and Photonics at the Kielce University of Technology in Kielce, Poland. She is currently lecturing on satellite and mobile communications, IP networks, multimedia technology, and Internet multimedia services. Her research interests lie in the area of telecommunication networks, with emphasis on Internet technology and multimedia transmission. She has authored many publications in these areas.

Robert R. Chodorek received an MS in electrical engineering from the Kielce University of Technology in Kielce, Poland (1990), and a PhD in computer sciences from the AGH University of Science and Technology in Krakow, Poland (1996). He is currently an assistant professor in the Department of Telecommunications at the AGH University of Science and Technology in Krakow, Poland. His current areas of research include performance evaluation of telecommunication networks, in particular broadband communications, IP multicasting, and multimedia communications. He is author or co-author of over 80 research papers and two books.

Justo Alcázar Díaz was born in Málaga, Spain, in 1978. He received an MSc in telecommunications engineering from Malaga University (Spain), and a master's degree in business administration from Clemson University (South Carolina, USA). The main focus of his MBA was in technology management and information technologies, and these are areas in which he has later pursued his professional career, working in the space sector for the European Space Agency. Currently he is pursuing a PhD at Polytechnic University of Madrid in the field of technology management.

Roberto Fernández received his degree in electrical engineering from Universidad Nacional del Sur in 1968 and is a full-time professor and researcher at the Engineering Faculty of the Universidad Nacional del Comahue in Neuquén, Argentina. He gives courses in digital design for electronic engineering and also assists at research projects. His papers have been presented and published in several congresses and meetings on electronics, control, and space science. Fernández has also been director of the Electrical Department, leading research and development projects and integrating many consulting commissions. He is currently developing his thesis for a master's degree in teaching physical and natural sciences.

Robert A. Goehlich was born in Berlin, Germany, in 1975. He studied aerospace engineering at the Technical University Berlin (1996-2000) and received a PhD in 2003. He has done research stays in the fields of reusable launch vehicles and space tourism at the Israel Institute of Technology (Haifa, Israel), at the University of Washington (Seattle, USA), at the National Aerospace Laboratory (Tokyo, Japan), and at EADS Space (Kourou Space-port, French Guiana). Dr. Goehlich has created and, since November 2003, lectured on the world's first "space tourism" master's/PhD courses as a visiting professor at Keio University (Yokohama, Japan). He is an author of three textbooks about the feasibility of space tourism and strategies to realize space tourism in near-, mid- and long-term.

Jeanne Holm is the chief knowledge architect at the Jet Propulsion Laboratory (JPL), California Institute of Technology. Holm leads NASA's knowledge management team, which helps to facilitate access to and reuse of NASA's knowledge and to support future missions and drive innovation. Her latest activities involve helping to transform NASA into a learning organization through innovative techniques in developing communities of practice and ensuring lessons are embedded across the organization. Holm chairs the U.S. Government Knowledge Management Working Group for Knowledge and Human Capital Retention. She chairs the IAA Knowledge Management in Aerospace Study Group. Her degrees are from UCLA and Claremont Graduate University in the management of information systems. She is an instructor at UCLA and Pepperdine University, and her courses focus on knowledge management, information design, and architectures for interoperability. She has been awarded numerous honors, including the NASA Exceptional Service Medal for leadership (twice), for her work on the Galileo and Voyager spacecraft, three Webby's from The International Academy of Digital Arts and Sciences, Competia's 2003 Champion of the Year, and the APQC award for "Using Knowledge Management to Drive Innovation."

Kay-Uwe Hörl, a German business lawyer, is heading the Contract Department of EADS ASTRIUM in Munich, Germany. He has worked in the aircraft and space divisions of the European Aeronautic Defence and Space Company (EADS). Dr Hörl's research interests are the interfaces between business law and space law.

James King is a director at Industry Canada's Communications Research Centre in Ottawa, where he manages a team of specialists who work on various Canadian and international satellite programs. He has over 25 years experience in Cospas-Sarsat, including 10 years as deputy head of the Cospas-Sarsat Secretariat in London. King participated in the development, launch, and testing of the first Sarsat Leosar satellites in the early 1980s, then the Geosar satellite enhancements and now the new Meosar enhancements to fly on future navigation satellites. King received a Bachelor of Science in electrical engineering in 1971, and a Master of Engineering in 1986, specializing in satellite communications, and has written numerous papers on Cospas-Sarsat and other satellite systems.

Dimitris J. Kraniou is a professor of international economics and management at the Graduate Business Programs, Point Park University. He holds a BA, MA, and PhD from the University of Pittsburgh. His areas of study were international economics and international administration. He also holds an advanced certificate in international political economy. Dr. Kraniou has functioned as chair and director of the Economic/Business Departments and of graduate business programs. Dr. Kraniou has been the VP and treasurer of International Professional Associations. He is the chair of the Greek Room, University of Pittsburgh, and the president of the Pan-Ionian Society. His continuous love affairs involve astrophysics and ancient Greek classics. Research interests include the management of extra-terrestrial resources, multinational corporate strategies, globalization, systems theory, and everything that is Greek.

Ikuko Kuriyama is special staff to the manager in the Japan Aerospace Exploration Agency. She is currently working for Office for Earth and Environmental Science, Ministry of Education, Culture, Sports, Science and Technology of Japan. Her professional expertise includes international cooperation and legal coordination related to bilateral and multilateral Earth observation programs. She now serves as a national focal point for the intergovernmental Group on Earth Observation (GEO) and a national delegate to the Conference of the Party to the UNFCCC, through which she promotes Earth observation activities. Kuriyama holds LLM from Keio University, Tokyo, Japan, and a Master of Space Studies (MSc) from International Space University, Strasbourg, France.

Jorge Lassig is a full-time professor with the Engineering Faculty, Comahue University in Neuquén Argentine. His research interests include atmospheric environment, space technology, and microgravity fluids research. Lassig is an aeronautic engineer with extensive experience in space project management. He carried out on board the NASA KC-135 airplane testing experiments in Zero-G environment. He was also PI of the GAS Canister G-761, the first Argentine-made payload to fly in the space shuttle. His research papers have been published in international journals and in three books. Lassig is a member of the editorial board for the *Latin-American Journal of Space Science and Technology.*

Paul Nelson is a recent graduate from Darien High School, in Darien, CT, USA. He will be attending the undergraduate engineering school at the University of Wisconsin-Madison. Nelson spent 3 years of his high school career in Authentic Science Research (ASR). The ASR program is a 3-year independent sciences research class in which students choose a science topic and work on it for 3 years. Nelson worked with the idea of a space elevator, during this period collaborating with the experts in the space elevator field. He attended international conferences presenting his papers and work with the space elevator in such places as Japan, Washington, DC, and Canada. He hopes to continue his research with the space elevator while he attends UW-Madison.

Pasquale Pace received an MS in computer engineering and a PhD in information engineering from the University of Calabria, Italy (2000 and 2005, respectively). From March 2005 to October 2005, he was a visiting researcher at the Centre for Communication Systems Research (CCSR) at the University of Surrey – UK, where he did research on multimedia satellite systems. In November 2005, he joined the D.E.I.S. Department, University of Calabria as research fellow. His research interests include multimedia satellite systems, DVB-RCS-satellite architectures, IP-satellite, mobility management, traffic & resource management, call admission control, and integration of satellite systems and high altitude platforms in heterogeneous communications networks.

Larry J. Paxton is a member of the principal professional staff at the Johns Hopkins University Applied Physics Laboratory. Dr. Paxton received a PhD from the University of Colorado at Boulder from the Department of Astrophysical, Planetary, and Atmospheric Sciences. His research interests include the atmospheres of the Earth, Mars, and Venus, and the application of instrumentation to remote sensing problems. He has published over 130 papers on planetary atmospheres, instrumentation, remote sensing, and space mission design. He is a member of the International Academy of Astronautics and he belongs to the American Geophysical Union, the American Astronomical Society, and the Society of Photo-optical Instrumentation Engineers. He is the project scientist on the Global Ultraviolet Imager on the NASA TIMED mission, and the principal investigator on the SSUSI instrument for DMSP and AURORA for NPOESS.

After obtaining his master's degree in aerospace engineering from the Technical University of Delft in The Netherlands (1998), **Michel van Pelt** started working as a cost and system engineer at the ESTEC centre of the European Space Agency. He is involved in many space projects, for which he prepares cost estimates and analyses of financial proposals. Furthermore, he regularly works in ESTEC's Concurrent Design Facility as cost engineer, system engineer, or team leader. He is also author of the book *Space Tourism: Adventures in Earth Orbit and Beyond*. This year, his new book *Space Invaders: How Robotic Spacecraft Explore the Solar System* will be published.

Juan Jorge Quiroga is a full-time associate professor and researcher with the Engineering Faculty, Universidad Nacional del Comahue in Neuquén Argentine. He gives courses in industrial electronics for electronic engineering and leads the project "Desarrollo de Aspectos Relevantes en la Tecnología Electrónica para Cargas Útiles Espaciales y Microsatélite." His research papers have been presented and published in a number of congress and meetings on electronics, control, and spatial science. Over the years, Quiroga has also provided consulting services to local companies in addition to university teaching. He is preparing his thesis for a master's degree in teaching physical and natural sciences. He is a senior member of the IEEE.

N. Raghavendra Rao is a professor at SSN School of Management & Computer Applications, Chennai, India. Dr. Rao has a master's degree in commerce from Osmania University, and a PhD in finance from the University of Poona. He has a rare distinction of having experience in the combined areas of information technology and business applications. His rich experience in industry is matched with a parallel academic experience in management & IT in business schools. He has over two decades of experience in the development of application software related to manufacturing, service-oriented organizations, financial institutions, and business enterprises. He presents papers related to information technology at conferences. He contributes articles on information technology to mainstream newspapers and journals. His area of research interest is mobile computing and space technology.

Chris Robinson is an associate professor of art at the University of South Carolina, whose work seeks cross-disciplinary relationships between the two cultures of art and science. Trained as a minimalist, he specializes in 3D imaging, art, and technology, and installations with mediums ranging from computer graphics to laser light. He was one of the early pioneers of the visual arts active role in zero gravity and space exploration, training and observing in the General Dynamics F-16 and NASA's Zero G aircraft, the Weightless Wonder. Robinson is currently part of a multidisciplinary team conducting National Science Foundation funded research on the societal implications and role of images in nanoscience/technology, and has been active in exhibitions and scholarly conferences in the U.S. and Europe.

Lesley Jane Smith, Scottish solicitor and member of the Hanseatic Bar of Bre-men, Germany, has long-standing experience as a teacher and professor of European and comparative law at the University of Lueneburg, Germany, and within various law programmes in Europe. She has worked for both public and private legal services, including a short period with the EC Commission. Dr Smith's consultancy work includes commercial consulting and latterly, law reform work in transformation states. Her research interests are in European competition and intellectual property law, European private law, and space law. She is a member of various international and academic lawyers' associations, including the International Institute of Space Law.

Tirso Velasco (Spain) graduated as telecommunications engineer in the Polytechnic University of Valencia (UPV) in 1998. Since then, he has developed his professional career in the aerospace field within academia (University of Valencia), space agencies (ESA, INTA), and the private sector (EADS Astrium, Indra). He has been involved as a systems engineer in different areas including planetary exploration, defence, and navigation projects, including the European Global Navigation Satellite System (GNSS), Galileo. He has 16 papers in international publications. His current research interest is the evolution of GNSS and space policies.

Stella Tkatchova was born in Sofia, Bulgaria. In 1996, she undertook a master's degree in international business from the University for National and World Economy in Sofia. She then worked as a financial expert in the Bulgarian National Bank (BNB). In 1999, she undertook a master's degree of space studies in the International Space University (ISU) in Strasbourg. Until recently, she worked at the ESTEC Centre of the European Space Agency as an ISS marketing analyst. At present, she is finalizing her PhD in space technology commercialization at the Faculty of Aerospace Engineering of the Technical University of Delft, The Netherlands.

Olga Zhdanovich has 16 years of experience working in Russian and international government institutions, academia, and private companies. She has a good mixture of experience from pure academic to private entrepreneurship. She received an MEng in cartography from Moscow University of Geodesy and Cartography (1990), and in 1996, she received an MSc in environmental science and policy from the Central European University. Zhdanovich is a faculty of the International Space University. Her major fields are environmental resources management, remote sensing, satellite navigation, future of space exploration, market research, and space education. She was coordinator of the "Space Programs in the 21st Century Forum" at UNISPACE III.

Index

D

E

F

J

L

M

N

O